AF772486

Springer Series in Electrophysics
Volume 19

Edited by Günter Ecker

Springer Series in Electrophysics

Editors: Günter Ecker Walter Engl Leopold B. Felsen

Volume 1 **Structural Pattern Recognition** By T. Pavlidis

Volume 2 **Noise in Physical Systems** Editor: D. Wolf

Volume 3 **The Boundary-Layer Method in Diffraction Problems**
By V. M. Babič, N. Y. Kirpičnikova

Volume 4 **Cavitation and Inhomogeneities** in Underwater Acoustics
Editor: W. Lauterborn

Volume 5 **Very Large Scale Integration (VLSI)**
Fundamentals and Applications
Editor: D. F. Barbe 2nd Edition

Volume 6 **Parametric Electronics** An Introduction
By K.-H. Löcherer, C. D. Brandt

Volume 7 **Insulating Films on Semiconductors**
Editors: M. Schulz, G. Pensl

Volume 8 **Theoretical Fundamentals of the Ocean Acoustics**
By L. Brekhovskikh, Y. P. Lysanov

Volume 9 **Principles of Plasma Electrodynamics**
By A. F. Alexandrov, L. S. Bogdankevich, A. A. Rukhadze

Volume 10 **Ion Implantation Techniques**
Editors: H. Ryssel, H. Glawischnig

Volume 11 **Ion Implantation: Equipment and Techniques**
Editors: H. Ryssel, H. Glawischnig

Volume 12 **VLSI Technology**
Fundamentals and Applications Editor: Y. Tarui

Volume 13 **Physics of Highly Charged Ions**
By R. K. Janev, L. P. Presnyakov, V. P. Shevelko

Volume 14 **Plasma Physics for Physicists**
By I. A. Artsimovich, R. Z. Sagdeev

Volume 15 **Relativity and Engineering** By J. Van Bladel

Volume 16 **Electromagnetic Induction Phenomena**
By D. Schieber

Volume 17 **Concert Hall Acoustics** By Y. Ando

Volume 18 **Planar Circuits** for Microwaves and Lightwaves
By T. Okoshi

Volume 19 **Physics of Shock Waves in Gases and Plasmas**
By M. A. Liberman, A. L. Velikovich

Volume 20 **Kinetic Theory of Particles and Photons**
Theoretical Foundations of Non-LTE Plasma Spectroscopy
By J. Oxenius

Volume 21 **Picosecond Electronics and Optoelectronics**
Editors: G. A. Mourou, D. M. Bloom, C.-H. Lee

M.A. Liberman A.L. Velikovich

Physics of Shock Waves in Gases and Plasmas

With 91 Figures

Springer-Verlag
Berlin Heidelberg New York Tokyo

Professor Michael A. Liberman
Dr. Alexander L. Velikovich

Institute for Physical Problems, Academy of Sciences of the USSR, ul. Kosygina, 2,
SU-117334 Moscow, USSR

Translated into English by A. S. Dobroslavsky (Moscow)
Copy-edited and upgraded by R. Breger (Bochum) and
H. Lotsch (Heidelberg)

Series Editors:

Professor Dr. Günter Ecker

Ruhr-Universität Bochum, Theoretische Physik, Lehrstuhl I,
Universitätsstrasse 150, D-4630 Bochum-Querenburg, Fed. Rep. of Germany

Professor Dr. Walter Engl

Institut für Theoretische Elektrotechnik, Rhein.-Westf. Technische Hochschule,
Templergraben 55, D-5100 Aachen, Fed. Rep. of Germany

Professor Leopold B. Felsen Ph.D.

Polytechnic Institute of New York, 333 Jay Street, Brooklyn, NY 11201, USA

ISBN 3-540-15605-4 Springer-Verlag Berlin Heidelberg New York Tokyo
ISBN 0-387-15605-4 Springer-Verlag New York Heidelberg Berlin Tokyo

Library of Congress Cataloging in Publication Data. Liberman, M. A. (Michael A.), 1942- Physics of shock waves in gases and plasmas. (Springer series in electrophysics; v. 19) Bibliography: p. Includes index. 1. Shock waves. 2. Gas dynamics. 3. Plasma dynamics. I. Velikovich, A. L. (Alexander L.), 1951-. II. Title. III. Series.
QC168.85.S45L53 1985 533'.293 85-12656

This work is subject to copyright. All rights are reserved, whether the whole or part of the material is concerned, specifically those of translation, reprinting, reuse of illustrations, broadcasting, reproduction by photocopying machine or similar means, and storage in data banks. Under § 54 of the German Copyright Law where copies are made for other than private use, a fee is payable to "Verwertungsgesellschaft Wort", Munich.

© Springer-Verlag Berlin Heidelberg 1986
Printed in Germany

The use of registered names, trademarks etc. in this publication does not imply, even in the absence of a specific statement, that such names are exempt from the relevant protective laws and regulations and therefore free for general use.

Offset printing: Beltz Offsetdruck, 6944 Hemsbach/Bergstr.
Bookbinding: J. Schäffer OHG, 6718 Grünstadt.
2153/3130-543210

Foreword

As Emile Jouguet remarked, "the shock wave flew off the tip of the pen of a theoretician for the first time" about a hundred years ago. The physics of shock waves has since grown into an independent branch of science closely linked with a wide range of research areas, from astrophysics and plasma physics to solid-state physics. Since the beginning, theoretical investigation has kept its leading role.

The present book is devoted to actual problems of the theory of shock waves in gases and plasmas, that are of general interest to physicists. It contains the results of studies on shock structure, stability, evolutionarity and dynamics. Of special interest is the theory of shock phenomena in magnetic fields, which is important for applied research on controlled nuclear fusion. A substantial contribution to this theory has been made by these authors.

This monograph is the first attempt in the literature to make a systematic presentation of the shock-structure theory.The theory is consistently substantiated by relevant experimental results obtained recently with the use of high-power electromagnetic shock tubes. The material contained here is applicable to the solution of a wide variety of problems arising in plasma physics, nuclear fusion and cosmic gasdynamics. I believe that this book will be of help and interest for a broad circle of research workers (physicists, astrophysicists and engineers concerned with energy accumulation, shock phenomena and other related problems of plasma hydrodynamics) as well as for university staff, post- and undergraduates.

Moscow — Academician Ya.B.Zel'dovich

Preface

The problem of shock waves pertains to many areas of physical science, from nuclear physics to astrophysics. Shock waves accompany the passage of meteoroids, a spacecraft and a supersonic aircraft through the atmosphere; and they arise in many laboratory experiments. Of special practical importance among the latter are studies of plasma heating by high-power electric discharges, or by laser, electron or ion beams interacting with targets, with a view to accomplishing controlled thermonuclear fusion.

This monograph deals with problems of production of shock waves and their structure, shock-shaping physical processes and mechanisms, and with the questions of dynamics and stability of shock waves. Special attention is paid to shock waves in a plasma, particularly, to the structures of such shocks.

We did not aim to produce another textbook on physical gasdynamics, since an excellent introduction to the subject already exists, namely *The Physics of Shock Waves and High Temperature Hydrodynamic Phenomena* by Ya.B. Zel'dovich and Yu.P. Raizer, the English translation of which was published by Academic Press in 1967. We rather develop a consistent treatment of those aspects of the shock-wave theory which totally or almost totally fell beyond the scope of that book, such as shock waves in plasmas, electromagnetic phenomena in shock waves, and ionizing shock waves in magnetic fields. We had to omit many important problems, which seem to pertain rather to the domain of fluid mechanics (reflection and diffraction of shock waves, propagation of shock waves in inhomogeneous fluids, etc.). We also consider mostly collision-dominated plasmas, otherwise referring the reader to the book by D.A. Tidman and N.A. Krall, *Shock Waves in Collisionless Plasmas* (Wiley, 1971). Furthermore, we tried to present our material so as to make it useful for both physicists and engineers, who have to deal with shock phenomena.

Chapter 1 is an introduction to the theory of gasdynamic shock waves. Here we introduce the concept of gasdynamic discontinuities, discuss the conditions of formation of shock waves, consider steady shock structures and the evolutionarity of shock waves. The basic concepts and techniques for analyzing the criteria of existence of steady shock structures and their linkage with the dynamics and stability of shock waves are illustrated by comprehensive examples from the theories of gasdynamic shock waves, and waves of detonation and deflagration.

Chapter 2 deals with shock waves in plasmas. Here we consider the ionization processes in the shock waves, the origins and structure of the region of precursor ionization, polarization of the plasma, radiative cooling of shock-heated plasmas, and the effects of a wall boundary layer in shock tubes.

Chapter 3 is entirely devoted to magnetohydrodynamic (MHD) shock waves (shock waves in a dense high-temperature plasma with a magnetic field).

In Chaps. 4,5 we consider one of the most interesting problems of the theory of shock waves: the problem of ionizing shocks in magnetic fields. Here we give the theory of ionizing shock waves, analyze their morphology for different orientations of the magnetic field relative to the shock plane, and investigate heating the plasma by ionizing shock waves in electromagnetic shock tubes. The dynamics of MHD and ionizing shock waves is analyzed in Chap. 5 on the basis of self-similar solutions and the results of numeric simulation.

Although this monograph is obviously mainly theoretical we discuss the operating principles of installations used for producing and studying the shock waves, and in most cases theoretical and experimental results. All mathematical expressions are presented in a closed form, convenient for practical use.

We are grateful to Professors S.I. Anisimov, A.V. Gurevich, A.A. Rukhadze, Yu.L. Klimontovich and Academician E.M. Lifshits, who have read parts or all of the manuscript, for useful discussions and criticism. Our thanks are due to Professor L.P. Pitaevskii for his most helpful comments. We wish to acknowledge the valuable collaboration of Professors I.I. Glass, J. Howard, J.A. Johnson III, S.S.R. Murty, M. Pinègre, P. Valentin, who made their original diagrams available for reproduction. We should like to express our special thanks for the material given to us before it came out of print: to Professor M. Pinègre for his doctoral thesis, which was useful in writing Chap. 2, to Professor C.D. Mathers for a technical description of the electromagnetic shock tube SUPPER VI. used in Sect.5.1, to Professors J. Howard, B.W. James, A.R. Low, and G.R. Fowles for the preprints of their papers, used in Sects.5.1.3 and 1.7.4.

We are most grateful to Academician Ya.B. Zel'dovich, whose help and constant attention we enjoyed during our work on the book. The discussions with him of the problems included in this book were most fruitful and stimulating.

We also acknowledge our gratitude to Dr. H.K.V. Lotsch of Springer-Verlag for his valuable assistance and kind cooperation.

We are very grateful to A. Dobroslavsky for his translation of the original Russian manuscript and help in preparing the final version.

To our grief, Academician I.M. Lifshits is no longer living to receive our heartiest thanks for his generous guidance, from which we benefited until his untimely death in 1982.

Moscow, July 1985

M.A. Liberman
A.L. Velikovich

Contents

1. Introduction to the Theory of Gasdynamic Shock Waves

The motion of a compressible fluid in a continuity approximation is described by the Navier-Stokes equations (Sect.1.1). In the kinetic theory from the microscopic point of view (Sect.1.2), this approximation is justifiable as long as the gasdynamic variations are low frequency and large scale, expressed as $Kn \ll 1$, Kn being a dimensionless parameter called the Knudsen number. Narrow shock fronts of strong gasdynamic shock waves are characterized by $Kn \sim 1$ (Sect. 1.3), so these shocks in gasdynamics should be viewed as infinitely thin discontinuities. The study of shock structures calls for a kinetic approach, and thus the Navier-Stokes equations may be used only as a model. However, shocks in plasmas are usually characterized by $Kn \ll 1$, and therefore the simple model, which in gasdynamics is mainly of methodological interest, is henceforth used consistently in calculations.

The conditions under which shock waves are produced in gasdynamic flows are analyzed in Sects.1.4,5. In Sect.1.4 we describe weak compression waves and the nonlinear expansion waves of arbitrary strength in a perfect fluid. In Sect.1.5 we demonstrate the forbiddenness of finite-amplitude compression waves: the nonlinear distortion of their profiles gives rise to discontinuities, or shock waves. Even with relatively low-intensity sound waves, profile steepening whereby the initially sine-wave flow assumes a saw-tooth appearance, containing weak shocks. This evolution is considered in detail in Sect.1.5 on the basis of Burgers' equation. The results of Sect.1.5 reveal that the presence of shock waves is a typical feature of gasdynamic flows in general.

Apart from shock waves, in gasdynamics discontinuities of a different kind may exist—tangential discontinuities, which are simply stepwise profiles of finite-amplitude vortex and entropy waves, traveling along with the medium and not experiencing nonlinear distortion (Sect.1.6.1). Dissipations included, these discontinuities, contrary to shock waves, are not associated with steady structures. The existence of the structure of a weak shock wave is proved in Sect.1.6.2 with the aid of Burgers' equation. In Sect.1.6.3 we obtain the boundary conditions on the gasdynamic shock front. The results obtained in Sects.1.4.5,1.6.2 are used in Sect.1.6.4 to explain the operation principle of the shock tube.

Section 1.7 is devoted to the questions of stability and evolutionarity of discontinuous solutions. The results of Sect.1.7 are not restricted to gasdynamic shock waves; they can also be applied to the waves of detonation, deflagration, laser radiation absorption waves, magnetohydrodynamic (MHD) shock waves, etc. In Sect.1.7.1 we introduce the basic concept of evolutionarity of a discontinuity; in Sect.1.7.2 we analyze the linkage between evolutionarity of a discontinuity and the existence of its structure. We also introduce the concept of the main and additional boundary conditions on the surface of discontinuity. In the low-frequency limit of the spectrum of waves propagating in a dissipative medium, we single out those waves which carry the disturbances from infinity to the shock front, and vice versa, and the shock-shaping dissipative waves. These concepts are illustrated in Sect.1.7.3 with the example of a fluid described by Burgers' equation. The three-dimensionality of the real flow rigorizes the conditions of stability of the discontinuity; in particular, the stability may come to depend on the equation of state.

In Sect.1.8 we analyze structures of gasdynamic shock waves on the basis of the Navier-Stokes equations for large and small Prandtl numbers Pr, the case typical of a plasma. In Sect.1.8.2 we calculate the viscosity-shaped structure of the shock wave ($Pr \gg 1$), which is shown to have no restrictions imposed on its strength. On the contrary, the high thermal conductivity ($Pr \ll 1$; Sect.1.8.3) is not capable of directly dissipating the momentum of the incident flow, and can therefore shape structures of only weak shock waves, whereas the strong shock waves contain viscous subshocks. The multiple-scale method, used for analyzing the structures of shock waves, is described in Sect.1.8.

In Sect.1.9 we consider structures of detonation and deflagration waves. Several types of discontinuity may propagate in the medium where exothermal reaction takes place, including (with the appropriate reaction kinetics) two types of detonation waves (strong and weak). This helps to illustrate the concept of additional boundary conditions and shows how boundary conditions help establish to establish the regime of propagation of the reaction under given circumstances.

Finally, in compliance with generally accepted practice, all the formulas throughout the book are written in the International System of Units (SI).

At the same time, certain physical quantities are traditionally expressed in conventional [eV, Torr], or in multiples of SI [cm, V/cm] units whenever convenient. Unless otherwise stated, the temperature is expressed in units of energy.

Dimensionless variables are used extensively. For convenience, the dimensionless variables are denoted by the same symbols as the respective dimensional variables. If necessary, however, distinction is made by placing a bar above the dimensional variables. All notation is explained when arising for the first time in the text.

1.1 Equations of Motion

1.1.1 Conservation Laws and the Euler Equation

The macroscopic behavior of a compressible continuous fluid is determined by the velocity distribution function $\mathbf{u}(\mathbf{r},t)$, which is a local function of coordinates and time and a pair of thermodynamic variables [density $\rho(\mathbf{r},t)$ and temperature $T(\mathbf{r},t)$, or density and pressure $p(\mathbf{r},t)$, etc.]. The values of the other thermodynamic variables are determined by those two chosen via the equation of state. Thus the macroscopic motion of a fluid is uniquely determined by the five gasdynamic variables, which must satisfy certain equations of motion. The basic equations of motion of a fluid follow from the conservation laws of mass, momentum and energy.

Consider an arbitrary volume element V in the fluid. The change in the mass of the fluid of density ρ in volume V is counterbalanced by the flow of fluid across the bounding surface. We can write the equation of conservation of mass in the form

$$\frac{\partial}{\partial t}\int_V \rho dV + \int_S \rho\mathbf{u}\cdot\mathbf{n}\, dS = 0 \quad , \tag{1.1.1}$$

where $\mathbf{n}$ is the outward normal to the bounding surface S of volume V. Going from the surface integral to the space integral according to the Gauss theorem, we get from (1.1.1)

$$\int_V \left[\frac{\partial\rho}{\partial t} + \nabla\cdot(\rho\mathbf{u})\right] dV = 0 \quad . \tag{1.1.2}$$

Since the integrand is continuous and (1.1.2) must be valid for any volume, we can write the continuity equation

$$\frac{\partial\rho}{\partial t} + \nabla\cdot(\rho\mathbf{u}) = 0 \quad . \tag{1.1.3}$$

Consider a volume element in the fluid. It experiences stress applied to the bounding surface and proportional to the area of the small surface element. The stress is characterized by the *stress tensor* Π_{ij}, whose components are represented by the ith component of the force acting on the unit surface element normal to the direction j. The symmetry of the stress tensor derives from the law of conservation of angular momentum:

$$\Pi_{ij} = \Pi_{ji} \quad . \tag{1.1.4}$$

For isotropic frictionless fluids the stress tensor is simply linked with the pressure p in the fluid:

$$\Pi_{ij}^{(0)} = p\delta_{ij} \quad . \tag{1.1.5}$$

The equation of conservation of the ith component of momentum of the fluid volume element can now be written as

$$\frac{\partial}{\partial t}\int_V \rho u_i \, dV + \int_S (\rho u_i n_j u_j + \Pi_{ij} n_j)\, dS = 0 \quad , \tag{1.1.6}$$

where the Cartesian components u_i of the velocity $\mathbf{u}(\mathbf{r},t)$ are introduced ($i,j = 1,2,3$ correspond to the symbols x,y,z, respectively; the Einstein summation rule for these indices is adopted). The first term here describes the rate of variation of the fluid's momentum in the volume, the second accounts for the transfer of momentum across the surface S of the volume V, and the third derives from the variation in momentum due to the tensile forces.

Converting the integral over the surface in (1.1.6) into the integral over the volume, we obtain, similar to (1.1.3),

$$\frac{\partial}{\partial t}(\rho u_i) + \frac{\partial}{\partial x_j}(\rho u_i u_j + \Pi_{ij}) = 0 \quad . \tag{1.1.7}$$

Substituting the stress tensor in the form of (1.1.5) into (1.1.7) and using the continuity equation, we obtain the equation of motion of frictionless fluid: the *Euler equation*

$$\frac{\partial u_i}{\partial t} + u_j \frac{\partial u_i}{\partial x_j} = -\frac{1}{\rho}\frac{\partial p}{\partial x_i} \quad . \tag{1.1.8}$$

The density of energy of the fluid is composed of the density of the kinetic energy $\rho u_i^2/2$ and the internal (thermal) energy $\rho\varepsilon$. The equation of conservation of energy for the volume element of frictionless fluid at zero heat exchange between individual fragments has the form

$$\frac{\partial}{\partial t}\int_V\left(\frac{1}{2}\rho u_i^2+\rho\varepsilon\right)dV+\int_S\left[\left(\frac{1}{2}\rho u_i^2+\rho\varepsilon\right)u_j+p\delta_{ij}u_i\right]n_j\,dS=0\quad. \tag{1.1.9}$$

From (1.1.9) follows the continuity equation for the density of energy of the fluid:

$$\frac{\partial}{\partial t}\left(\frac{1}{2}\rho u_i^2+\rho\varepsilon\right)+\frac{\partial}{\partial x_j}\left[\left(\frac{1}{2}\rho u_i^2+\rho\varepsilon+p\right)u_j\right]=0\quad. \tag{1.1.10}$$

Let us transform (1.1.10), making use of the continuity equation (1.1.3) and Euler's equation (1.1.8), and taking advantage of thermodynamic equality

$$T\,dS=d\varepsilon-\frac{p}{\rho^2}\,d\rho\quad, \tag{1.1.11}$$

where T is the temperature, and S the density of the fluid's entropy.

After some straightforward algebra (1.1.10) can be written as

$$T\,\frac{dS}{dt}=0\quad, \tag{1.1.12}$$

where, as usual, the total time derivative denotes the entropy variation of the given moving fluid particle, which enables us to express (1.1.12) in the form

$$\frac{dS}{dt}=\frac{\partial S}{\partial t}+\mathbf{u}\nabla S=0\quad. \tag{1.1.13}$$

Equation (1.1.13) is a direct implication of the energy conservation law. In the flow of frictionless fluid in the absence of dissipation and heat exchange between individual portions of the fluid, the entropy of each fragment remains constant along the path of its motion. Such flow is called *isentropic* or *adiabatic*.

Let us introduce a thermal function, defined as the enthalpy of unit mass of the fluid W by a thermodynamic equality

$$dW=T\,dS+\frac{1}{\rho}\,dp\quad. \tag{1.1.14}$$

For an adiabatic flow (1.1.14) becomes

$$dW=\frac{1}{\rho}\,dp\quad, \tag{1.1.15}$$

and hence the Euler equation for an adiabatic flow can be written as

$$\frac{\partial\mathbf{u}}{\partial t}+(\mathbf{u}\nabla\mathbf{u})=-\nabla W\quad. \tag{1.1.16}$$

The enthalpy is related to the internal energy of the fluid:

$$W = \varepsilon + p/\rho \quad . \tag{1.1.17}$$

The same quantity is easily detected in (1.1.10) in the term accounting for the energy flux, enabling us to write the equation of energy balance as

$$\frac{\partial}{\partial t}\left(\frac{1}{2}\rho\mathbf{u}^2 + \rho\varepsilon\right) + \nabla\left[\rho\mathbf{u}\left(W + \frac{\mathbf{u}^2}{2}\right)\right] = 0 \quad . \tag{1.1.18}$$

As we have seen, (1.1.3,16,18) describe the adiabatic flow of a fluid, disregarding viscosity and heat transfer. The flow may be considered adiabatic only if its nonuniformity is not too high, that is, all the pertinent hydrodynamic variables change in a rather smooth and slow way. In formal language this means that all hydrodynamic functions can be differentiated a sufficient number of times. When we come to dealing with shock waves, where the gradients are high and discontinuities arise, these equations are no longer valid and must be modified. This is especially true for (1.1.12). The entropy in the shock wave is not constant, but rather suffers an abrupt change across the shock front.

For convenient reference let us complete this section with a list of thermodynamic relations for an ideal gas [1.1].

The equation of state for an ideal gas is

$$p = R\rho T \quad . \tag{1.1.19}$$

The internal energy and enthalpy ε and W, respectively, can be expressed via the temperature and the specific heat capacities at constant volume c_V and constant pressure c_p:

$$\varepsilon = c_V T \quad ; \quad W = c_p T \quad . \tag{1.1.20}$$

The ratio of specific heat capacities $\gamma = c_p/c_V$ is called the adiabatic exponent. Thus, we can write

$$\varepsilon = \frac{1}{\gamma - 1}\frac{p}{\rho} \quad , \quad W = \frac{\gamma}{\gamma - 1}\frac{p}{\rho} \quad . \tag{1.1.21}$$

Using (1.1.21) and the thermodynamic relation (1.1.11), we get

$$S - S_0 = c_V \ln(p/\rho^\gamma) \quad . \tag{1.1.22}$$

For an isentropic flow $S = \text{const}$, and therefore p and ρ are linked via the adiabatic equation

$$p/p_0 = (\rho/\rho_0)^\gamma \quad . \tag{1.1.23}$$

For an ideal monoatomic gas $\gamma = 5/3$.

1.1.2 Viscosity and Heat Transfer in a Fluid. The Navier-Stokes Equation

If the flow of fluid is nonuniform and the gradients of thermodynamic variables become rather high, then the exchange of heat and momentum between individual fragments of the fluid traveling at different speeds can no longer be neglected. The existence of viscosity and heat transfer leads to irreversible redistribution of heat and momentum among the individual fragments of the fluid, so that the entropy is no longer conserved. Such flow cannot be considered adiabatic, and the equations of motion and energy balance require modification. The continuity equation (1.1.3), which derives from the law of conservation of mass, remains valid.

The simplest modified form of the Euler equation, accounting for the viscosity of the fluid, is the Navier-Stokes equation, corresponding to the first-order approximation with respect to velocity gradients. The second term in the Euler equation (1.1.7) represents the divergence of the density tensor of the flux of momentum deriving from the reversible (entropy-conserving) processes. In the viscous first-order approximation this must be supplemented by the tensor of viscous stress, linear in the velocity gradients:

$$\Pi_{ij} = p\delta_{ij} - \sigma_{ij} \quad . \tag{1.1.24}$$

In the general form the viscous stress tensor can be presented as

$$\sigma_{ij} = \eta\left(\frac{\partial u_i}{\partial x_j} + \frac{\partial u_j}{\partial x_i} - \frac{2}{3}\,\delta_{ij}\,\frac{\partial u_k}{\partial x_k}\right) + \zeta\delta_{ij}\,\frac{\partial u_k}{\partial x_k} \quad , \tag{1.1.25}$$

where η and ζ are the viscosity coefficients.

Inasmuch as energy dissipation leads to an increase in entropy, the coefficients of viscosity must be positive [1.2]:

$$\eta > 0 \quad , \qquad \zeta > 0 \quad .$$

With the aid of (1.1.24,25) the equation of motion for viscous fluids can be written in the form

$$\frac{\partial u_i}{\partial t} + u_j\,\frac{\partial u_i}{\partial x_j} = \frac{1}{\rho}\,\frac{\partial}{\partial x_j}\left[-\,p\delta_{ij} + \eta\left(\frac{\partial u_i}{\partial x_j} + \frac{\partial u_j}{\partial x_i} - \frac{2}{3}\,\delta_{ij}\,\frac{\partial u_k}{\partial x_k}\right) + \zeta\delta_{ij}\,\frac{\partial u_k}{\partial x_k}\right] \quad . \tag{1.1.26}$$

or in vector form

$$\rho\left[\frac{\partial \mathbf{u}}{\partial t} + (\mathbf{u}\nabla)\mathbf{u}\right] = -\,\nabla p + \eta\Delta\mathbf{u} + \left(\zeta + \frac{\eta}{3}\right)\nabla(\nabla\cdot\mathbf{u}) \quad . \tag{1.1.27}$$

Similar changes must be made in the equation of energy balance (1.1.18) by supplementing the energy-flux-density term with the viscous-dissipation-heat flux and the heat flux due to thermal conduction, giving

$$\frac{\partial}{\partial t}\left(\frac{1}{2}\rho u^2 + \rho\varepsilon\right) = -\nabla\cdot\left[\rho\mathbf{u}\left(\frac{\mathbf{u}^2}{2} + W\right) - (\hat{\sigma}\mathbf{u}) - \kappa\nabla T\right] \quad , \tag{1.1.28}$$

where κ is the coefficient of thermal conductivity of the fluid, $(\hat{\sigma}\mathbf{u}) \equiv \sigma_{ik}u_k$, the tensor σ_{ik} being the viscous contribution to the stress tensor (1.1.24). By virtue of the continuity and Navier-Stokes equations, and taking into account the thermodynamic relations (1.1.11,14), Eq. (1.1.28) can be rewritten in the form

$$T\frac{dS}{dt} = \frac{1}{\rho}\sigma_{ij}\frac{\partial u_i}{\partial x_j} + \frac{1}{\rho}\nabla\cdot(\kappa\nabla T) \quad . \tag{1.1.29}$$

From (1.1.29) it is clear that in the absence of dissipation the flow is adiabatic, and the production of entropy is associated with viscosity and thermal conduction. The term T(dS/dt) in (1.1.29) accounts exactly for the rate of change of the heat content of the unit mass of the fluid due to the heat dissipation by viscous friction and heat transfer by thermal conduction.

1.2 Kinetic Theory and Gasdynamic Equations

1.2.1 Kinetic Equations for a Gas

To elucidate the physical meaning of each term and establish the limits of applicability of the hydrogynamic equations, let us analyze these equations from the viewpoint of the kinetic theory of gases. In the kinetic theory the state of a gas particle (atom or molecule) is characterized by the distribution function $f(\mathbf{r},\mathbf{v},t)$, understood as the probability of the particle to be found, at instant t, at the point $\mathbf{r},\mathbf{v}$ of six-dimensional phase space. In other words, the quantity

$$f(\mathbf{v},\mathbf{r},t)d\mathbf{v}\,d\mathbf{r} \tag{1.2.1}$$

determines the number of particles whose velocities $\mathbf{v}$ and coordinates $\mathbf{r}$ fall within the interval $d\mathbf{v}\,d\mathbf{r}$ at the time t. Normalization of the distribution function follows from the condition that the integral over the six-dimensional phase space must be equal to the total number of particles

$$\int f(\mathbf{v},\mathbf{r},t)\,d\mathbf{v}\,d\mathbf{r} = N \quad . \tag{1.2.2}$$

The number density of particles is given by

$$n(\mathbf{r},t) = \int f(\mathbf{v},\mathbf{r},t)\,d\mathbf{v} \quad , \tag{1.2.3}$$

[the mass density being $\rho(\mathbf{r},t) = mn(\mathbf{r},t)$], and the mean velocity by

$$\mathbf{u}(\mathbf{r},t) = \frac{1}{n(\mathbf{r},t)} \int \mathbf{v} f(\mathbf{v},\mathbf{r},t)\, d\mathbf{v} \quad . \tag{1.2.4}$$

In the kinetic theory the distribution function is a solution of the kinetic Boltzmann equation

$$\frac{\partial f_\alpha}{\partial t} + \mathbf{v}\,\frac{\partial f_\alpha}{\partial \mathbf{r}} + \mathbf{F}_\alpha\,\frac{\partial f_\alpha}{\partial \mathbf{v}} = \left\{\frac{\partial f_\alpha}{\partial t}\right\}_{st} \quad , \tag{1.2.5}$$

where $\mathbf{F}_\alpha$ is the external force acting upon a particle of sort α; for charged particles $\mathbf{F}_\alpha$ is the Lorentz force

$$\mathbf{F}_\alpha = q_\alpha(\mathbf{E} + \mathbf{v} \times \mathbf{B}) \quad . \tag{1.2.6}$$

On the right-hand side of (1.2.5) is the collision integral, which determines the change in the distribution function as a result of interactions between particles interpreted as collisions. For a sufficiently rarefied gas, where the collisions may be considered pairwise, the Boltzmann collision integral can be written in the form

$$\left\{\frac{\partial f_\alpha}{\partial t}\right\}_{st} = \sum_\beta \int d\mathbf{v}_\beta d\sigma_{\alpha\beta} v_{\alpha\beta}(f'_\alpha f'_\beta - f_\alpha f_\beta) \quad , \tag{1.2.7}$$

where $v_{\alpha\beta}$ is the relative velocity of colliding α and β particles; $d\sigma_{\alpha\beta}$ is the differential cross section of scattering.

At equilibrium (1.2.5) becomes

$$\left\{\frac{\partial f_\alpha}{\partial t}\right\}_{st} = 0 \quad , \tag{1.2.8}$$

whose solution is a Maxwellian distribution function for the α component of the fluid

$$f_\alpha(\mathbf{v}) = n_\alpha\left(\frac{m_\alpha}{2\pi T}\right)^{3/2} \exp\left[-\frac{m_\alpha}{2T}(\mathbf{v} - \mathbf{u})^2\right] \quad , \tag{1.2.9}$$

where n_α is the number density of α particles, T and **u** are, respectively, the temperature and the mean velocity, being equal for all sorts of particles in equilibrium.

1.2.2 Obtaining the Gasdynamic Equations

The gasdynamic equations give a simpler – if less accurate – description of the dynamic behavior of a fluid compared with the kinetic equations. Consequently, the gasdynamic approximation proves satisfactory only under certain conditions, to be discussed. Here we suppose that the fluid contains only one sort of particle.

At equilibrium the distribution function has the Maxwellian form (1.2.9):

$$f(\mathbf{v},\mathbf{r},t) = f^{(0)}(\mathbf{v}) = n\left(\frac{m}{2\pi T}\right)^{3/2} \exp\left[-\frac{m}{2T}(\mathbf{v}-\mathbf{u})^2\right] \quad , \tag{1.2.10}$$

where n, $\mathbf{u}$, T are constants. The Maxwell distribution function (1.2.10) brings the Boltzmann collision integral to zero. This function is a steady-state solution of the kinetic equation in the absence of external forces and spatial gradients, exactly when the operator on the left-hand side of the kinetic equation is identically zero. Should the initial and/or boundary conditions give rise to spatial and/or temporal nonuniformity, i.e., explicit dependence of gasdynamic variables on coordinates and time, the fluid may no longer be considered in equilibrium, and the distribution (1.2.10) becomes invalid. Let us investigate the limits to which the distribution function (1.2.10) — the so-called local Maxwellian distribution function, whose parameters now depend on coordinates and time [$n = n(\mathbf{r},t)$, $\mathbf{u} = \mathbf{u}(r,t)$, $T = T(\mathbf{r},t)$] — will nevertheless be a fair approximation, without being the exact solution of the kinetic equation.

If (1.2.10) really is a fair approximation, then at a given point $(\mathbf{r}_0,t_0)$ in the center of mass coordinate system, the characteristic velocity of a particle is $v_T = (2T/m)^{1/2}$, m being the mass of the particle, and the distribution function is essentially nonzero within the volume v_T^3 in the phase space of velocities. Let the scale of spatial nonuniformity, determined by the initial and boundary conditions, be L, and the scale of the external force be F_0. The Boltzmann equation can be reduced to the dimensionless form by introducing the dimensionless variables

$$\mathbf{v}' = \mathbf{v}/v_T \quad , \quad \mathbf{r}' = \mathbf{r}/L \quad , \quad t' = v_T t/L \quad , \quad \mathbf{F}' = \mathbf{F}/F_0 \quad , \quad f' = nf/v_T^3 \quad , \tag{1.2.11}$$

where $v_T = v_T(r_0,t_0)$, $n = n(r_0,t_0)$; below we shall write the dimensionless variables without primes. The left-hand side of the kinetic equation takes the form

$$\frac{v_T}{L} K(f) \quad , \tag{1.2.12}$$

where the dimensionless operator K(f) is defined by

$$K(f) \equiv \left(\frac{\partial}{\partial t} + \mathbf{v}\frac{\partial}{\partial \mathbf{r}} + \frac{1}{Fr}\mathbf{F}\frac{\partial}{\partial \mathbf{v}}\right)f$$

($Fr = mv_T^2/F_0L$ is the dimensionless Froude number, which characterizes the relative contribution of the external forces). From the Boltzmann collision integral on the right-hand side of the kinetic equation we single out the dimensionless collision operator I(f),

$$\left\{\frac{\partial f}{\partial t}\right\}_{st} = n\sigma v_T I(f) = n\sigma v_T \int (f'f_1' - ff_1)\, u \frac{h\,dh}{\sigma} d\varphi d\mathbf{v} \quad , \tag{1.2.13}$$

where h is the impact parameter, φ is the scattering angle, u is the relative velocity [1.3]; u and **v** in the integrand are dimensionless and have the order of unity. The characteristic transverse scattering cross section is denoted by σ. From (1.2.12,13) we get the sought-for dimensionless kinetic equation:

$$K(f) = \frac{1}{Kn} I(f) \quad , \tag{1.2.14}$$

where $Kn = 1/n\sigma L$ is the dimensionless Knudsen number. According to a trivial gaskinetic estimate, $\lambda = 1/n\sigma$ is the mean free path of the particle, and so the Knudsen number can be presented as

$$Kn = \lambda/L \quad . \tag{1.2.15}$$

The Knudsen number characterizes the relative importance of the discreteness of the particles which make up the fluid. In the limit $Kn \to \infty$ the discreteness is decisive, since the collisions leading to "Maxwellization" of the distribution function are rare, and the statistically averaged gasdynamic variables make little sense: it is the distribution function which describes the behavior of a separate particle under given circumstances that matters in this case (the so-called free-molecular flow). On the contrary, if Kn is small, then Maxwellization is essential: the local Maxwellian distribution function (1.2.10) will give a fair approximation of the solution of (1.2.14), and the medium may be considered continuous.

Under the condition

$$Kn \ll 1 \quad , \tag{1.2.16}$$

the distribution function can be expanded in a power series of this parameter:

$$f = f_0 + Kn \cdot f_1 + (Kn)^2 f_2 + \ \dots \quad , \tag{1.2.17}$$

f_0 being the dimensionless form of (1.2.10). Having substituted (1.2.17) into (1.2.14), one may follow the iteration procedure to determine step by step the functions f_1, f_2, etc. In the zero-order approximation in Kn $f = f_0$; in the first-order approximation f_1 can be found from the integral equation

$$I(f_0 + Kn \cdot f_1) = Kn \cdot K(f_0) \tag{1.2.18}$$

and so on.

Since f_0 is the function of only the local values of the gasdynamic variables n, **u**, T, the expressions for f_1 and all the higher corrections, in accordance with (1.2.14,17), include only these variables. By substituting the obtained distribution function into the expressions for the components of the stress tensor and heat flux vector, we may relate these quantities to the hydrodynamic variables and their derivatives, thus obtaining a closed set of gasdynamic equations. This method of obtaining the gasdynamic equations from the Boltzmann equation was developed by Chapman and Enskog.

In the zero-order approximation (small Kn) we obtain the Euler equation. The first-order approximation in Kn leads to the Navier-Stokes equations (Sect.1.1). The transport coefficients are no longer phenomenological parameters, as in Sect.1.1: their values are calculated in the course of obtaining the gasdynamic equations from the potential of the interaction between the particles. If, for instance, the molecules are considered as rigid spheres of radius a, the coefficients of (first) viscosity (η) and thermal conductivity (κ) are

$$\eta = \frac{5}{64a^2}\left(\frac{mT}{\pi}\right)^{\frac{1}{2}} \quad ; \quad \kappa = \frac{75}{256a^2}\left(\frac{T}{\pi m}\right)^{\frac{1}{2}} \quad , \tag{1.2.19}$$

while the coefficient of second viscosity ζ vanishes.

The still higher approximation is represented by the so-called Burnett equations. The corrections of the order of $(Kn)^2$ in the expressions for the stress-tensor and heat-flux vector components include terms linear with respect to the second derivatives of gasdynamic variables and terms quadratic with respect to the first derivatives, the coefficients being determined by the potential of intermolecular interaction. The awkwardness of these equations is a good excuse for not reproducing them here; the interested reader is referred to [1.4,5]. It should be noted that the Burnett equations not only improve the accuracy of the Navier-Stokes equations, but also describe a number of new nonuniform, nonisothermal phenomenona, e.g., previously unknown types of convective flows [1.6,7].

Further pursuance of the Chapman-Enskog method with a view to taking into account the corrections of the order of $(Kn)^n$, $n \geqslant 3$, leads to equations containing n^{th}-order derivatives of the hydrodynamic variables and n^{th}-order nonlinearities. The task of constructing and solving such equations even for $n = 3$ becomes quite formidable.

Grad proposed a method allowing gasdynamic equations to be obtained, which correspond to high-order expansions in small Kn, but in a form more suitable for analytical and numerical calculations [1.8]. The basic variables adopted in this technique are the centered moments of the distribution function, i.e., the tensor components

$$M^{(n)}_{\alpha_1\ldots\alpha_k}(\mathbf{r},t) = \int \left(\prod_{k=1}^{n} V_{\alpha_k}\right) f\, d\mathbf{v} \quad , \quad \alpha_k = 1,2,3 \quad ; \quad 0 \leqslant n < \infty . \quad (1.2.20)$$

There are $(n+1)(n+2)/2$ independent moments of the nth order, and the total number of independent moments of 0th, 1st,...n^{th} order is equal to

$$\frac{1}{6}(n+1)(n+2)(n+3) \quad .$$

Since the centered first-order moments are identically equal to zero, the set (1.2.20) is complemented by three noncentral first-order moments: the three components of the gasdynamic velocity **u** (1.2.4). The gasdynamic variables and the components of the stress tensor and heat-flux vector are obviously moments themselves, or can be expressed in terms of moments:

$$n(\mathbf{r},t) = M^{(0)}(\mathbf{r},t) \quad ; \quad \Pi_{ij}(\mathbf{r},t) = mM^{(2)}_{ij}(\mathbf{r},t) \quad ,$$

$$q_i(\mathbf{r},t) = \frac{m}{2}\sum_{j=1}^{3} M^{(3)}_{ijj}(\mathbf{r},t) \quad . \quad (1.2.21)$$

All the moments in Grad's equations, including Π_{ij}, q_i, in contrast to the equations of Navier-Stokes, Burnett, etc., are considered as independent variables, rather than being expressed via the derivatives of gasdynamic variables. Consecutive calculation of the moments on the right-hand and left-hand sides of the Boltzmann equation proves the latter's equivalence to an infinite chain of equations in increasing order of moments: the continuity equation (1.1.3) links the zeroth and first order moments, the equation of motion (1.1.7) employs the moments from zeroth to second order, the energy balance equation (1.1.28) involves the moments of up to third order, and so on. A finite set of equations may be obtained by truncating the chain at some point, i.e., disregarding all the moments above the preset order.

The equations of the five-moment approximation with respect to n, **u**, and $p = \frac{1}{3}\Sigma_{j=1}^{3}\Pi_{jj}$ coincide with Euler's equations, not taking the heat flux and viscous stress into account. The corresponding ten-moment approximation for a two-component fluid (plasma) is, however, capable of describing dissipative phenomena such as the heat exchange between the components and intercomponent friction. Used most commonly are the thirteen-moment Grad equations: a set of thirteen equations in the physically meaningful independent variables n,

$\mathbf{u}$, Π_{ij}, $\mathbf{q}$. Taking into account the remaining seven moments of third order, M_{ijk} will result in the twenty-moment set of equations. Without writing them out explicitly, let us only observe that Grad's equations are quasilinear (linear with respect to the derivatives of the pertinent moments) and thus have a relatively simple structure. For this reason the 13-moment equations (or the 26-moment equations for a two-component plasma) are usually preferred to the Burnett equations [1.9].

Obtaining Grad's equations from the kinetic Boltzmann equation is based on expanding the distribution function in Hermite polynomials:

$$f = f^{(0)}\left(a^{(0)}H^{(0)} + a^{(1)}_{\alpha_1}H^{(1)}_{\alpha_1} + \ldots + a^{(n)}_{\alpha_1\ldots\alpha_n}H^{(n)}_{\alpha_1\ldots\alpha_n} + \ldots\right) , \quad (1.2.22)$$

where $f^{(0)}$ is the local Maxwellian distribution function; $H^{(n)}_{\alpha_1\ldots\alpha_n}$ are the Hermite polynomials in the three independent components of the dimensionless vector

$$\xi = \mathbf{v}/v_T \quad , \quad (1.2.23)$$

defined by

$$H^{(n)}_{\alpha_1\ldots\alpha_n} = (-1)^n \exp[\xi^2/2](\partial^n/\partial\xi_{\alpha_1}\ldots\partial\xi_{\alpha_n}) \exp(-\xi^2/2) \quad ,$$

$\alpha_k = 1,2,3$; note the summation convention in (1.2.22). The coefficients $a^{(n)}_{\alpha_1\ldots\alpha_n}$ are functions of coordinates and time and are expressed via the moments (1.2.20). If $Kn \ll 1$, then the nonequilibrium contribution to the random velocity component, which is ultimately responsible for the difference between f and $f^{(0)}$, is of the order of $v_T\lambda/L$, and hence [in accordance with (1.2.23)], the expansion (1.2.22) is essentially an expansion in powers of Kn. The more or less arbitrary truncation of the chain of equations in moments by considering a fixed number of moments usually results in that all the terms up to a certain order in Kn, but only some terms of higher order are accounted for. For example, the 13-moment and 20-moment Grad equations at small Kn are accurate to second order in Kn, and equivalent, to this order, to Burnett's equations. The third-order terms in Kn are accounted for only partly. For this reason an attempt to employ the 13-moment or 20-moment equations to explore the effects not described by the Burnett equations would be unjustified.

We see that the applicability of gasdynamic equations has a sound basis under the condition (1.2.16). In this case also corrective procedures are available for improving the Navier-Stokes approximation in fluid dynamics.

1.3 Limits of Applicability of the Gasdynamic Equations in Studying Shock-Wave Structure

In the present study the gasdynamic equations are consistently employed for calculating the structure of shock waves. Let us examine the validity of this approach. Here, we assume that the reader has mastered the basic concepts of shock waves (e.g., on the level of [1.2]); otherwise we suggest first reading the remaining part of this chapter.

The transport processes in a simple gas, made up of structureless molecules of the same kind, have only the characteristic length: the mean free path λ. Since the width L of the shock front is determined by the transport processes (and not by the external initial and boundary conditions), we have

$$L \sim \lambda \ , \qquad (1.3.1)$$

or

$$\mathrm{Kn} \sim 1 \ . \qquad (1.3.2)$$

Estimate (1.3.2) does not fit in with condition (1.2.16), which is necessary for deriving the gasdynamic equations from the Boltzmann equation. The violation of this condition in shock waves is quite logical: in the shock-front region the distribution function does not originate locally, in the neighborhood of a given point, as assumed in the expansions (1.2.17,22), but rather is shaped by the regions upstream and downstream of the front, where the gasdynamic variables assume highly different values. [The exception is the range of weak shock waves, where these values differ little. Then it is possible to introduce a small parameter ε (generally unavailable), which characterizes their difference (the "weakness" of the shock; $\varepsilon = M - 1$, M being the Mach number). The estimates (1.3.1,2) are then replaced by $L \sim \lambda/\varepsilon$ and $\mathrm{Kn} \sim \varepsilon$, and condition (1.2.16) is met.] This is directly emphasized by (1.3.1), according to which the molecules, without colliding, may come to a given point of the front from both upstream and downstream regions.

We see that no basis exists for employing the gasdynamic equations in this case: rigorous calculation of the structure of the gasdynamic shock wave can be carried out only within the framework of kinetic theory. This means getting a solution of the integrodifferential Boltzmann equation with a realistic potential of intermolecular interaction. In a unidimensional steady-state problem the sought-for distribution function will depend on the three arguments x, v_x and $v_\perp = (v_y^2 + v_z^2)^{1/2}$, x being the direction of propagation of the shock. This problem is so complicated that it defies even very rough analytic approximations. The most promising way of tackling it seems to be the Monte Carlo method; the calculations involved are extremely formidable even considering today's computational facilities.

The difficulties in getting straightforward solutions force one to seek detours from ad hoc principles. The most well-known of such approaches was developed independently by *Mott-Smith* [1.10] and *Tamm* [1.11]. It consists of postulating the distribution function at the front in the form of a superposition of equilibrium distribution functions downstream and upstream of the front, with the respective coefficients $\alpha(x)$ and $1-\alpha(x)$ switching from 0 to 1 in the front region. The profile $\alpha(x)$, which determines the structure of the front, is found by substituting the distribution function into the Boltzmann equation. Neither being an exact solution of the kinetic equation, nor even warranting an established degree of accuracy, the bimodal Mott-Smith distribution function nevertheless correctly renders the basic property of the actual distribution function at the front of the shock wave, which is shaped simultaneously by the flows ahead of the front and behind it. The structures of the gasdynamic shock waves calculated thus fit well with the experimental findings [1.12,13]. The quantitative agreement with experiment can be further improved by increasing the number of components which build up the superposition of the same form, or by adopting a different way of separating the particles into groups, each having its own distribution function [1.14,15].

The Navier-Stokes equations with the transport coefficients either calculated conventionally under the assumption (1.2.16) (Sect.1.2.2) or empirically, give a good description of the structure of relatively weak shock waves ($M \lesssim 2$). For stronger shocks these equations also supply solutions describing the structure of the front, although the calculated front width usually differs appreciably from the values obtained from experiment or Mott-Smith calculations. If condition (1.2.16) is not satisfied, the Burnett and Grad equations cannot provide a better approximation to improve on the Navier-Stokes description. For weak shock waves ($M \lesssim 2$) the solutions of the Burnett equations are no better (and those of Grad's equations poorer) fit with the experiment than the solutions of the Navier-Stokes equations. For strong shock waves ($M \gtrsim 2$) the Burnett and Grad equations have no solutions to represent the structure of the shock waves [1.16]; the higher approximations also cannot be used, since, for instance, the series (1.2.22) in Hermite polynomials, representing the distribution function in a strong shock wave, is divergent [1.17]. There is obviously nothing surprising in that the rather rough Navier-Stokes model is useful for at least the qualitative description of shock-wave structures, when the more elegant Burnett and Grad models prove entirely inadequate. The latter are constructed under certain assumptions regarding the form of the distribution function (Sect.1.2.2) and fail once the necessary conditions are not satisfied, whereas the Navier-Stokes equations can be derived phenomenologically, on the basis of laws of mechanics and thermodynamics ([1.2] and Sect.1.1). The inadequacy of the Navier-Stokes equations in describing the actual structure of strong shock waves can also be attributed to the inappropriate choice of the transport coefficients, computed or measured under condition (1.2.16), serving for the phenomenological parameters η, ζ, κ in the Navier-Stokes model. The agreement between the Navier-Stokes calculations and experiment could be improved considerably by carefully choosing these coefficients (considered as functions of density and temperature) [1.18]. Nevertheless, the description of the gasdynamic shock structures is essentially a kinetic problem, and effective improvements can be made only within the framework of kinetic theory.

These problems, however, fall beyond the scope of this book which is dedicated to quite a different task. We intend to study the shock waves in a plasma, a fluid highly different from a simple gas in that it comprises two (or three) components, one of which (the electrons) is much lighter than the other(s) (atoms and/or ions). The particles in a plasma may possess inner degrees of freedom (excitation and relaxation of atoms), engage in reversible reactions (ionization and recombination), interact with the external electromagnetic field (Joule heating). Consequently, instead of one characteristic length λ we shall be dealing with a hierarchy of scales, each corresponding to a separate physical process, among which λ is usually the smallest. By virtue of this, the shock profiles in a plasma appear as successive or isolated layers, differing in the scales of nonuniformity.
The spatial scales, which characterize these layers (the regions of the temperature and ionization relaxation, precursor ionization, Joule and heat-conductive heating, radiative cooling) are large compared to λ, so that condition (1.2.16) is satisifed, thus justifying the use of the Navier-Stokes equations.

One of these layers (and, in most cases, the thinnest) may be the shock transition with the characteristic length λ, which is termed a gasdynamic or viscous jump (or subshock). Its structure, not described by the Navier-Stokes equations, is not important for us in view of the smallness of its width on the scale of other physical processes. They determine the structure and the dynamic behavior of shock waves in a plasma. Usually it is permissible either to treat the viscous jump as an infinitely narrow discontinuity, or to smooth it out by formal application of the Navier-Stokes equations: the experimentally measured parameters of the shock waves will not be affected. Even when the length λ dominates in the overall structure and can be detected (as for a strong shock in a completely ionized magnetized plasma, Sect.3.4.2), the considerable increase (by several orders of magnitude) of this length within the limits of the front violates condition (1.2.16) in only a narrow forefront region, not affecting the agreement between the Navier-Stokes calculations and experiment.

The asymptotic approach of the Navier-Stokes shock profile to the exact solution of the model kinetic equation was also found for gasdynamic shock waves in the high-pressure region [1.19]. Here, however, the path length across the entire front varies within an order of magnitude, and the discrepancy between the two solutions in the low-pressure region leads to a considerable difference in the estimated widths of the front.

The arguments developed above are confirmed not only by the good agreement between the gasdynamic (and magnetohydrodynamic) calculations of the shock-wave structures in a dense plasma and the relevant experimental results, but also by numerous examples of successful numeric simulations of diverse phenomena in collisive plasmas, where the shocks play an important role (e.g., the dynamics of Z and θ pinches, the compression of targets in laser thermonuclear fusion). Starting in Chap.2 we shall therefore use these equations freely, without questioning their applicability. Let it be emphasized that further treatment of the gasdynamic shock waves in Sect.1.8 on the basis of the Navier-Stokes equations is entirely modelwise. We do not aim at giving a description of the real gasdynamic shock-wave structures, but rather wish to illustrate the computational techniques to be used in the following chapters.

1.4 Linear and Nonlinear Waves

1.4.1 Linear and Sonic Waves

Small perturbations of pressure, density and velocity propagate in a fluid in the form of sonic waves. It follows from the equation of energy balance that they travel through the fluid at the speed of sound. Indeed, let us consider the small perturbations p', ρ' and $u \ll a$ in an initially homogeneous and isotropic gas, the equilibrium values of pressure, density and velocity being p_0, ρ_0 and $\mathbf{u}=0$, respectively. We further assume (on the grounds given below), that p'/p_0, ρ'/ρ_0, $\mathbf{u}/a$ are of the first order of smallness, while the order of smallness of the dissipative terms in (1.1.26) and the change in the entropy due to the deviation of p, ρ and $\mathbf{u}$ from their equilibrium values according to (1.1.29), are higher. Then, in a first approximation, the process may be considered adiabatic, the small variation in the pressure being uniquely linked with the small variation in the density via the equation of state

$$p = p(\rho) \quad , \; S = \text{const} \quad . \tag{1.4.1}$$

The speed of sound is given by

$$a^2 = \left(\frac{\partial p}{\partial \rho}\right)_S \quad . \tag{1.4.2}$$

Retaining only the first nonvanishing terms, we get

$$a^2 = a_0^2 = \left(\frac{\partial p(\rho)}{\partial \rho}\right)_S \Bigg|_{\rho=\rho_0} \quad . \tag{1.4.3}$$

Up to the first nonvanishing terms (1.4.3) is equivalent to the adiabatic equation (1.1.13), i.e.,

$$\frac{dp'}{dt} = a^2 \frac{d\rho'}{dt} \quad . \tag{1.4.4}$$

In this approximation the equations for perturbations of density, velocity and pressure are represented by the continuity equation and Euler's equation. Linearizing (1.1.3,8), we obtain the following equations for small perturbations ρ', $\mathbf{p}'$, $\mathbf{u}$:

$$\frac{\partial \rho'}{\partial t} + \rho_0 \operatorname{div} \mathbf{u} = 0 \quad , \tag{1.4.5}$$

$$\frac{\partial u}{\partial t} = -\frac{1}{\rho_0} \operatorname{grad} p \quad . \tag{1.4.6}$$

The Helmholtz theorem in vector calculus states that the field of velocities $\mathbf{u}$ is expressible as the sum of an irrotational (potential) field with the divergent-free (solenoidal) field:

$$\mathbf{u} = \mathbf{u}_1 + \mathbf{u}_2 \quad ; \quad \operatorname{curl} \mathbf{u}_1 = 0 \quad , \quad \operatorname{div} \mathbf{u}_2 = 0 \quad .$$

It is clear that (1.4.5,6) relate the perturbations p' and ρ' to the potential component of velocity only. For waves with $p' \neq 0$ (sonic waves) we may assume $\mathbf{u} = \mathbf{u}_1$ and introduce the velocity potential φ by

$$\mathbf{u} = \operatorname{grad}\varphi \quad . \tag{1.4.7}$$

From (1.4.6) we get

$$p = -\rho_0 \frac{\partial\varphi}{\partial t} \quad . \tag{1.4.8}$$

In the same order, and taking (1.4.4) into account, we obtain from (1.4.8)

$$\rho = -\frac{p_0}{a_0^2}\frac{\partial\varphi}{\partial t} \quad . \tag{1.4.9}$$

Substituting (1.4.7,9) into (1.4.5) yields the wave equation

$$\frac{\partial^2\varphi}{\partial t^2} - a_0^2\Delta\varphi = 0 \quad , \tag{1.4.10}$$

whose solutions may be viewed as a superposition of traveling plane waves, propagating with the speed a_0. For plane monochromatic waves of the form $\varphi = \exp[i(kx - \omega t)]$, a dispersion equation evolves from (1.4.10):

$$\omega^2 = a_0^2 k^2 \quad . \tag{1.4.11}$$

Using (1.4.5,6,11) we proceed to find the order of smallness of the perturbed amplitudes of the pressure p_m' and the density ρ_m', expressing it via the Mach number of the sonic wave, $M = u_m/a_0$ (u_m being the amplitude of the velocity perturbation):

$$\frac{p_m}{p_0} = \gamma M \quad , \quad \frac{\rho_m}{\rho_0} = M \quad . \tag{1.4.12}$$

The equations can be linearized only if

$$M \ll 1 \quad . \tag{1.4.13}$$

Neglecting the dissipative terms on the right-hand sides of (1.1.26,29) is based on the smallness of another dimensionless parameter, the Knudsen number, see (1.2.15), which here can be defined by $Kn = k\lambda$, λ being the path length. The condition of applicability of the gasdynamic equations (1.2.16) assures the smallness of the dissipative terms: being of the order $M \cdot Kn$, they are small in Kn compared with the terms retained.

Note, that the condition of applicability of (1.4.10) does not really reduce to the requirement of (1.2.16,1.4.13) being simultaneously satisfied, but it is more restrictive, see inequality (1.5.21).

The dispersion equation (1.4.11) represents the first term in the expansion of $\omega(k)$ with respect to Kn. The dissipations due to viscosity and heat conduction require corrections to account for the damping

$$\omega = a_0 k - i\mu k^2 \quad , \tag{1.4.14}$$

where

$$\mu = \frac{1}{2\rho_0}\left[\frac{4}{3}\eta + \zeta + \kappa\left(\frac{1}{c_V} - \frac{1}{c_p}\right)\right] \quad . \tag{1.4.15}$$

Condition (1.2.16) ensures the relative smallness of damping of the sonic wave: the length of damping, $L \sim \bar{\lambda}/Kn \gg \bar{\lambda}$, $\bar{\lambda} = 2\pi/k$ being the sonic wavelength.

Note that (1.4.5,6) also describe linear waves which cannot be reduced to sonic waves. Their solutions obviously are the arbitrary time-constant perturbations of the rotational velocity component or the entropy (and density) perturbations at constant pressure. Neglecting dissipation, such perturbations (e.g., in the unidimensional case, the profiles $u_y = u_y(x)$, $u_z = u_z(x)$, $S = S(x)$ at $p(x) = \text{const}$) are stationary, i.e. the phase velocity of the corresponding wave is zero. Expansion of the dispersion equation in powers of Kn, similar to (1.4.14), for the rotational and entropy disturbances in a nonmoving medium starts with the second term; it accounts for the diffusive smearing of the profile.

1.4.2 Nonlinear Plane Waves

Consider the exact planar solutions for the nondissipative flow, described by the Euler equations (1.1.3,8,13), which for the unidimensional flow ($\mathbf{u} = \{u,0,0\}$) have the form

$$\frac{d\rho}{dt} + \rho\frac{\partial u}{\partial x} = 0 \quad , \tag{1.4.16}$$

$$\frac{du}{dt} + \frac{1}{\rho}\frac{dp}{dx} = 0 \quad , \tag{1.4.17}$$

$$\frac{dS}{dt} = 0 \quad , \tag{1.4.18}$$

recalling the total time derivative

$$\frac{d}{dt} \equiv \frac{\partial}{\partial t} + u\frac{\partial}{\partial x} \quad .$$

As hyperbolic equations [1.20], Eqs. (1.4.16-18) must have characteristics, which cannot be directed along the x axis, and thus can be of the form $x = x(t)$. The characteristics are curves in the plane (t,x), along which the corresponding gasdynamic variables remain constant.

Equation (1.4.18) by itself describes a characteristic with the characteristic velocity u, namely

$$\frac{dS}{dt} = 0 \quad \text{along} \quad C_0: \frac{dx}{dt} = u \quad . \tag{1.4.18a}$$

It follows that in the adiabatic flow the flow lines are the characteristics per se, namely C_0.

To reduce the remaining two equations to the characteristic form, let us take advantage of the adiabatic equation (1.4.18), rewriting it by virtue of $\rho = \rho(p,S)$ in the form

$$\frac{d\rho}{dt} = \left(\frac{\partial\rho}{\partial P}\right)_S \frac{dp}{dt} + \left(\frac{\partial\rho}{\partial S}\right)_p \frac{dS}{dt} = \frac{1}{a^2}\frac{dp}{dt} \quad . \tag{1.4.19}$$

In (1.4.16) expressing $d\rho/dt$ via dp/dt according to (1.4.19), we get

$$\frac{\partial p}{\partial t} + u\frac{\partial p}{\partial x} + a^2\rho\frac{\partial u}{\partial x} = 0 \quad .$$

Adding this equation to and subtracting from (1.4.17) yields the two equations in the characteristic form:

$$\left[\frac{\partial p}{\partial t} + (u + a)\frac{\partial p}{\partial x}\right] + \rho a\left[\frac{\partial u}{\partial t} + (u + a)\frac{\partial u}{\partial x}\right] = 0 \tag{1.4.20}$$

$$\left[\frac{\partial p}{\partial t} + (u - a)\frac{\partial p}{\partial x}\right] - \rho a\left[\frac{\partial u}{\partial t} + (u - a)\frac{\partial u}{\partial x}\right] = 0 \quad . \tag{1.4.21}$$

Thus (1.4.16-18), expressed along the respective characteristics, assume the form[1]

$$\frac{dS}{dT} = 0 \qquad \text{along } C_0 \ ; \ \frac{dx}{dt} = u \tag{1.4.22}$$

$$\frac{dp}{dt} + \rho a\frac{du}{dt} = 0 \quad \text{along } C_+ \ ; \ \frac{dx}{dt} = u + a \tag{1.4.23}$$

$$\frac{dp}{dt} - \rho a\frac{du}{dt} = 0 \quad \text{along } C_- \ : \ \frac{dx}{dt} = u - a \quad . \tag{1.4.24}$$

---

1 If the rotational components of velocity, $u_y(x,t)$ and $u_z(x,t)$, are nonzero (which amounts to taking the nonunidimensionality of the plane flow into account), (1.1.8) yields additional equations of the form (1.4.18 or 20) with S replaced by u_y or u_z. They could have been written down directly by analogy with (1.4.18,20), knowing (Sect.1.4.1) that the rotational and the entropy perturbations travel along with the flow

As follows from (1.4.23,24), the perturbations in a gas propagate along the characteristics C_+ and C_- (respectively, parallel and antiparallel to the direction of the x axis) with the velocities $u+a$ and $u-a$ in the laboratory frame of reference.

Linearizing (1.4.22-24) brings us back to the equations of sonic waves, discussed in Sect.1.4.1. Indeed, in this case the equations are

$$\frac{dS}{dt} = 0 \qquad \text{along } C_0 : \frac{dx}{dt} = 0 \tag{1.4.25}$$

$$\frac{dp}{dt} + \rho_0 a_0 \frac{du}{dt} = 0 \qquad \text{along } C_+ : \frac{dx}{dt} = a_0 \tag{1.4.26}$$

$$\frac{dp}{dt} - \rho_0 a_0 \frac{du}{dt} = 0 \qquad \text{along } C_- : \frac{dx}{dt} = -a_0 \quad , \tag{1.4.27}$$

whence

$$p + \rho_0 a_0 u = f_1(x - a_0 t) \tag{1.4.28}$$

$$p - \rho_0 a_0 u = f_2(x - a_0 t) \quad . \tag{1.4.29}$$

We see that in the acoustic approximation the perturbations of the gasdynamic variables travel along the characteristics at the constant speed of sound a_0. In the general case, the small perturbation (which, incidentally, is not too arbitrary) propagates along all three characteristics C_0, C_+, C_-, which come out from the given point of the plane (x,t). It turns out, however, that the perturbation can be decomposed into components which travel along characteristics of the same family. [Since the characteristic then separates the perturbed part of the (x,t) plane from the unperturbed part, the smoothness condition of the flow variables on the characteristic may be violated. The inverse is also true: the locus of discontinuities of the spatial derivatives of any of the hydrodynamic variables (weak jump) in the (x,t) plane must coincide with one of the characteristics [1.20,21].] This circumstance leads to the most important solutions of (1.4.16-18): simple traveling waves, representing the extension of linearized acoustic waves to the general case of waves of arbitrary amplitude.

1.4.3 The Riemann Invariants

In the adiabatic flow the entropy is constant throughout, so that (1.4.22) is an identity, while the pressure p and the speed of sound a are some functions of ρ according to the equation of state $p = p(\rho)$. For an isentropic flow we can introduce new functions, called the *Riemann invariants*:

$$I_+ = u + \int \frac{dp}{\rho a} \quad ; \quad I_- = u - \int \frac{dp}{\rho a} \quad . \tag{1.4.30}$$

With the aid of these functions the characteristic equations (1.4.20,21) can be presented in the form

$$\left[\frac{\partial}{\partial t} + (u + a)\frac{\partial}{\partial x}\right] I_+ = 0 \quad , \tag{1.4.31}$$

$$\left[\frac{\partial}{\partial t} + (u - a)\frac{\partial}{\partial x}\right] I_- = 0 \quad , \tag{1.4.32}$$

or, alternatively,

$$I_+ = u + \int \frac{dp}{\rho a} = \text{const} \quad \text{along } C_+ : \frac{dx}{dt} = u + a \tag{1.4.33}$$

$$I_- = u - \int \frac{dp}{\rho a} = \text{const} \quad \text{along } C_- : \frac{dx}{dt} = u - a \quad . \tag{1.4.34}$$

We see that along the characteristics C_+, C_- in the (x,t) plane the respective variables I_+, I_- remain constant. Consequently, the small perturbation I_+ travels along C_+, while the small perturbation I_- travels along C_-.

The equation of state for an ideal gas with constant heat capacity and an adiabatic exponent $\gamma = c_p/c_V$ has the simple form of (1.1.23). The Riemann invariants then have the form (up to an arbitrary additive integration constant):

$$I_+ = u + \frac{2}{\gamma - 1} a \quad , \quad I_- = u - \frac{2}{\gamma - 1} a \quad . \tag{1.4.35}$$

The Riemann invariants I_+, I_- may be considered the new flow variables. Since they are uniquely related to u and a, by solving (1.4.33,34) with respect to u and a, we may revert from I_+, I_- to u, a. From (1.4.35), for the ideal gas we have

$$u = \frac{I_+ + I_-}{2} \quad ; \quad a = \frac{\gamma - 1}{4} (I_+ - I_-) \quad . \tag{1.4.36}$$

The characteristic equations for an isentropic gas flow take the form

$$C_+ : \frac{dx}{dt} = \frac{\gamma + 1}{4} I_+ + \frac{3 - \gamma}{4} I_- \quad , \tag{1.4.37}$$

$$C_- : \frac{dx}{dt} = \frac{3 - \gamma}{4} I_+ + \frac{\gamma + 1}{4} I_- \quad . \tag{1.4.38}$$

In the general case of nonisentropic flow (1.4.33,34) become invalid, $dp/\rho a$ no longer being a total differential. It is always possible, however, to single out the perturbations which travel along their own characteristics.

According to (1.4.22-24), these are the entropy perturbation traveling along C_0, and the perturbations of the kind $\delta u \pm \delta p/\rho a$, which travel along C_+ and C_-.

1.4.4 Simple Waves

When the Riemann invariant I_+ is constant along the characteristic C_+, the slope of the characteristic at a given point of the (x,t) plane is determined by I_-. Conversely, the slope of C_- is determined by I_+.

If one of the Riemann invariants for the isentropic flow remains constant throughout (for all x,t), this implies the possible existence of a wave traveling in the same direction. In the acoustic approximation (1.4.28,29) directly indicate that for the wave traveling out $I_- = \text{const}$, and $f_2 = 0$. This property, however, is more universal: one of the Riemann invariants remains constant for a traveling wave having an arbitrary (not necessarily small) amplitude.

Assume for definiteness $I_- = \text{const}$. Since I_+ is constant along C_+, it follows from (1.4.38) that the characteristics C_+ are represented by a family of straight lines in the (x,t) plane. The equation for the characteristics C_0 can be written as

$$x = f_+(I_+,I_-)t + \varphi(I_+) \quad , \tag{1.4.39}$$

where $f_+(I_+,I_-) = \text{const}$ is a known thermodynamic function, and $\varphi(I_+)$ is an arbitrary integration constant. Equation (1.4.39) together with the condition $I_-(x,t) = \text{const}$ furnishes a general solution of the equations for simple waves of an isentropic flow, implicitly defining $I_+(x,t)$.

Let us now obtain the solution for the simple wave in a different—and occasionally more convenient—form, the variables being the velocity of the gas, u, and the speed of sound, a. Note that for isentropic flow the condition

$$I_- = u - \int \frac{dp}{\rho a} = \text{const} \tag{1.4.40}$$

implies that the speed of sound and the density are single-valued functions of u, and do not depend explicitly on the (independent) variables x and t, i.e., $\rho = \rho(u)$, $a = a(u)$. Then from (1.4.16,17) it follows that

$$a(u)\,\frac{d\rho}{du} = \rho(u) \quad . \tag{1.4.41}$$

Substituting (1.4.41) into (1.4.17) and making use of the condition $a^2(\rho) = (\partial p/\partial\rho)_{S=\text{const}}$, we get the following equation for a simple wave traveling in the x-direction:

$$\frac{\partial u}{\partial t} + [u + a(u)]\frac{\partial u}{\partial x} = 0 \quad . \tag{1.4.42}$$

The general solution of (1.4.42) is nothing else but the characteristic C_+, i.e., (1.4.23 or 39). The implicit form of the solution of (1.4.42) is

$$x = [u + a(u)]t + f(u) \quad , \tag{1.4.43}$$

where f(u) is an arbitrary function, and a(u) is found from (1.4.40).

The solution (1.4.43) was originally obtained by Riemann in 1860 [1.2]. The equivalence of Riemann's simple-wave solutions (1.4.43,39) is self-evident. The sense in which the solution represents a traveling wave is also clear from (1.4.43): a point in which the velocity u has a given value travels in space at a constant speed.

From (1.1.23,1.4.2) for adiabatic flow of ideal gas we have

$$a^2 = a_0^2\left(\frac{\rho}{\rho_0}\right)^{\gamma-1} \quad , \tag{1.4.44}$$

where $a_0 = a(\rho_0)$ is the speed of sound in the linear (acoustic) approximation.

Substituting (1.4.44) into (1.4.41), we get

$$a(u) = a_0 + \frac{\gamma - 1}{2} u \quad . \tag{1.4.45}$$

Thus, the solution (1.4.43) now has the form

$$x = \left(a_0 + \frac{\gamma + 1}{2} u\right)t + f(u) \quad . \tag{1.4.46}$$

For the simple wave traveling in the (-x) direction ($I_+ =$ const, characteristic C_-) the respective solution is given by

$$x = \left(-a_0 + \frac{\gamma + 1}{2} u\right)t + f(u) \quad . \tag{1.4.47}$$

It is often convenient to invert (1.4.46,47), presenting the solution in the form

$$u = F\left[x - \left(\pm a_0 + \frac{\gamma + 1}{2} u\right)t\right] \quad , \tag{1.4.48}$$

where F is an arbitrary function defined by the form of the simple wave at $t = 0$:

$$u(x,0) = F(x) \quad . \tag{1.4.49}$$

The characteristics allow some important conclusions regarding the nature of the flow to be drawn directly. For example, the linearity of the characteristics of simple waves implies that a simple wave borders on the region of constant flow. Indeed, in the region of constant flow the two Riemann invariants I_+, I_- are constants, and the two families of characteristics C_+, C_- are linear. One of the characteristics C_+ separates the region of constant flow from the simple-wave region, and therefore the characteristics C_+ do not pass from one region to the other, whereas the characteristic C_- with the constant value of I_- cross the border line. This interpretation proves the validity of the above statement.

The boundary in the (x,t) plane between two regions of the same plane where the flow is described by two different analytical solutions of the equations of motion is a locus of weak discontinuity, and thus also a characteristic.

1.4.5 Expansion Waves

Let us start our study of the flow in a simple wave by considering the motion of a gas in a tube, caused by the movement of a piston at the end of the tube.

Consider the gas initially at rest, the pressure, density and speed of sound being constant: p_0, ρ_0, a_0. The gas in the tube at rest ($u=0$) occupies the region $x>0$, limited, on the left, by the piston, which at $t=0$ starts moving to the left according to the law $x=X(t)$. At $t=0$ the gas to the right of the piston ($x>0$) is at rest, and the paths of all particles, including those in the immediate vicinity of the piston, originate on the x axis in the homogeneous region. The entropy is constant along each characteristic C_0 in accordance with (1.4.25), and since the initial value of entropy is the same for all the characteristics C_0, the entropy is constant throughout the entire flow (isentropic flow).

The value of dx/dt is smaller on the characteristics C_- than on the characteristics C_0, and all the characteristics C_-, originating on the x axis, carry along the constant value of the Riemann invariant $I_-(x,t)$, defined by the initial values of the velocity of the flow and the speed of sound at $t=0$. In other words, for an ideal gas along all the characteristics C_- we have

$$I_- = u - \frac{2}{\gamma - 1} a = - \frac{2}{\gamma - 1} a_0 \quad . \tag{1.4.50}$$

Since the initial value of $I_-(x,t) = I_-(x,0)$ is the same for all the characteristics C_-, the Riemann invariant (1.4.50) is constant throughout. Hence it follows that the flow in front of the piston is described by a simple wave; the speed of sound can be found from (1.4.50):

$$a(u) = a_0 + \frac{\gamma - 1}{2} u \quad . \tag{1.4.51}$$

The characteristics C_+, originating on the x axis ($x>0$), carry along the initial velocity $u=0$ and are thus parallel to one another: $dx/dt=a_0$, down to the characteristic C_+^0, which starts at $x=0$ and is defined by

$$x = a_0 t \quad . \tag{1.4.52}$$

Thus the entire region to the right of C_+^0 is covered by a grid of parallel straight-line characteristics C_+ and C_-.

For the other characteristics C_+ we have

$$I_+ = u + \frac{2}{\gamma - 1} a = \text{const} \quad \text{along } C_+ \; ; \frac{dx}{dt} = u + a \quad . \tag{1.4.53}$$

Hence, taking (1.4.51) into account,

$$u = \text{const} \quad \text{along } C_+ : \frac{dx}{dt} = a_0 + \frac{\gamma + 1}{2} u \quad . \tag{1.4.54}$$

The velocity of the gas at the edge of the piston coincides with the velocity of the piston $-V(t) = \dot{X}(t)$ and has a negative value (the piston moves to the left; the simple expansion wave travels to the right). Accordingly, the speed of sound $a(u)$, as well as the pressure and the density near the piston, are lower than the initial values. The value of u on each characteristic C_+, being constant, is defined by the intercept of the characteristic with the path of the piston. Hence, we have the boundary condition

$$u = -V(t) \quad \text{at} \quad x = X(t) \quad . \tag{1.4.55}$$

The characteristic C_+ which intersects with the piston path at $t=\tau$ is defined by

$$x = X(\tau) + \left[a_0 - \frac{\gamma + 1}{2} V(\tau)\right](t - \tau) \quad . \tag{1.4.56}$$

The instant at which the characteristic intercepts with the path of the piston $\tau(x,t)$ is implicitly defined by (1.4.56). Ultimately, the complete solution for the expansion wave is

$$\tau = \tau(x,t) \; , \quad u = -V(\tau) \; , \quad a = a_0 - \frac{\gamma - 1}{2} V(\tau) \quad . \tag{1.4.57}$$

Since the piston is moving to the left only, and its motion is either uniform or accelerated, the characteristics C_+ outgoing from the piston path are either divergent or parallel, with the inclination

$$\frac{dx}{dt} = a_0 - \frac{\gamma + 1}{2} V(t) \quad . \tag{1.4.58}$$

An important particular case of the expansion waves is the centered expansion wave, which occurs when at $t=0$ the piston abruptly starts moving to

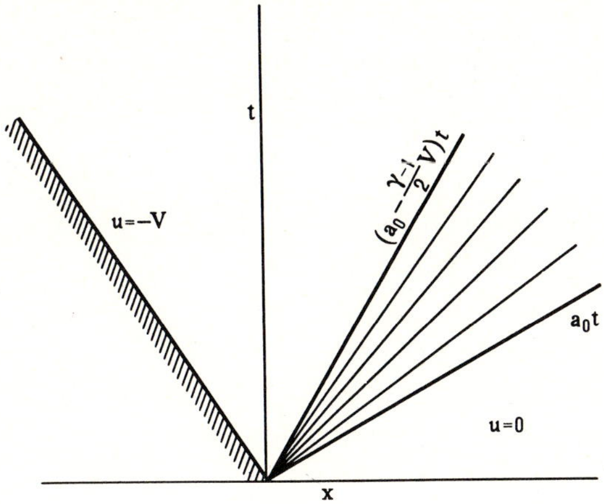

Fig.1.1. Fanning characteristics C_+ in a centered expansion wave. At $t=0$ the piston starts moving to the left with velocity V

the left, having instantaneously assumed the velocity -V. In this case, according to (1.4.46), all the characteristics originate in the point $x=0$, $t=0$ and are defined by

$$x = [u + a(u)]t \quad , \tag{1.4.59}$$

where all the values in the range $(-V \leqslant u \leqslant 0)$ are achieved at the origin of the coordinate system instantaneously. Each value of u defines a characteristic of the "fan", see Fig.1.1. Substituting a(u) from (1.4.51) into (1.4.59), we obtain the equation for the fan-shaped family of characteristics in the form

$$x = \left(a_0 + \frac{\gamma + 1}{2} u\right)t \quad . \tag{1.4.60}$$

Solving (1.4.51,60) with respect to u and a, we get the expressions for the velocity of the flow and the speed of sound in the region within the fan of characteristics. The complete solution is given by

$$u = 0 \quad , \quad a = a_0 \quad \text{at} \quad x/t > a_0 \quad ; \tag{1.4.61}$$

$$u = \frac{2}{\gamma + 1}\left(\frac{x}{t} - a_0\right) \quad ; \quad a = a_0\left(\frac{2}{\gamma + 1} + \frac{\gamma - 1}{\gamma + 1}\frac{x}{a_0 t}\right)$$
$$\text{at} \quad a_0 - \frac{\gamma - 1}{2}V < x/t < a_0 \quad ; \tag{1.4.62}$$

$$u = -V \quad , \quad a = a_0 - \frac{\gamma - 1}{2}V$$
$$\text{at} \quad x/t < a_0 - \frac{\gamma - 1}{2}V \quad . \tag{1.4.63}$$

The solutions (1.4.61-63) describe, respectively, the region of unperturbed gas ahead of the expansion wave, the profile of the expansion wave itself, and the region of the constant flow next to the piston. The head of the centered expansion wave travels at the speed of sound (for the unperturbed gas) a_0, while the tail flows with the gas at the speed $-V$. The pressure and density profiles in the centered expansion wave can easily be found with the aid of (1.1.23,1.4.44).

The centered expansion wave is formed if a density jump initially separates two homogeneous masses of gas in a tube. Suppose, for instance, that a tube is filled with a gas which has different densities on either side of a partition $\rho_1 < \rho_0$. As soon as the partition is removed, a centered expansion wave starts running towards the higher-density region.

1.5 Origins of Discontinuities

1.5.1 Profile Distortion of a Running Wave

We have considered the problem of simple waves: the expansion wave, traveling towards the higher-density regions and formed by the density jump or by instantaneously pulling the piston out of the tube (the centered expansion wave, Fig.1.1), or the expansion wave which is formed when the piston is drawn out of the tube according to the prescribed law: $x = X(t)$.

It might seem that the same solution would equally apply when the piston is pushed into the tube. We shall see, however, that this is not true. Let us try to set up a formal solution: just replace $-V$ by V in the formulas of the preceding section, because the piston is now moving to the right. As the piston is driven into the tube with velocity V (or, which is the same, a flow arises from the initial discontinuity, represented by a step in density and velocity), the head of the wave travels in the gas with the speed of sound a_0 along the characteristic $x = a_0 t$, while the piston abuts on the region of constant flow, in which the velocity of the gas is governed by the velocity of the piston V, and the speed of sound, in accordance with (1.4.63), is $a = a_0 + \frac{1}{2}(\gamma - 1)V$. The velocity of flow between head and tail of the wave assumes the distribution (1.4.62). It appears that the tail of the wave travels faster than its head. In the (x,t) plane this situation is depicted by an "upside-down" characteristics fan: the characteristics in the region corresponding to the simple wave intersect with one another. The corresponding

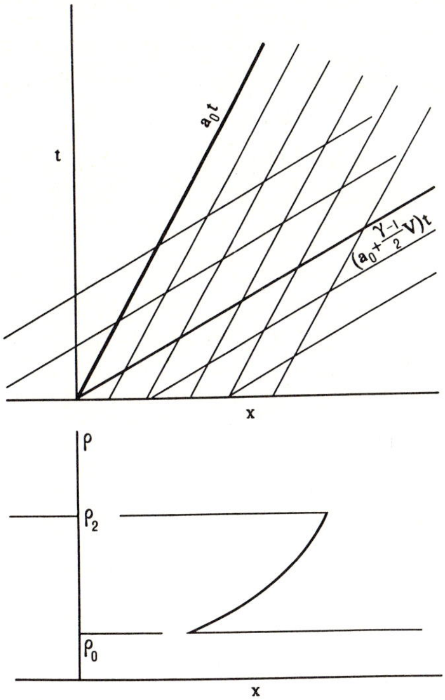

Fig.1.2. Characteristics C_+ and the density profile in a "centered compression wave", arising when the piston is instantaneously accelerated to the right with velocity V at t = 0

profile of density (Fig.1.2) has a kink, which indicates that the solution is ambiguous and lacks physical sense. A similar ambiguity arises when the piston is gradually driven into the tube.

The distortion of the profile, resulting in discontinuities, is a general property of the running wave. In contrast to the linear approximation, the velocity of a point of the wave profile (to be specific, running to the right), equals to

$$u + a(u) \quad , \tag{1.5.1}$$

is composed of the velocity of perturbation (the speed of sound a_0) and the velocity of the gas itself u. Since this velocity depends on the density, it will differ for different points of the profile, and so the profile will gradually change its shape. Using (1.4.41) we can easily check that

$$\frac{d(u + a)}{d\rho} > 0 \quad . \tag{1.5.2}$$

Hence it follows that the velocity of a given point of the profile is greater, the higher the density. The compression portions of the wave gain on the others, distorting the profile, until it finally becomes so bent up that the curve, say, $\rho(x)$, becomes ambiguous at some t. The ambiguities indicate breaks; the wave ceases to be a simple wave. The expansion waves do not break, since the density, as we have seen, exhibits a monotonic rise in the direction of propagation of the wave.

The point of discontinuity (the origin of the shock wave) is easily determined by simultaneously solving two trivial equations

$$\left(\frac{\partial x}{\partial u}\right)_t = 0 \quad ; \quad \left(\frac{\partial^2 x}{\partial u^2}\right)_t = 0 \quad , \tag{1.5.3}$$

which define the condition of self-overlap for u(x). For an ideal gas, using (1.4.46), we get

$$\begin{aligned} t &= -\frac{2}{\gamma + 1} f'(u) \quad , \\ f''(u) &= 0 \quad . \end{aligned} \tag{1.5.4}$$

1.5.2 Breakdown of the Sonic Wave Front

Let us now take advantage of the results just obtained and investigate some properties of sonic waves in a second-order approximation with respect to the Mach number. In the linear (acoustic) approximation all the points of the wave profile travel at the same speed a_0. In the next approximation the speed v of a point of the wave profile is given by

$$v = a_0 + \frac{\partial u}{\partial \rho} (\rho_0)\rho \quad . \tag{1.5.5}$$

For the ideal gas then follows the (exact) relation, see (1.4.46),

$$v = a_0 + \frac{\gamma + 1}{2} u \quad . \tag{1.5.6}$$

Thus the profile of the sonic wave distorts, which is manifested in a number of ways. First, the steepening of the profile may result in breaks in each period of the wave, so that in the long run the initially sine wave will assume a saw-tooth shape. Secondly, the steepening of the profile, while preserving the periodicity of the wave, changes its harmonic composition. The initially monochromatic wave with frequency ω exhibits a build-up of harmonics, the upper overtones $n\omega$ with large n gaining strength as the profile becomes steeper. The energy of the fundamental frequency is constantly pumped into the higher harmonics. As the damping of sound is proportional to the squared frequency, this results in additional attenuation of the wave. The steepening of the front will therefore go on until checked by the dissipative processes. Consequently, the profile of the wave will ultimately depend on the interrelation between the nonlinear and the dissipative effects and the initial intensity of the wave. If the wave amplitude is sufficiently large, the nonlinear effects dominate and are capable of producing shocks. Otherwise the wave will fade out before the nonlinear effects gain strength.

Let us consider the evolution of a plane harmonic sonic wave, generated in the ideal gas by the plane oscillator at $x = 0$, i.e.,

$$u = u_0 \sin\omega t \quad \text{at} \quad x = 0 \quad . \tag{1.5.7}$$

From (1.4.46) with the boundary condition (1.5.7) we get the following solution:

$$x = \left(a_0 + \frac{\gamma + 1}{2} u\right)t - \frac{\lambda}{2\pi}\left(1 + \frac{\gamma + 1}{2} \frac{u}{a_0}\right) \text{arc sin} \frac{u}{u_0} \quad , \tag{1.5.8}$$

where $\lambda = 2\pi/k = 2\pi a_0/\omega$ is the sonic wavelength.

The instant t_s and location x_s of the break in the sonic wave are defined by

$$\left(\frac{\partial x}{\partial u}\right)_{u=0} = 0 \quad ; \quad x = a_0 t \quad , \tag{1.5.9}$$

whence

$$t_s = \frac{\lambda}{u_0} \frac{1}{\pi(\gamma + 1)} \quad ; \quad x_s = \frac{a_0}{u_0} \frac{\lambda}{\pi(v + 1)} \quad . \tag{1.5.10}$$

The jump amplitude u_s is found from the transcendent equation (1.5.8) at $t = t_s$, $x = x_s$.

Let us estimate the lowest amplitude of the sonic wave necessary for the break to occur. Damping taken into account, the amplitude of the wave fades out as $u_0 \exp(-\mu t)$, where the damping decrement μ is given by (1.4.15). Now (1.5.10) for the time of break becomes

$$t_s = \frac{\lambda}{u_0 \pi(\gamma + 1)} \exp(\mu t_s) \quad . \tag{1.5.11}$$

The highest value of μ for which (1.5.11) still has a solution is $\mu = 1/t_s = \pi u_0(\gamma + 1)(\lambda e)^{-1}$ ($e = 2.72\ldots$). Hence the lowest oscillation amplitude of the velocity in the sonic wave for the break to occur is

$$u_0 = \frac{\mu e \lambda}{\pi(\gamma + 1)} \quad . \tag{1.5.12}$$

Estimating $\mu \sim \nu/\lambda^2$, [where $\nu \sim \ell a_0$ is the kinematic viscosity, see (1.4.15)] we get

$$u_0 \gtrsim \frac{\ell}{\lambda} a_0 \tag{1.5.13a}$$

which can also be expressed via the Mach and Knudsen numbers ($M_0 = u_0/a_0$; Sect. 1.4.1):

$$M_0 \gtrsim Kn \quad . \tag{1.5.13b}$$

1.5.3 Burgers' Equation. Evolution of Spectral Composition of the Sonic Wave

As demonstrated in Sect.1.4.1, the dispersion equation in the linear approximation (dissipation taken into account) has the form (1.4.14). Hence it follows that the linearized equation for a weakly damped sonic wave is

$$\frac{\partial u}{\partial t} + a_0 \frac{\partial u}{\partial x} = \mu \frac{\partial^2 u}{\partial x^2} \quad . \tag{1.5.14}$$

This allows one to construct an equation which would also account for nonlinear effects (different velocities for different portions of the wave):

$$\frac{\partial u}{\partial t} + [u + a(u)] \frac{\partial u}{\partial x} = \mu \frac{\partial^2 u}{\partial x^2} \quad , \tag{1.5.15}$$

where for an ideal gas $a(u) = a_0 + \frac{1}{2}(\gamma - 1)u$. Substituting $a(u)$ into (1.5.15) and using the designation

$$v = a_0 + \frac{\gamma + 1}{2} u \quad ,$$

we arrive at the well-known Burgers' equation

$$\frac{\partial v}{\partial t} + v \frac{\partial v}{\partial x} - \mu \frac{\partial^2 v}{\partial x^2} = 0 \quad , \tag{1.5.16}$$

which is the simplest equation accounting for both the nonlinear distortion of the wave profile and the dissipative damping [1.20,22]. The remarkable property of this equation is its analytic solvability when reduced to the linear "heat conduction" equation [1.23,24]. We shall not use this exact solution [1.20]; it suffices to say that all the results obtained in this section can be rigorously deduced from this solution.

Let us examine the distortion of the initial sine-wave profile, described by (1.5.16). Assume that at $t = 0$

$$v(x,0) = a_0 + v_0 \sin kx \quad . \tag{1.5.17}$$

The measure of the intensity of initial perturbation is the Mach number $M_0 = v_0/a_0$; the adopted approximation requires that

$$M_0 \ll 1 \quad . \tag{1.5.18}$$

In dimensionless variables

$$\tau = M_0\omega t \quad , \quad \xi = kx \quad , \quad u = \frac{v - a_0}{a_0} \quad , \quad \text{where} \quad k = \omega/a_0 \quad ,$$

the Burgers' equation takes the form

$$u_\tau + uu_\xi - \frac{1}{R}u_{\xi\xi} = 0 \quad . \tag{1.5.19}$$

The dimensionless parameter

$$R = \frac{v_0}{\mu k} = M_0 \frac{a_0}{\mu k} \tag{1.5.20}$$

characterizes the relative role of the nonlinear and dissipative effects of the wave in question, and by hydrodynamic analogy it can be termed the acoustic Reynolds number.

From (1.5.15) it becomes clear that damping of the sine wave in the linear approximation is low when the second factor in (1.5.20) is large. This condition, as demonstrated in Sect.1.4.1, amounts to the requirement that

$$\ell \ll \lambda \quad , \quad \text{or} \quad Kn \ll 1 \quad ,$$

which, as a matter of fact, is a general condition of applicability of the equations we are using, see (1.2.16). However, in nonlinear theory – in contrast to the linear approximation – the separate fulfillment of the inequalities (1.2.16,1.5.18) does not ensure the applicability of the acoustic approximation of "infinitesimal perturbations". The steady domination of the dissipative effects over the nonlinear ones is ensured only by the more rigid condition $R \ll 1$, equivalent to

$$M_0 \ll Kn \ll 1 \quad . \tag{1.5.21}$$

Unless this condition is fulfilled [even providing that (1.2.16,18) are satisfied], the wave must be viewed as a wave of finite amplitude, dominated by nonlinear effects. At the other extreme ($R \gg 1$) the initial dissipation can be neglected, and in place of (1.5.11) we get the dissipationless equation

$$u_\tau + uu_\xi = 0 \quad ,$$

whose solution with the initial condition (1.5.17) has the form

$$u = \sin(\xi - u\tau) \quad . \tag{1.5.22}$$

Since by virtue of (1.5.22)

$$u_\xi = \frac{\cos(\xi - u\tau)}{1 + \cos(\xi - u\tau)} \quad ,$$

at $\tau = 1$ the value of u_ξ tends to infinity, i.e., breaks occur in those points where $\xi = \pi(2n+1)$, $u = 0$. Therefore, the profile (1.5.22) approximates the solution of (1.5.19) at $\tau < 1$, and the dissipative effects become more pronounced at τ close to unity the larger the value of R is. In the region of applicability of this solution the generation of harmonics due to nonlinearity is described by the Bessel-Fubini expansion [1.25]

$$u = \sum_{n=1}^{\infty} \frac{2(-1)^n}{n\tau} J_n(n\tau) \sin n\xi \quad . \tag{1.5.23}$$

With small τ the amplitude of the nth harmonic grows as τ^{n-1}. As τ approaches unity (the closer to unity, the larger R), the dissipative effects start counterbalancing the nonlinear effects. The amplitudes of harmonics, having reached their maximum values (which for the lower harmonics are of the order of unity), start going down in accordance with *Fay*'s solution [1.26]

$$u = \sum_{n=1}^{\infty} \frac{2}{R \sinh\left[\frac{n}{R}(1+\tau)\right]} \sin n\xi \quad . \tag{1.5.24}$$

The linear theory predicts a more rapid fading of harmonics, proportional to $\exp(-n^2\tau/R)$, than does (1.5.24). Actually, the damping is weaker because of the nonlinear interaction between harmonics, which sustains the higher harmonics at the expense of energy pumped from the lower ones, against the background of the dissipation of the total energy of the wave. In the end, the dimensionless amplitude M of the wave decreases from M_0 down to the values which ensure the fulfilment of (1.5.21), and the wave becomes a linear acoustic wave.

1.6 Discontinuities and Shocks

1.6.1 Discontinuous Solutions

The physically meaningless ambiguities in the solutions for simple compression waves indicate that the basic assumptions for obtaining continuous solutions become invalid. We must therefore expand the class of solutions of the initial equations by allowing for discontinuous solutions. The ambiguity can be eliminated by allowing for the ordinary discontinuity (discontinuity of the first kind). The expansion of the class of solutions necessitates revising the initial assumptions. At the point of discontinuity the gradients are large, and therefore the energy dissipation by viscosity and heat conduction is high; consequently, the flow at the point of discontinuity cannot be considered isentropic. As a matter of fact, the real flow is fairly well approximated by the introduction of a jump, satisfying certain conditions, together with the initial equations for the continuous flow.

In this approach (a extended class of dissipationless solutions) the shock is represented by the discontinuity surface of zero thickness. The boundary conditions on the discontinuity surface must be chosen so as to characterize the flow on either side of the jump. Formal mathematics offers a choice of various classes of discontinuous solutions, of which the adequate ones must be sorted out from physical reasons. The differential equations, obtained for continuous flow, now become invalid. Their integral prototypes (i.e., the conservation laws), however, remain valid, and they determine the boundary conditions on the surface of discontinuity.

This formulation of the problem brings us to the concept of the extended class of weak solutions, consisting of the continuous part, described by differential equations, and the discontinuity surfaces which satisfy the appropriate boundary conditions. Subsequent approximations may improve the accuracy of the solution by considering the discontinuity as a narrow region with large (but finite) gradients and calculating the structure of this region.

The importance of weak solutions derives from the fact that the extended class of "piecewise-continuous" solutions of Euler's equations can give a good description of a large number of real flows. In particular, the weak solutions prove indispensable for the self-similar problems whose initial and boundary conditions contain no length or time parameters. As the dimensionalities of the variables p, ρ, u are not independent (the combination $\rho u^2/p$ is dimensionless), the solution of a self-similar problem can depend on x and t only in the velocity-like combination $\eta = x/t$. An example of a self-similar flow is the centered expansion wave, see (1.4.61-63). Any solution of the self-similar problem is also centered (all variables are constant along rays $x/t = \mathrm{const}$ fanning out from the origin of coordinates), although it generally includes —apart from centered expansion waves —discontinuities, such as shock waves. In many cases the self-similar solutions are asymptotic to the solutions of a more general form, when the particulars of the initial conditions become irrelevant (for more detail see [1.27]). That is why the self-similar solutions are important for studying the dynamics of shock waves.

Flows containing shock waves can often be correctly described by weak solutions of the equations of the ideal fluid, Sect.3.3.5. [A fluid motion is called self-similar if the profiles of all the fluid variables (density, velocity, etc.) are conserved; a problem is self-similar if it has a self-similar solution. The evolution of profiles for self-similar solution is reduced just to a change of scales with time. This indicates that a self-similar solution can contain the independent variable x only in some combination $\eta = x/R(t)$, R(t) being the time-dependent length scale. Here $R(t) \propto t$.]

Consider the element of a discontinuity surface, moving at speed D; the x axis is directed normal to the surface. From (1.1.1), which expresses the condition of conservation of the flux of mass, follows

$$(\rho_0 - \rho_2)D = \rho_0 u_0 - \rho_2 u_2 \quad .$$

Designating the discontinuous jump of a variable by bracketing it, e.g., $[\rho] = \rho_0 - \rho_2$, we can rewrite this equation in the form

$$D[\rho] = [\rho u] \quad . \tag{1.6.1}$$

Hereafter, the subscript 0 denotes the values of variables upstream of the discontinuity, and the subscript 2 those downstream.

Similarly, from (1.1.6) of continuity for the flux of momentum, we get the equations of conservation for the x, y, z components of the flux of momentum:

$$D[\rho u_x] = [p + \rho u_x^2] \quad , \tag{1.6.2}$$

$$[\rho u_x u_y] = 0 \quad , \quad [\rho u_x u_z] = 0 \quad . \tag{1.6.3}$$

From the continuity condition for the energy flux (1.1.10) it follows

$$D\left[\frac{1}{2}\rho \mathbf{u}^2 + \rho\varepsilon\right] = [\rho u_x(w + \mathbf{u}^2/2)] \quad . \tag{1.6.4}$$

Observe that consistent with the arguments developed above, there is no such equation for the continuity of the entropy flux; moreover, the entropy production at the discontinuity depends on the amplitude of the jump.

The boundary conditions are conveniently written in the coordinate system tied to the discontinuity surface. Denoting by v_0 and v_2 the gas velocities before and after the jump, respectively, in the discontinuity-surface coordinate system, we get the following complete set of boundary conditions on the discontinuity surface:

$$[\rho v_x] = 0 \quad , \tag{1.6.5}$$

$$[p + \rho v_x^2] = 0 \quad , \tag{1.6.6}$$

$$[\rho v_x v_y] = 0 \quad , \quad [\rho v_x v_z] = 0 \quad , \tag{1.6.7}$$

$$[\rho v_x(w + \mathbf{v}^2/2)] = 0 \quad . \tag{1.6.8}$$

The boundary conditions (1.6.5-8) indicate the possible existence of two types of discontinuities. The first type, which exhibits zero mass flux across the discontinuity surface, is called the tangential discontinuity:

$$\rho_0 v_{0x} = \rho_2 v_{2x} = 0 \quad . \tag{1.6.9}$$

Hence it follows that $v_{0x} = v_{2x} = 0$, and from (1.6.6) that

$$p_0 = p_2 \quad . \tag{1.6.10}$$

We see that the tangential discontinuity affects only the tangential components of velocity, density, etc., while the normal component of velocity and the pressure preserve their continuity.

With the second type of discontinuities – the shock waves – the flow of gas across the discontinuity surface is nonzero. Since in the shock waves $\rho v_x \neq 0$, (1.6.7) implies the continuity of the tangential components of velocity:

$$[v_y] = 0 \quad , \quad [v_z] = 0 \quad . \tag{1.6.11}$$

By virtue of (1.6.11) the boundary conditions (1.6.5-8) for the shock waves can be written in the form

$$[\rho v_x] = 0 \quad , \tag{1.6.12}$$

$$[\rho v_x^2 + p] = 0 \quad , \tag{1.6.13}$$

$$[w + v_x^2/2] = 0 \quad . \tag{1.6.14}$$

Further below we shall be concerned with the second type of discontinuous solutions, the shock waves. The problems of tangential discontinuities were dealt with in [1.2].

1.6.2 The Solution of Burgers' Equation for the Profile of a Weak Shock Wave

As already indicated (Sect.1.5.3), Burgers' equation provides a good approximation for nonlinear effects in low-amplitude perturbations, accounting also for dissipation. The exact solution of Burgers' equation is a good illustration of the concept of weak solutions. The exact solution of (1.5.15) for $\mu \to 0$ can be demonstrated to approach the discontinuous solution of the Euler equation.

Let us use Burgers' equation for calculating the structure of a weak shock wave. For this purpose we introduce in (1.5.15) a new function $c = u(\gamma + 1)/2$ and consider a stationary solution, which describes the wave traveling at a constant speed $a_0 + D$, i.e., $c = c(x - Dt)$. We get

$$- Dc' + cc' = \mu c'' \quad . \tag{1.6.15}$$

Integrating (1.6.15) once, we get

$$- Dc + c^2/2 + \text{const} = \mu c' \quad . \tag{1.6.16}$$

The boundary conditions are

$$c = c_0 \qquad \text{at} \quad x - Dt \to \infty$$

$$c = c_2 > c_0 \quad \text{at} \quad x - Dt \to - \infty \quad .$$

Hence we find

$$D = \frac{c_0 + c_2}{2} \quad , \quad c_2 > D > c_0 \quad , \; \text{const} = c_0 c_2/2 \quad . \tag{1.6.17}$$

Substituting (1.6.17) into (1.6.16) and integrating once again, we get the solution for the profile of the shock front

$$c = c_0 + \frac{c_2 - c_0}{1 + \exp[(x - Dt)/\Delta + \text{const}]} \quad , \tag{1.6.18}$$

where $\Delta = 2\mu/(c_2 - c_0)$ is the width of the transition region, which can be viewed as the shock-front width. Observe that $\Delta \to 0$ for $\mu \to 0$, i.e., in the dissipation-free limit, the solution describes a jump of zero width, traveling at the speed D defined by (1.6.17). Note that the profile (1.6.18) is quite general and describes weak shocks of arbitrary nature.

1.6.3 The Shock Adiabat

Let us examine more closely the boundary conditions for the shock wave. Since the tangential components of velocity are continuous, it is always possible to go to the intrinsic coordinate system, where the tangential components of gas variables are zero, and the gas flows in and out in the direction normal to the shock-front surface. Then (1.6.12-14) comprise a set of three algebraic equations, which connect the unperturbed values of the gas parameters upstream of the front (u_0, ρ_0, p_0, w_0) with the values downstream of the front

(u_2, ρ_2, p_2, w_2). Supplementing the boundary conditions with the equation of state, e.g., $w = w(p, \rho)$, assumed to be known, we can calculate the parameters of the flow ahead of the front on the basis of the known amplitude of the shock wave. The "amplitude" can be represented by the speed of the shock front, or the speed of the piston generating the shock wave, or the pressure behind the front, etc.

Let us rewrite (1.6.12-14), expressing the enthalpy via the internal energy of the gas, $w = \varepsilon + p/\rho$:

$$\rho_0 u_0 = \rho_2 u_2 \quad , \tag{1.6.19}$$

$$\rho_0 u_0^2 + p_0 = \rho_2 u_2^2 + p_2 \quad , \tag{1.6.20}$$

$$\varepsilon_0 + p_0/\rho_0 + u_0^2/2 = \varepsilon_2 + p_2/\rho_2 + u_2^2/2 \quad . \tag{1.6.21}$$

The first two equations can be used to obtain the velocity of the downstream flow relative to the velocity of the upstream flow:

$$u_0 - u_2 = \sqrt{\frac{(p_2 - p_0)(\rho_2 - \rho_0)}{\rho_0 \rho_2}} \quad . \tag{1.6.22}$$

From (1.6.21,22) we derive the following expression, which links the pressures and densities on either side of the front:

$$\varepsilon_2(p_2,\rho_2) - \varepsilon_0(p_0,\rho_0) = \frac{1}{2}(p_2 + p_0)(\rho_2 - \rho_0)/\rho_2\rho_0 \quad . \tag{1.6.23}$$

By analogy with the equation of adiabatic gas compression, (1.6.23) is termed the *shock adiabat*, or the Hugoniot-Rankine adiabat.

Consider the shock adiabat for an ideal gas with constant heat capacities and adiabatic exponent γ. Substituting (1.1.21) for the ideal gas into the shock adiabat equation (1.6.23), we get the ideal-gas shock adiabat equation in the explicit form

$$\frac{p_2}{p_0} = \frac{(\gamma + 1)\rho_2 - (\gamma - 1)\rho_0}{(\gamma + 1)\rho_0 - (\gamma - 1)\rho_2} \quad . \tag{1.6.24}$$

From (1.6.24) it is clear that the shock adiabat passes through the point (ρ_0, p_0), and the compression downstream of the front ρ_2/ρ_0 exhibits a monotonic increase as the strength of the shock increases, and tends to the limit

$$\frac{\rho_2}{\rho_0} = \frac{\gamma + 1}{\gamma - 1} \tag{1.6.25}$$

as the strength of the shock tends to infinity, $p_2/p_0 \gg 1$. The limit of compression equals 4 for a monatomic gas ($\gamma = 5/3$), 6 for a diatomic gas ($\gamma = 7/5$), etc. In reality, as shown in Chap.2, the adiabatic exponent does not remain constant in a strong shock wave, because the gas particles dissociate and ionize. Then the compression can become arbitrarily high.

The relationships between the variables in the shock wave are conveniently expressed via the Mach number (the ratio of velocity of the flow to the local speed of sound, $M = u/a$), which characterizes the intensity of the flow. The "upstream" Mach number is $M_0 = u_0/a_0$; the "downstream" Mach number is $M_2 = u_2/a_2$. We shall also introduce the dimensionless velocity, density and temperature, defined as the ratio of a given quantity to its value upstream or downstream. For example, the dimensionless density $\rho(2) = \rho_2/\rho_0$ is the compression, or the density jump across the shock front. The dimensionless variables will be denoted by the same letters as the dimensional quantities, but without the subscript.

For the dimensionless variables the boundary conditions (1.6.19-21) assume an especially simple form:

$$\rho(2)u(2) = 1 \quad , \tag{1.6.26}$$

$$u(2) - 1 + \frac{1}{\gamma M_0^2}\left(\frac{T(2)}{u(2)} - 1\right) = 0 \quad , \tag{1.6.27}$$

$$u^2(2) - 1 + \frac{2}{(\gamma - 1)M_0^2}[T(2) - 1] = 0 \quad , \tag{1.6.28}$$

where

$$\rho(2) = \rho_2/\rho_0 \quad ; \quad u(2) = u_2/u_0 \quad ; \quad T(2) = T_2/T_0 \quad .$$

If we choose to use the dimensionless variables $\rho(0) = \rho_0/\rho_2$, etc., (1.6.26-28) remain the same, with the transcription $0 \leftrightarrow 2$.

By solving the (1.6.26-28), we obtain the following expressions for the jumps of variables across the shock front as functions of Mach number, which characterizes the intensity of the shock wave:

$$\rho(2) = \frac{M_0^2(\gamma + 1)}{2 + (\gamma - 1)M_0^2} \quad ; \quad u(2) = 1/\rho(2) \quad ; \tag{1.6.29}$$

$$T(2) = \frac{[2 + (\gamma - 1)M_0^2][2\gamma M_0^2 - (\gamma - 1)]}{M_0^2(\gamma + 1)^2} \quad ; \tag{1.6.30}$$

$$p(2) = \frac{2\gamma M_0^2 - (\gamma - 1)}{\gamma + 1} \quad . \tag{1.6.31}$$

Let us also write the expression for M_2:

$$M_2^2 = \frac{2 + (\gamma - 1)M_0^2}{2\gamma M_0^2 - (\gamma - 1)} \quad . \tag{1.6.32}$$

In the high-intensity shock wave $M_0 \gg 1$, which allows some useful estimates to be made:

$$\rho(2) \simeq (\gamma + 1)/(\gamma - 1) \quad ;$$

$$p(2) \simeq \frac{2\gamma}{\gamma + 1} M_0^2 \quad ; \tag{1.6.33}$$

$$T(2) \simeq \frac{2\gamma(\gamma - 1)}{(\gamma + 1)^2} M_0^2 \quad .$$

As we shall see below, the shock waves are always supersonic ($1 < M_0 < \infty$); at the same time $1 > M_2 > \gamma - 1/2\gamma$, i.e., the downstream flow is subsonic with respect to the local speed of sound.

As we have already indicated, the shock wave considered as a discontinuity complies with three out of the four conservation laws, corresponding to the differential equations. The condition of adiabaticity, i.e., the continuity of entropy, is violated on the jump. Moreover, the entropy production in the shock wave is uniquely determined by the intensity of the shock. For the ideal gas, the entropy (up to an additive constant) is $S = c_V \ln(p/\rho^\gamma)$. Accordingly, the change in the entropy across the shock front, with due account taken of (1.6.29,31), can be expressed as

$$S_2 - S_0 = c_V \ln\left\{p(2) \left[\frac{(\gamma - 1)p(2) + (\gamma + 1)}{(\gamma + 1)p(2) + (\gamma - 1)}\right]^\gamma\right\} \quad . \tag{1.6.34}$$

Expression (1.6.34) implies that the entropy in the shock wave can only increase ($\gamma > 1$). This indicates directly that the shock wave in the ideal gas (under the condition of increasing entropy) is a compression wave, i.e., $p(2) = p_2/p_0 > 1$. Using (1.6.29-31), we immediately find that

$$M_0 > 1 \ , \ M_2 < 1 \ , \ \rho(2) = \rho_2/\rho_0 > 1 \ , \ u(2) = u_2/u_0 < 1 \quad . \tag{1.6.35}$$

This statement is proved in hydrodynamics for a large class of media, whose equations of state are classified as normal, i.e., they satisfy the conditions

$$\left(\frac{\partial^2 p}{\partial \rho^2}\right)_S > 0 \quad , \quad \left(\frac{\partial p}{\partial S}\right)_p > 0$$

(Zemplen's theorem [1.2]). As shown in Sect.1.7.1, the subsonic-supersonic condition $M_0 > 1$, $M_2 < 1$ is valid irrespective of the equation of state. At the same time the conclusion that the shock wave is a compression wave [$\rho(2) > 1$] relies heavily on the assumption of normality of thermodynamic properties of the medium. As demonstrated by *Zel'dovich* [1.28], thermodynamically anomalous media may exhibit expansion shock waves. This theoretical prediction has been recently verified by observing the expansion shock waves in a gas near the critical point, where $(\partial^2 p/\partial \rho^2)_S < 0$ [1.29].

The entropy increase in the shock wave does not depend on the mechanism of dissipation and is determined only by the intensity of the shock wave, in accordance with the laws of conservation of fluxes of mass, momentum and energy. As long as the consideration does not involve any parameters of length, the shock is represented by a zero-thickness discontinuity surface. When the dissipative processes are taken into account, the length parameters are brought into the consideration, the shock front has a finite thickness, and the profiles of variables are shaped by the dissipative mechanisms. However, the dissipative mechanisms govern only the profiles of variables within the front, whereas the jumps of these variables are determined only by the conservation laws, that is, by the boundary conditions (1.6.19-21).

1.6.4 Production of Shock Waves. Elementary Theory of a Shock Tube

A commonly employed tool for producing gasdynamic shock waves is a shock tube. Let us briefly describe its principles of operation. The shock tube is usually a circular or rectangular tube, length L of which is much larger than its diameter D in order to ensure the unidimensionality of the gas flow (as a rule, L is several meters, and D is several centimeters). The tube is divided into two parts by a diaphragm; the volume on one side of the diaphragm constitutes the compression (high-pressure) chamber; the other side is the expansion (low-pressure) chamber. The low-pressure chamber contains the test gas, used to study the characteristics of the shock wave. The high-pressure chamber contains the driver gas, which acts as a piston to excite the shock wave in the test gas. The diaphragm instantaneously ruptures, and the driver gas starts expanding into the low-pressure chamber, producing a shock wave in the test gas. At the same time, the expansion wave starts running, in the driver gas, in the opposing direction.

Let us express the parameters of the shock wave in the test gas via the ratios of density and temperature of the test gas (subscript 0) to those of the driver gas (subscript 3), ρ_0/ρ_3 and T_0/T_3, assuming the flows of both gases to be unidimensional and adiabatic (the adiabatic exponents being γ_0 and γ_3), and the shock wave to be strong ($M_0 \gg 1$). After the shock wave and the expansion wave have traveled a distance equal to several diameters of the tube from the location of the diaphragm, the flow can be considered to be self-similar and the expansion wave centered. The pressures and velocities of the driver gas and the test gas on either side of the interface (which in the self-similar problem is a contact discontinuity) must be the same. The values pertaining to the interface — the region behind the shock wave and ahead of the expansion wave — are denoted by the subscript 2. Making use of (1.6.33,1.4.62), for the region downstream of the shock we obtain

$$u_2 = \frac{2}{\gamma_0 + 1} u_s \quad , \qquad p_2 = \frac{2}{\gamma_0 + 1} \rho_0 u_s^2 \quad ,$$

where u_s is the velocity of the shock; or, in dimensionless form (a being the speed of sound)

$$\frac{u_2}{a_0} = \frac{2}{\gamma_1 + 1} M_0 \quad , \tag{1.6.36}$$

$$\frac{p_2}{\rho_0 a_0^2} = \frac{2}{\gamma_1 + 1} M_0^2 \quad . \tag{1.6.37}$$

Simultaneously, for the region upstream of the expansion wave we have

$$u_2 = \frac{2}{\gamma_3 - 1} (a_3 - a_2) \quad , \qquad p_2 = p_3 (a_2/a_3)^{2\gamma_3/\gamma_3 - 1} \quad ,$$

or

$$\frac{u_2}{a_0} = \frac{2}{\gamma_3 - 1} \frac{a_3}{a_0} \left[1 - (p_2/p_3)^{\gamma_3 - 1/2\gamma_3} \right] \quad . \tag{1.6.38}$$

Substituting the value of u_2 from (1.6.36) into the left-hand side of (1.6.38), and the value of p_2 from (1.6.37) into the right-hand side, after some trivial transformation we obtain the equation for M_0:

$$M_0 = \frac{(\gamma_0 + 1) a_0}{(\gamma_3 - 1) a_3} \left[1 - \left(\frac{2\gamma_0}{\gamma_0 + 1} \frac{p_0}{p_3} M_0^2 \right)^{\gamma_3 - 1/2\gamma_3} \right] \tag{1.6.39}$$

Evidently, (1.6.39) has a single positive root, which is smaller than the term in front of the square bracket. This term gives a good approximation to the sought-for solution, if only its substitution into the term in parentheses on the right-hand side of (1.6.39) makes this term much smaller than unity. This amounts to satisfying the inequality

$$\frac{\rho_0}{\rho_3} \ll \frac{(\gamma_3 - 1)^2}{2\gamma_3(\gamma_0 + 1)} \quad . \tag{1.6.40}$$

If (1.6.40) is satisfied—which requires that the density of the test gas be lower by at least two orders of magnitude than the density of the driver gas—the driver gas expands into the low-pressure chamber at the highest speed possible, as though this chamber were evacuated. Then the dimensionless parameters of the shock wave depend only on the ratio between the temperatures of the driver and test gases, and do not depend on the ratio of their densities:

$$M_0 = \frac{(\gamma_0 + 1)a_3}{(\gamma_3 - 1)a_1} = \frac{(\gamma_0 + 1)}{(\gamma_3 - 1)} \left(\frac{\gamma_3\mu_0 T_3}{\gamma_0\mu_3 T_0}\right)^{\frac{1}{2}} \quad , \tag{1.6.41}$$

$$\frac{p_2}{p_0} = \frac{2\gamma_3(\gamma_0 + 1)\mu_0}{(\gamma_3 - 1)^2\mu_3} \frac{T_3}{T_0} \quad , \tag{1.6.42}$$

$$\frac{T_2}{T_0} = \frac{2\gamma_3(\gamma_0 - 1)\mu_0}{(\gamma_3 - 1)^2\mu_3} \frac{T_3}{T_0} \quad , \tag{1.6.43}$$

where μ_0 and μ_3 are the molecular masses of the test gas and the driver gas, respectively.

Expressions (1.6.41-43) indicate that the best way to obtain strong shock waves consists in making the ratio μ_0/μ_3 as large as possible, by choosing a heavy test gas and a light driver gas. The shock wave also becomes stronger the higher the ratio T_3/T_0; and the temperature downstream is proportional to T_3. The intensity of the shock wave and the shock heating downstream can be raised by heating the driver gas: for example, sparking off the explosive gas mixture before rupturing the diaphragm, or energizing the driver gas by an electric discharge. An alternative way of increasing the Mach number of the shock wave consists in cooling the test gas. This method is realized in cryogenic shock tubes [1.30,31], where cooling the test gas (helium) down to 2 K makes it possible to produce shock waves with Mach numbers as high as 50.

In reality, the velocities of shock waves and the temperatures of shock heating obtainable in the shock tubes are somewhat lower than predicted by (1.6.41,43). One of the reasons seems the difficulty to satisfy the inequality

(1.6.40) with a good margin, which, as follows from (1.6.39), results in a reduction of M_0: the inertia of the test gas hampers the expansion of the driver gas, thus weakening the shock wave. In addition, heating to several thousand kelvin causes extra energy losses due to dissociation and ionization of the downstream gas, thus reducing the velocity and temperature downstream of the shock wave (Sect.2.2.1).

More efficient ways of producing strong shock waves also exist, using a high-pressure gas stream as a "piston". Shock waves can be reinforced by driving them through convergent channels, see (1.7.25) below. Some shock tubes use an intermediate chamber between the high-pressure and low-pressure ones: after the rupture of the first diaphragm the driver gas generates a shock wave in the intermediate chamber, which is used for exciting the final shock wave (after the rupture of a second diaphragm, separating the intermediate chamber from the low-pressure part) in the test gas. This allows one to obtain much higher Mach numbers with the same pressure of the driver gas. Finally, the strongest shock waves are obtained in explosion shock tubes, which use the blasts of solid explosives, cumulative jets and explosion-propelled metallic plates.

1.7 Criteria of Stability and Evolutionarity of Discontinuities

1.7.1 Evolutionarity

Let us now discuss the physical reasons for singling out from the great variety of formal discontinuous solutions those solutions which describe physically realizable flows. Such conditions are, as a rule, formulated in the form of requirements for the existence and stability of continuous solutions for the structures with jumps. It was found that these properties of discontinuities are determined by the spectrum of linear waves, which may exist in the medium under investigation with due account taken of the dissipative and other physical processes which shape the structure of the front. When the thickness of the front region is small compared with the characteristic scale of the problem (which justifies considering the discontinuity as a zero approximation), then in the first-order approximation the structure of the front can be considered stationary. Then we are concerned with the low-frequency limit $\omega \to 0$, where the spectrum breaks up into two classes of waves: undamped waves characterizing the ideal (dissipationless) medium, and the so-called dissipative waves, which fade out over a distance of the order of the front width. The former are responsible for the stability of the front (being the adequate response to the external disturbances), and the latter for the structure of the front. The necessary existence of both types of divergent waves on either side of the front determines the relationship between the stability and the existence of the front structure.

Our treatment of the problem will be sufficiently general, unrestricted by the framework of gasdynamic theory of ideal gas with constant heat capacity. We set up a vector W, whose components represent n independent variables of the flow; for the one-dimensional problem $W = \{\rho, u_x, S\}$. It would be natural to demand that (with due account taken of the dissipative effects) the solution, discontinuous in the idealized description, should display a certain structure. In other words, in the shock-front-fixed coordinate system the appropriately expanded set of equations accounting for dissipations must have a solution of the form (Sect.1.6)

$$W = W_0(x) \quad . \tag{1.7.1}$$

As $x \to \mp \infty$, this solution must asymptotically approach $W_0(\mp \infty)$, the components of the vectors $W_0(-\infty)$ and $W_0(+\infty)$ being the values of the flow variables upstream and downstream, respectively.

Furthermore, the solution (1.7.1) must be stable. While remaining within the framework of the one-dimensional approximation, let us impose a small perturbation on this solution

$$W = W_0(x) + \varepsilon e^{-i\omega t} W_1(x) \quad , \quad \varepsilon \to 0 \quad . \tag{1.7.2}$$

The substitution of (1.7.2) into the set of equations describing the nonstationary flow in the first-order approximation in ε gives us a set of linear equations with respect to the components of the vector $W_1(x)$. By solving these equations with appropriate boundary conditions (corresponding, for instance, to a particular type of wave incident on the discontinuity surface, or to the localization of perturbation in the vicinity of the front, $W_1(\mp \infty) = 0$, and so on), we can obtain the set of eigenfrequencies of the problem $\{\omega_i\}$. If none of the eigenfrequencies belong to the upper half of the complex plane, i.e., $\mathrm{Im}\{\omega_i\} < 0$ for all i, then the solution is stable, at least with respect to perturbations of this particular type; otherwise the solution is unstable.

In practice, however, realization of this procedure encounters major difficulties even with the most simple functions $W_0(x)$ [1.32]. In most cases the explicit form of $W_0(x)$ is not known. So, even though the nature of the spectrum $\{\omega_i\}$ can sometimes be determined from general considerations [1.33], as a rule, one needs more practical criteria of stability. The most convenient and physically clear such criterion is the condition of evolutionarity of the discontinuity, which we are going to formulate.

When considering the shock front as a discontinuity, we assume that its stationary thickness Δ and the time of flow across the discontinuity τ are much smaller than any of the characteristic times or lengths of the problem. In particular, this applies to frequencies ω and wave vectors k of simple linear waves, assumed to be propagating against the background of the constant flow upstream and downstream of the front,

$$\omega\tau \ll 1 \quad , \quad k\Delta \ll 1 \quad . \tag{1.7.3}$$

In this extreme the flow is described by idealized equations, those of the dissipationless medium (as $\omega \to 0$, $k \to 0$ the damping can be neglected: $\mathrm{Im}\{k\}/\mathrm{Re}\{k\} = O(k)$ with real-valued ω). Consequently, the problem contains no parameters of length and time, and the spectrum of perturbations corresponding to the prescribed boundary conditions cannot be determined. It is possible, however, to explore the linear response of the discontinuity to small perturbations, presented in the form of superpositions of a finite number of linear simple waves. The requirement of unambiguous definiteness of this linear response constitutes the criterion of evolutionarity of the discontinuity.

In this case the perturbation can be presented in the form

$$W - W_0(x) = \sum_i \varepsilon_i W_i f_i[x - (c_i + u_x)t] \quad , \tag{1.7.4}$$

where u_x is the flow velocity in the present equilibrium state; the summation is carried out over all types of waves which possibly exist against the background of the given equilibrium state [$W(-\infty)$ or $W(+\infty)$] in the limit (1.7.1). Here $f_i(x)$ is an arbitrary scalar function, whose scale of variation is superior to Δ, $\varepsilon_i \to 0$ is the amplitude of the ith wave, and the relationship between the components of vector W_i is uniquely determined by the type of the wave. Note that waves of the same type, propagating in the opposing direction (opposite sign of the phase velocity c_i), are described by different terms in the summation (1.7.4).

In the zeroth approximation with respect to small parameters (1.7.3)

$$\delta_t = \omega\tau \quad , \quad \delta_x = k\Delta = \frac{\omega\Delta}{c_i + u_x} \quad , \tag{1.7.5}$$

the set of equations describing the structure of the shock reduces to the set of algebraic relations (boundary conditions at the front) between the perturbations of the upstream and downstream values of variables. Here the smallness of parameters δ_t and δ_x (the existence of different length and time scales) is essential. In terms of the external scales k^{-1} and ω^{-1}, the front is an infinitesimally thin discontinuity, so that the perturbations of the flow parameters on one side of the front are instantaneously transmitted across, and the values of these parameters on either side of the front are connected via the boundary conditions. In terms of the internal scales Δ and τ, the perturbations represent extremely slow, small variations of the equilibrium values in the upstream and downstream infinities ($x \to \mp \infty$), which adjust the steady-state structure of the front in accordance with the changing boundary conditions.

The waves propagating on either side of the front can be divided into waves incident on the front and waves outgoing from the front. The outgoing waves are those for which $c_i + u_x < 0$ upstream, and those for which $c_i + u_x > 0$ downstream; all the rest are classified as incident. The special case, when at least on one side of the front waves with $c_i + u_x = 0$ exist, must be considered separately (Sect.3.3.5). Obviously, the present theory is incapable of handling this case, since the parameter δ_x, defined by (1.7.5), cannot be small if $c_i = -u_x$.

For the amplitudes of outgoing waves we get a set of linear equations, one less than the number r of the boundary conditions at the front. Indeed, the perturbation δW_i of each component of the vector W is a sum over all types of waves which contribute to the change of this variable. At the same time, the perturbation also affects the speed of the shock front. Having written the continuity equation (1.6.1) in the form

$$\rho_0(u_{x0} - D) = \rho_2(u_{x2} - D) \quad ,$$

where D is the speed of the shock front (in the front-fixed system $D = 0$, and the subscripts 0 and 2 denote the states upstream and downstream, respectively), we can express the perturbation of the front speed via the perturbation of the flow variables:

$$\delta D = \left(u_{x0}\frac{\delta\rho_0}{\rho_0} - u_{x0}\frac{\delta\rho_2}{\rho_2} + \delta u_{x0} - \frac{\rho_2}{\rho_0}\delta u_{x2}\right)\left(1 - \frac{\rho_2}{\rho_0}\right)^{-1} \quad . \tag{1.7.6}$$

Eliminating δD with the aid of (1.7.6) leads to the set of $(r-1)$ linear equations, connecting the amplitudes of the incident and outgoing waves. Recall once again that the coefficients of this set do not depend on the frequency ω, and so these equations by themselves do not allow one to determine the spectrum corresponding to the given boundary conditions. [As follows from (1.7.3), perturbations of this particular type correspond to the real-valued frequencies near $\omega = 0$.]

The existence and uniqueness of the solution for the interaction between discontinuity and small perturbations in the form of sufficiently long-period waves, incident on the front from upstream and downstream infinities (that is, the solution of the set of equations for the amplitudes of outgoing waves), constitutes the condition of evolutionarity of discontinuity, which is the condition of stability. Indeed, if the solution of this set of linear equations is not unique, then nontrivial solutions exist with zero amplitude for the incident waves. Accordingly, the front is capable of spontaneously emitting waves and thus is unstable. On the other hand, if the system of equations is overdetermined (no solution), this indicates that the response of the discontinuity to a small perturbation is not small: such, for instance, is the case when the discontinuity shows a tendency of breaking up into several finite-amplitude discontinuities.

The condition of evolutionarity requires that the rank of the matrix A for the coefficients of the unknown amplitudes of outgoing waves be equal to the number of these amplitudes:

$$\mathrm{rank}\{A\} = s_1 + s_2 \quad , \tag{1.7.7}$$

where s_1 and s_2 are the number of the outgoing waves upstream and downstream, respectively. In the most important nondegenerate case, to which we presently restrict our attention, the number of equations equals the number of unknowns, that is, the matrix A is square, $\mathrm{rank}\{A\} = \dim\{A\} = r - 1$, and from (1.7.7) we obtain the condition of evolutionarity of the discontinuity:

$$r = s_1 + s_2 + 1 \quad . \tag{1.7.8}$$

When using the criterion of evolutionarity (1.7.8) one has to ensure that the matrix A does not have a block structure, i.e., that the set of equations in the amplitudes of the outgoing waves does not fall apart into several independent sets. Such would be the case if the perturbations of certain variables are exclusively transmitted by certain types of waves. Then (1.7.8) ought to be satisfied for each separate block of the matrix A. A relevant nontrivial example is encountered in magnetohydrodynamics (Chap.3).

The condition of evolutionarity of discontinuity, (1.7.8), can also be explained in a different way, recalling that the discontinuous solutions are typical of hyperbolic sets of equations [1.20,21]. Going over to the characteristic form of equations describing the nonstationary flow, we note at once that the values of $n - s_1$ flow parameters upstream and $n - s_2$ parameters downstream can be changed arbitrarily: their perturbations are transmitted to the front along the characteristics, corresponding to the incident waves. The description of the dynamic behavior of the front uses $2n + 1$ variables (n parameters of dissipationless flow on either side of the front, plus the speed of the front). The values of $2n - s_1 - s_2$ of these parameters are determined by the boundary conditions in the upstream and downstream infinities. Accordingly, to give an unambiguous description of the flow in terms of a discontinuity moving in an ideal medium, we need another r relations, connecting the values of flow variables on either side of the front,

$$r = 2n + 1 - (2n - s_1 - s_2) = s_1 + s_2 + 1 \quad .$$

Let us use the criterion (1.7.8) for determining the evolutionarity of gasdynamic shock waves. In the one-dimensional problem with $n = 3$ variables (ρ, u_x, S) we have $r = 3$ boundary conditions at the front, expressing the continuity of the fluxes of mass, momentum and energy. According to (1.7.8), the number of the outgoing waves must be $r - 1 = 2$.

Since both the sonic and the entropy waves do perturb the density ρ, the matrix A does not separate into blocks. This argument can easily be extrapolated to the three-dimensional case for $n = 5$ variables (ρ, u_x, u_y, u_z, S). Here we must observe that the rotational perturbations of the transverse velocity components u_y and u_z travel along with the flow. We additionally get two outgoing waves downstream (incidentally, their coefficients form separate unidimensional blocks in A), together with two boundary conditions, expressing the continuity of transverse momentum flux, which implies the continuity of the tangential velocity components at the front. Undoubtedly, the evolutionarity condition remains the same.

Returning to the one-dimensional case, two outgoing waves downstream of the front always exist: the entropy wave, traveling along with the flow, and the sonic wave, whose wave vector is directed downstream. It follows that the evolutionarity condition reduces to the requirement of absence of one outgoing sonic wave upstream (i.e., $-a_0 + u_{x0} > 0$, or $M_0 > 1$), and another downstream ($-a_2 + u_{x2} < 0$, or $M_2 < 1$), or to the familiar subsonic-supersonic condition.

In the present case the evolutionarity condition permits the following obvious physical interpretation [1.34]. The condition $M_0 > 1$ ensures that the perturbation of pressure does not overtake the front, destroying its structure (the condition of mechanical stability). The condition $M_2 < 1$ implies the possibility to control the motion of the front by changing the downstream pressure by means of, say, a piston, the pressure variations being transmitted to the front by sonic waves. (If the medium exhibits several nonzero characteristic velocities, then the flow in the evolutionary shock wave can be "subsonic" with respect to some of these velocities upstream, and "supersonic" downstream.) Notice that obtaining these conditions did not involve specifying the equation of state of the medium.

In the present case the number of the boundary conditions on the front equals the number of variables which characterize the nondissipative flow ($r = n$), and therefore the speed of the shock may assume any value within certain limits (namely, $1 < M_0 < \infty$), and for every fixed value of velocity the parameters of state downstream are uniquely determined by the parameters of the unperturbed state. Such is the case with many types of shock waves, although this should not be taken for universal truth. In Sect.1.9 and Chap.4 we shall encounter examples of evolutionary discontinuities with $r \neq n$, when, for instance, either a degree of freedom is lost, associated with the arbitrariness of the shock velocity, or, on the contrary, extra degrees of freedom arise, resulting in the ambiguity of the state downstream even though the front velocity is fixed.

1.7.2 Evolutionarity Condition and Existence of the Shock Structure. Basic and Additional Relations on the Front

The condition of evolutionarity, which is a necessary condition of stability of the shock front, is formulated in terms of time and length scales "external" with respect to the front, the latter being treated as an infinitely thin discontinuity, see (1.7.3). There exists also an alternative necessary condition of stability, formulated in terms of "internal" scales: the solution of type (1.7.1) is stable as long as a unique solution exists [up to the shift $W(x) \to W(x + C)$] at the prescribed boundary conditions at $\mp \infty$. The solution must persist against changes in the steady-state front structure, caused by slight variations of the boundary conditions, transmitted to the front by incident waves (Sect.1.7.1).

It would be natural to expect these two conditions to be in some way connected with one another. Indeed, the evolutionarity condition turns out to be necessary for the existence of the uniquely determined steady-state structure of the front [1.35,36]. In other words, before undertaking the complicated task of determining the structure of a given type of discontinuity, one would be wise to check first its evolutionarity; at the same time, the evolutionarity is ensured by the existence and uniqueness of the structure (1.7.1).

The statement formulated above is valid for dissipative media, in which any space-periodic perturbation against a uniform background tends to fade out with time. The damping is due to the same dissipative mechanism, which takes part in shaping the structure of the front. Taking dissipation into account, the set of N equations, describing a nonstationary dissipative flow, can be written in the sufficiently general form

$$\tilde{A}(\tilde{W}) \frac{\partial \tilde{W}}{\partial t} + \tilde{B}(\tilde{W}) \frac{\partial \tilde{W}}{\partial x} + \tilde{C}(\tilde{W}) + \tilde{D}(\tilde{W}) \frac{\partial^2 \tilde{W}}{\partial x^2} = 0 \quad , \tag{1.7.9}$$

where $\tilde{A}(\tilde{W})$, $\tilde{B}(\tilde{W})$, $\tilde{C}(\tilde{W})$, $\tilde{D}(\tilde{W})$ are $N \times N$ matrices, depending on the parameters of the flow. The idealized picture of dissipationless flow is obtained by nullifying the terms of the lowest order with respect to derivatives of $\tilde{W}$ in (1.7.9). From the condition

$$\tilde{C}(\tilde{W}) = 0 \tag{1.7.10}$$

we get the relations imposed on the parameters of the vector $\tilde{W}$, which describes the uniform background, namely the solutions of (1.7.9) of the form $\tilde{W}(x,t) = \text{const}$. If we use (1.7.10) to express some variables via the rest, we come to the n-dimensional vector $W(n < N)$ whose components represent the parameters of dissipationless flow. The set of hyperbolic equations for the components of this vector can be obtained by substituting the variables, expressed with the aid of (1.7.10), into the first two terms of (1.7.9):

$$A(W) \frac{\partial W}{\partial t} + B(W) \frac{\partial W}{\partial x} = 0 \quad , \tag{1.7.11}$$

where $A(W)$, $B(W)$ are $(n \times n)$ matrices.

For instance, in magnethydrodynamics (under the assumption of finite conductivity) the electric field $\mathbf{E}$, the magnetic field $\mathbf{B}$ and the velocity $\mathbf{u}$ are independent variables, but in a uniform flow they are bound by the relation of type (1.7.10)

$$\mathbf{E} + \mathbf{u} \times \mathbf{B} = 0 \quad . \tag{1.7.12}$$

If we demand that (1.7.12) be satisfied identically, we obtain the set of equations of ideal magnetohydrodynamics in a smaller number of variables, since $\mathbf{E}$ is expressed via $\mathbf{u}$ and $\mathbf{B}$ by (1.7.12).

The small perturbation of the uniform background, according to (1.7.11), is represented by a linear combination of n undamped traveling waves with fixed phase velocities [the transition from (1.7.9 to 11) is correct in the limit $\omega \to 0$, $k \to 0$, see (1.7.3)].

The involvement of m dissipative mechanisms [second-order derivatives in (1.7.9)] may be viewed as the extension of the respective system to the space of $(n+m)$ dimensions, treating some of the derivatives $\partial W_i/\partial x$ as independent variables. Then the number of roots of the dispersion equations, which connect k and ω for the perturbations of the uniform background W_0 and have the form

$$W(x,t) - W_0 = \varepsilon W_1 \exp(-i\omega t + ikx) \qquad (1.7.13)$$

(where W_1 is a constant vector, $\varepsilon \to 0$), becomes $n+m$. If ω is in the upper half of the complex plane ($\mathrm{Im}\{\omega\} > 0$), which corresponds to the time-increasing amplitude of disturbance of type (1.7.13), then none of the solutions $k(\omega)$ can lie on the real axis, as follows from the presence of the dissipative terms in the system (1.7.9). Therefore, the waves pertaining to different branches of the dispersion curve die away as x increases (decreases), i.e., propagate in the positive (negative) direction if $k(\omega,W_0)$ lies in the upper (lower) half-plane. Let the lower half-plane contain $p_1(\omega,W_0)$ roots of the dispersion equation, and the upper half-plane contain $p_2(\omega,W_0)$ roots.

On each branch of the dispersion curve, $k(\omega,W_0)$ is an analytic function of ω and W_0. In particular, $k(\omega,W_0)$ as a function of complex-valued ω, $\mathrm{Im}\{\omega\} > 0$, cannot display a singularity, because in the vicinity of a singularity real values of k are always to be found. Fixing any value of ω in the upper half-plane and varying W_0, we see that the roots initially belonging to a certain half-plane (either the upper or the lower one) cannot go to the other half-plane: they cannot cross the real axis, and the transition via infinity would imply a singularity for $\mathrm{Im}\{\omega\} > 0$. Thus the numbers $p_1(\omega,W_0)$ and $p_2(\omega,W_0)$ are the invariants of the dissipative set of equations. In other words, there are always p_1 waves traveling in the negative direction, and p_2 waves traveling in the positive direction. Recall that

$$p_1 + p_2 = n + m \quad . \qquad (1.7.14)$$

For studying the stationary solutions of type (1.7.1) we go to the limit $\omega \to 0 + i0$. In this limit some of the roots of the dispersion equation tend to $k = 0$; these roots presently correspond to undamped waves propagating in a dissipationless medium. The numbers of roots tending to $k = 0$ from the upper and lower half-planes are s_2 and s_1, respectively, in accordance with the number of undamped waves traveling downstream and upstream from the front. The number of roots remaining in the lower half-plane is $q_1 = p_1 - s_1$; and

$q_2 = p_2 - s_2$ in the upper half-plane. The strongly damped waves described by these solutions are called dissipative waves. The asymptotic solution of type (1.7.1) as $x \to \mp \infty$ is readily represented in the form of an expansion over the dissipative waves:

$$x \to -\infty \ , \quad W_0(x) - W_0(-\infty) \sim \sum_{i=1}^{q_1} C_i^- a_i \ \exp(ik_i x) \quad , \quad \mathrm{Im}\{k_i\} < 0 \quad , \tag{1.7.15}$$

$$x \to +\infty \ , \quad W_0(x) - W_0(+\infty) \sim \sum_{j=1}^{q_2} C_j^+ b_j \ \exp(ik_j x) \quad , \quad \mathrm{Im}\{k_j\} > 0 \quad , \tag{1.7.16}$$

where a_i, b_j are the eigenvectors of the set of equations (1.7.9), linearized near $W = W_0(-\infty)$ and $W = W_0(+\infty)$, respectively; C_i^- and C_j^+ are constants. The $(n+m)$ boundary conditions of the continuity of the flow variables and their spatial derivatives must be satisfied at the point $x = 0$, where the solutions join whose asymptotics as $x \to \mp \infty$ are described by (1.7.15,16). One more condition is required for eliminating the ambiguity arising from the possible shift of argument in (1.7.1). This condition can be obtained by equating one of the components of vector W at $x = 0$ to some average value between the two extremes:

$$W_i(0) = W_i' \quad ,$$

where i is any whole number from 1 to n; W_i' is assigned any fixed value from $W_i(-\infty)$ to $W_i(+\infty)$. Thus we have a total of $n + m + 1$ boundary conditions.

The solution of type (1.7.1) is entirely determined by its asymptotics (1.7.15,16); that is, by the set of $q_1 + q_2$ constants C_i^- and C_j^+. Since $q_1 + q_2 < n + m + 1$, the existence of the solution implies that the asymptotic values of W(x) — the components of vectors $W_0(-\infty)$ and $W(+\infty)$ — are not independent, but rather are bound by certain relations, whose number

$$r = n + m + 1 - q_1 - q_2 = n + m + 1 - (p_1 - s_1) - (p_2 - s_2)$$
$$= s_1 + s_2 + 1$$

coincides with the number indicated in (1.7.8), and thus complies with the evolutionary conditions. This was to be proved.

If, however, (1.7.8) is violated, then $(q_1 + q_2)$ parameters C_i^-, C_j^+ have to satisfy $\ell = n + m - s_1 - s_2 \neq q_1 + q_2$ boundary conditions. In other words, either the prescribed boundary conditions allow the existence of a multitude of different structures (if $\ell < q_1 + q_2$), or a stationary structure (1.7.1) exists only with certain fixed values of $W_0(-\infty)$, $W_0(+\infty)$, lacking the capability of conforming to their perturbations (if $\ell > q_1 + q_2$); such discontinuities are nonevolutionary, and there is little sense in considering them.

The above-mentioned r boundary conditions should, in general, depend on the properties of the dissipative set of equations, (1.7.9). In most cases of practical interest, however, the boundary conditions which reflect certain physical realities on the front, such as the continuity of the fluxes of mass, momentum, energy, the tangential component of the electric field, the normal component of the magnetic field, do not depend on the nature of dissipations, and can be written down directly. These conditions may be called basic. In gasdynamics and magnetohydrodynamics the number of basic boundary conditions equals the number of parameters of dissipationless flow, and conforms with the conditions of evolutionarity of the shock. Accordingly, the dissipative qualities of the system affect only the characteristics of the dissipative waves, which determine the structure of the front. For example, the transition from the Navier-Stokes gasdynamics to the more accurate approximations (Sect.1.2) may result in a more detailed description of the front structure (1.7.1), but will not require modification of the Rankine-Hugoniot conditions, i.e., will not entail the possible discovery of new classes of gasdynamic shock waves.

In the theory of detonation and deflagration (Sect.1.9) and in the theory of ionizing shock waves (Chap.4), the basic conditions are sometimes insufficient; in other words, additional boundary conditions must exist, associated with the front-shaping physical processes, i.e., with the nature of dissipation. The above-named theories are concerned mainly with obtaining these additional boundary conditions.

For example, if $r = s_1 + s_2 + 1 > n$, then the existence of the front structure is made possible only by satisfying $(r - n)$ relations between the front velocity and the parameters of the unperturbed state. Occasionally, the front velocity can be fixed, as in the case of weak detonation or deflagration (Sect.1.9). If, on the other hand, $r = s_1 + s_2 + 1 < n$, then a given set of initial conditions corresponds to a number of finite $(n-r)$ parametric states, each of which is linked to the unperturbed state via the unique structure (1.7.1) and can be physically realized by the appropriate choice of the initial and boundary conditions. Examples of the ionizing shock waves of types 3 and 4 can be found in Chap.4 (dealing with the existence of structures) and Chap.5 (treating the realization of different structures in the piston problem).

Finally, let us observe that the existence of the stationary structure (1.7.1), postulated above, is not guaranteed by the prerequisite of evolutionarity alone. Such a solution can be constructed only if the set of equations (1.7.9) accounts for a sufficient number of dissipative mechanisms, whose exact number and nature should be determined in accordance with each particular case. As a rule, the inability to construct a solution representing the structure of the evolutionary shock wave simply indicates that some important dissipative mechanisms have been overlooked. In any case, we never have encountered any evidence violating this statement.

1.7.3 Spectra of Dissipative Waves, Corresponding to Shock-Wave Structure Described by Burgers' Equation

As we have seen in Sect.1.7.2, in the limit $\omega \to 0 + i0$ the waves describing the perturbations against the uniform background of the form (1.7.13) fall into two classes: the undamped waves ($k \to 0$), and the dissipative waves ($k \neq 0$). The undamped waves, incident on the front, transmit perturbations from the upstream and downstream infinities, corresponding to the varying boundary conditions; the outgoing undamped waves transmit the variations of variables, caused by the adjustment of the front, to the upstream and downstream infinities. The dissipative waves are responsible for the structure of the front, which is determined by the nonlinear interaction of the finite-amplitude dissipative waves. In the most simple case ($r = n$) we have: $q_1 + q_2 = m + 1$. The solution of the problem of the front structure can then be reduced, as a rule, to the m-dimensional phase space, in which the manifolds of integral curves have the dimensions q_1 and q_2, respectively, these manifolds being formed by the profiles of the dissipative waves dying away towards the upstream and downstream infinities, going out of point 0 and into point 2. The existence of the front structure implies that these manifolds intersect with one another. When these manifolds transversally intersect, the dimensionality of the intersection is $q_1 + q_2 - m = 1$. In other words, in the nondegenerate case the structure of the front is unique (the illustrations for $m = 2$ and $m = 3$ can be found in Sects.1.8 and 3.5.3, respectively). The shock structure is in a one-to-one correspondence with the spectrum of dissipative waves (the set of constants C_i^- and C_j^+). In accordance with (1.7.15,16), the front width is estimated as

$$\Delta \sim \left[\min_{\substack{1 \leqslant i \leqslant q_1 \\ C_i^- \neq 0}}(-\mathrm{Im}\{k_i\})\right]^{-1} + \left[\min_{\substack{1 \leqslant j \leqslant q_2 \\ C_j^+ \neq 0}}(\mathrm{Im}\{k_j\})\right]^{-1} . \tag{1.7.17}$$

Observe, finally, that the strict discrimination between the dissipative (front-shaping) waves and the undamped waves, which transmit perturbations, is based on the adopted hydrodynamic approximation, the equations of type (1.7.9). In reality this question is more complicated. The flow parameters always fluctuate about the hydrodynamic averages. The spectra of fluctuations are in equilibrium in the regions far upstream and downstream. The equilibrium fluctuations arrive at the front as incident waves; however, after having been reflected or refracted by the front, the outgoing fluctuations are no longer in equilibrium: in the second-order approximation with respect to the amplitude they affect the flow parameters near the front [1.37,38]. It is not hard to find the law describing the decrease of the fluctuations' contribution with distance. The damping coefficient for the long-wavelength fluctuations with the frequency ω is proportional to ω^2; hence at a given distance $|x|$ from the front, those fluctuations are important whose frequency is below $\mathrm{const}|x|^{-\frac{1}{2}}$. Their effect is proportional to their number, i.e., to the occupied portion of the phase space. The phase volume has the order of the third power of the maximum wave number (i.e., of the third power of frequency), and thus is found to decrease as $|x|^{-3/2}$. An expression for the coefficient in this law can be found in [1.38].

Let us use the simplest Burgers' equation for illustrating the aforesaid. In the coordinate system traveling at speed D, (1.5.15) has the form

$$c_t - Dc_x + cc_x - \mu c_{xx} = 0 \quad . \tag{1.7.18}$$

The stationary solution of (1.7.18), which describes the shock transition $c_0 \to c_2$, is given by (1.6.18), the front velocity being $D = (c_0 + c_2)/2$, see (1.6.17). The dissipationless flow is described by (1.7.18) with $\mu = 0$, i.e., by a single variable c ($n = 1$). The waves against the uniform background $c = c_0$ propagate at the speed $c_0 - D$; in other words, there are no undamped outgoing waves ($s_1 = s_2 = 0$) due to the condition (1.6.17). Consequently, the evolutionarity of the shock wave guaranteed by the existence of the structure (1.6.18) depends on the existence of a single condition (1.6.17): $r = s_1 + s_2 + 1 = 1$.

The exact equation (1.7.18) [following Sect.1.7.2, the last term in (1.7.18) implies that $m = 1$], corresponds with the following dispersion equation for the perturbations against the uniform background c_0:

$$k^2 + ik\frac{c_0 - D}{\mu} - i\frac{\omega}{\mu} = 0 \quad . \tag{1.7.19}$$

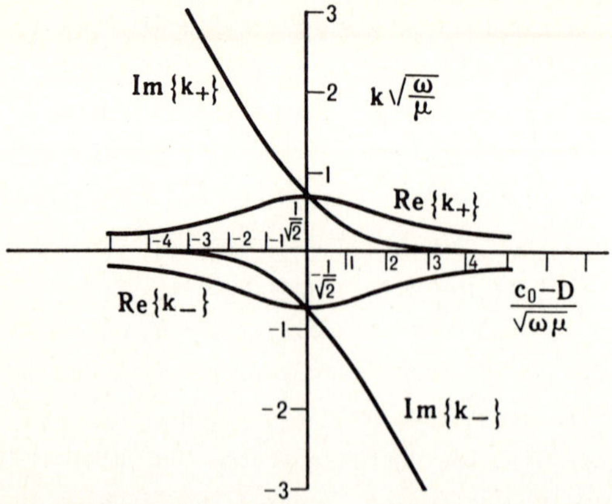

Fig.1.3. The dispersion curves for a fluid described by Burgers' equation

The change of variables $\zeta = (c_0 - D)/\sqrt{\omega\mu}$ and $\kappa = k\sqrt{\omega/\mu}$ illustrates the form of the solutions of (1.7.19) for any values of the parameters c_0, D, ω, μ. Figure 1.3 shows the branches of the dispersion curves for real ω. In accordance with Sect.1.7.2, one wave always exists traveling in the positive direction and one wave traveling in the negative direction (in Fig.1.3 they are denoted by + and -, respectively). Depending on the relation between the components c_0 and D of the vector W_0 (i.e., on the sign of ζ), the same branches of the dispersion curve may correspond to either dissipative or undamped waves. In the limit $\omega \to 0$ (i.e., $\zeta \to \pm\infty$ for $c_0 \neq D$), for $c_0 > D$ the undamped wave runs in the positive direction ($k_+ \to 0$, $\mathrm{Im}\{k_+\}/\mathrm{Re}\{k_+\} \to 0$), and the dissipative wave in the negative direction ($\mathrm{Im}\{k_-\} \to -(c_0 - D)/2\nu$). The situation is reversed with $c_0 < D (\mathrm{Im}\{k_+\} \to -(c_0 - D)/2\nu)$.

Setting in (1.6.18) const = 0, we get the asymptotics of the form (1.7.15, 16):

$$c \sim c_0 - (c_2 - c_0) \exp\left(-\frac{c_2 - c_0}{2\mu} x\right) \quad \text{as} \quad x \to -\infty \ , \tag{1.7.20}$$

$$c \sim c_2 + (c_2 - c_0) \exp\left(-\frac{c_2 - c_0}{2\mu} x\right) \quad \text{as} \quad x \to +\infty \ . \tag{1.7.21}$$

The spectrum of dissipative waves, corresponding to the structure of the shock wave of the form (1.6.18) (const = a) is thus determined by the values of two constants, $C^+ = -C^- = c_2 - c_0$. The shifting of the structure, i.e., the different choice of the constant, const = a, would correspond to

$$C^+ \to C^+ \exp\left(-\frac{c_2 - c_0}{2} a\right) \ ,$$

$$C^- \to C^- \exp\left(\frac{c_2 - c_0}{2\mu} a\right) .$$

Using the two constants C^- and C^+ allows the location of the front to be fixed [i.e., the value of the constant in (1.6.18)], and the continuity of c and c_x to be secured (a total of $n+m+1=3$ conditions), since the initial and the final values of $c-c_0$ and c_2^- are bound by one ($r=1$) condition (1.6.17). For the interaction of shock waves with perturbations of finite amplitude, described by Burgers' equation, see [1.20,39].

The right-hand side of (1.7.17) estimates the width of the shock front Δ exactly coinciding with the Prandtl estimate:

$$\Delta \sim \frac{1}{|\mathrm{Im}\{k_-\}|} + \frac{1}{|\mathrm{Im}\{k_+\}|} = \frac{2\mu}{|c_0 - D|} + \frac{2\mu}{|c_2 - D|}$$

$$= \frac{8\mu}{c_2 - c_0} = \frac{c_2 - c_0}{\max|dc/dx|} . \qquad (1.7.22)$$

Let us also observe the peculiar form of the solutions of the dispersion equations (1.7.19) at $c_0 = D$:

$$k_\pm = \pm \frac{i+1}{2}\sqrt{\frac{\omega}{2\mu}} . \qquad (1.7.23)$$

The dispersion law (1.7.23) corresponds to the heat-conduction or diffusion type parabolic equation, which describes the diffusive blurring of the running wave in the wave-fixed coordinate system, due to the last term in (1.7.9). The dispersion law for the rotational and the entropy perturbations has exactly the same form (Sect.1.4.1). The waves of the form (1.7.23) are strongly damped ($\mathrm{Im}\{k_\pm\} = \mathrm{Re}\{k_\pm\}$), and cannot exist in a dissipationless medium, which makes them similar to the dissipative waves. On the other hand, at $\omega \to 0$ we have $k_+ \to 0$, $k_- \to 0$ and thus a certain similarity with the undamped waves as well. With regard to the front of any finite width the waves of sufficiently low frequency can be assumed to transmit the perturbations from the front to the upstream and downstream infinities [the criterion (1.7.3) is here replaced by the condition $\omega \ll \Delta^2/\mu$]. It would be natural to call these waves the diffusion waves; they constitute a class of waves intermediate between undamped and dissipative waves. The diffusion waves must be taken into account when checking the evolutionarity of singular shock waves (Chap.3).

1.7.4 Stability and Evolutionarity of Plane Discontinuities in Three Dimensions

So far we have dealt only with plane discontinuities. The perturbations also had the form of plane waves whose wave vectors were normal to the front and thus did not violate the unidimensionality of the problem. This approach is based on the same approximation which allows the concept of a discontinuity to be introduced (the smallness of the front width on the characteristic scale of the problem). It is assumed that as soon as the reciprocal curvature of the front is small on the characteristic scale, the front in the vicinity of any given section can be considered plane in a first approximation. This assumption is justified if (a) the front shape is a local characteristic of the flow, i.e., there is no long-range nonhydrodynamic interaction between different regions of the front, and (b) the plane front is stable, i.e., tends to resume its shape after the removal of perturbations. If either or both of these conditions are violated, the nonunidimensionality of the discontinuity becomes important. In Chap.3 we shall discuss the rigorousness of condition (a) for the shocks in a plasma with long-range electromagnetic interactions. At present we shall restrict our attention to condition (b).

The formal analysis of stability in the three-dimensional case may be carried out along the same lines as described in Sect.1.7.1. The greater diversity of solutions of the dispersion equation is due to the appearance of a new length-like variable, namely, the perturbation wavelength along the front plane. For this reason the spectrum of even an ideal medium may display frequencies with $\mathrm{Im}\{\omega\} \neq 0$ (in particular, $\mathrm{Im}\{\omega\} > 0$, which indicates instability). Such treatment of shock waves in a one-component medium in the absence of external fields was performed in [1.40-42]. Our treatment will be based on a somewhat different approach, which allows a vivid description of nonunidimensional shock waves: the geometrical dynamics of the shock waves [1.20].

The principal variable used in this approach is the area A of the so-called ray tube, formed by the normals to the front in its successive positions. The equation which connects A with the Mach number M_0 of the shock wave for the near-unidimensional flow (Fig.1.) has the form

$$\frac{dM_0}{dx} = -\frac{1}{g(M_0)A}\frac{dA}{dx} \quad , \tag{1.7.24}$$

where the unperturbed discontinuity plane is normal to the x axis, and the function $g(M_0)$ is determined by the medium's equation of state and can be

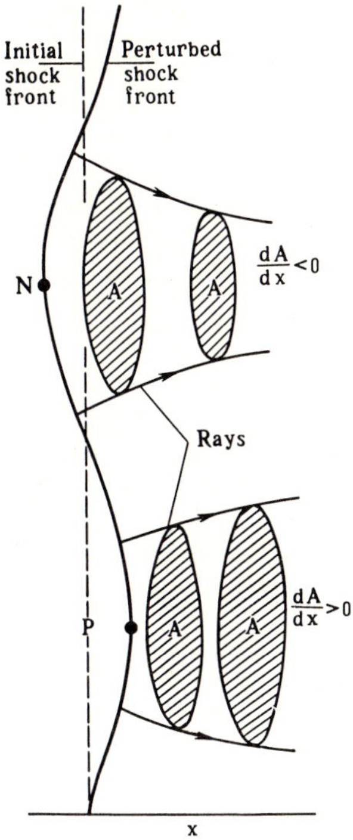

Fig.1.4. Variations of area of ray tubes upon disturbing the planarity of the shock front

presented in the form [1.43]

$$g(M_0) = \frac{1}{M_0}\left[M_2\left(\frac{\rho_2}{\rho_0} - 1\right) + 1\right]\frac{1 + 2M_2 - \eta}{1 + \eta} , \tag{1.7.25}$$

where ρ_2/ρ_0, M_2 and the parameter

$$\eta = M_2^2 \frac{(dp_2/d\rho_2)_S}{(dp_2/d\rho_2)_H} \tag{1.7.26}$$

are with a given equation of state the known functions of M_0 (the derivative in the numerator corresponds to the adiabatic speed of sound downstream; the derivative in the denominator is taken along the shock adiabat). The expression for the ideal gas (γ = const), corresponding to (1.7.25), can be found in [1.20].

As becomes clear from Fig.1.4, the derivative dA/dx is positive for the leading portions of the front (near point P) and negative for the lagging portions (near point N). If $g(M_0) > 0$, then in accordance with (1.7.24) the lagging portions will be accelerated, and the leading retarded. The perturbation of the front (dissipations disregarded) will be propagated along the

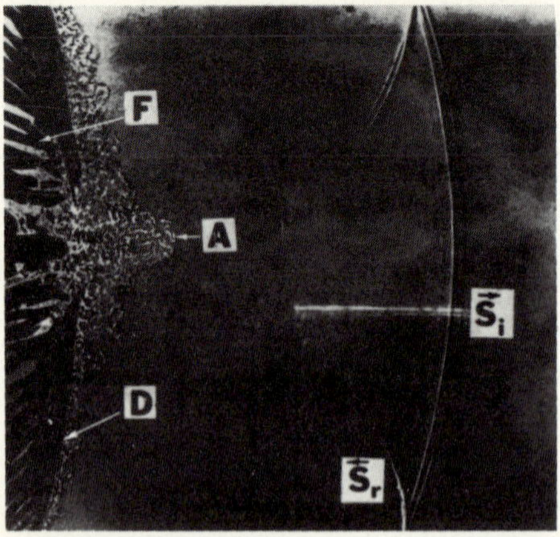

Fig.1.5. Production of a shock wave in a shock tube. When diaphragm D is ruptured, its fragments F let through a jet of compressed air A, driving the shock wave $\vec{S}_i$ · $\overleftarrow{S}_r$ denotes the shock reflected from tube wall [1.44]

front in the form of running linear waves [1.20], while the front as a whole remains locally plane. In particular, $g(M_0) > 0$ for an ideal gas, which is why the fronts of gasdynamic shock waves display a very symmetric shape even when their movement is maintained by quite complicated flows (Fig.1.5).

On the contrary, if $g(M_0) < 0$, then the perturbations of the initially plane front tend to build up: the lagging portions lose and the leading portions gain. The corrugation of the front becomes more and more pronounced; hence the term corrugation instability. Consequently, the shock front cannot remain plane, and assumes a more intricate shape. This instability is evidently similar to the Rayleigh-Taylor instability: the shock-compressed medium acts as a heavy fluid driven by a light fluid (the unperturbed flow). This instability arises in a thermodynamically normal medium ($\rho_2/\rho_0 > 1$) if either of the following conditions is satisfied:

$$\eta < -1 \quad , \tag{1.7.27}$$

or

$$\eta > 1 + 2M_2 \quad . \tag{1.7.28}$$

Then the spectrum displays frequencies with $\mathrm{Im}\{\omega\} > 0$. Together with the corrugation instability, the possible existence of branches of the dispersion curve with real-valued ω, k, corresponding to the stationary undamped perturbations, not tending to infinity, was discovered quite early [1.40,41]. This instability arises if an inequality

$$\eta > 1 - \frac{2M_2^2(\rho_2/\rho_0)}{1 + M_2^2(\rho_2/\rho_0 - 1)} \quad , \tag{1.7.29}$$

weaker than (1.7.28), is satisfied.

The physical meaning of this latter kind of instability was elucidated in [1.42,45]: when the perturbations taken into consideration include plane sonic waves with arbitarily directed wave vectors, then the plane discontinuity is nonevolutionary. The condition of evolutionarity of a plane discontinuity in three dimensions can be formulated in terms of the requirement that the coefficients of reflection and transmission of linear waves should take on finite values for all types of waves and all angles of incidence (in the one-dimensional case we consider only the normal incidence, $\theta = 0$). The inequality (1.7.29) implies that for a certain angle of incidence $\theta = \theta_{cr}$ of the sonic waves coming from downstream, the coefficient of reflection goes to infinity. Consequently, the front becomes capable of emitting spontaneously the sonic waves in certain directions, which leads to the splitting of the initially plane discontinuity [1.42,43,45]. Near the instability threshold (1.7.29) the wave vectors of emitted waves are parallel to the discontinuity plane.

In an ideal gas with $\gamma = \text{const}$ the inequality (1.7.29) cannot be satisfied, and the splitting instability does not arise. The estimates indicate that this kind of instability can be manifested by the shock waves in solids, and by ionizing shock waves (Chap.2).

We see that including three-dimensional perturbations into the consideration results in a much more complicated relationship between the conditions of stability (or evolutionarity) and the existence of the stationary structure of the front of type (1.7.1), than in a purely one-dimensional problem (Sect.1.7.2). As a matter of fact, an ingenious choice of transfer coefficients allows one to construct one-dimensional structures of type (1.7.1) also for the shock fronts in such media, where the instability conditions (1.7.27-29) are satisfied. Consequently, the existence and uniqueness of the structure (1.7.1) no longer ensure the evolutionarity of the front. The nonunidimensionality of the flow in the present case may possibly entail spontaneous violation of the symmetry of the problem imposed by the boundary conditions. In other words, not only may many nonunidimensional stationary (or nonstationary) shock structures exist of the form $W_0(x,y,z)$ [or $W_0(x,y,z,t)$], which, as $x \to \mp\infty$, tend to the same asymptotic values $W_0(\mp\infty)$, but they also may be more energetically advantageous than the higher-symmetry solution (1.7.1) [1.46]. The question regarding the form of the flow when the conditions of stability of the plane shock front are violated is as yet unsettled. There is little doubt, however, that near the initial discontinuity the flow must become turbulent. The study of these problems, which are of great importance for the theory of collisionless shock waves and the theory of detonation, is one of the major directions of future development of shock-wave science.

1.8 Structures of Gasdynamic Shock Waves

1.8.1 Equations of the Shock Layer

Now we obtain the equations which describe the structures of gasdynamic shock fronts in the Navier-Stokes approximation. This approximation works fairly well with shocks of moderate intensity ($M_0 \gtrsim 1$), while stronger shock waves are more adequately described by the Mott-Smith approximation [1.34]. At the moment, however, we are concerned with the computative techniques used for analyzing the properties of shock waves in plasmas, rather than with the actual structures of gasdynamic shock fronts. Accordingly, we shall base our discussion on a number of instructive examples, without going into particulars and experimental results [1.12-15,18,47,48].

The equations for the stationary shock layer in the present case are represented by the first integrals of the equations of continuity, motion and energy balance of the gas flow, which reflect the conservation laws of mass, momentum and energy. Hereafter it will be convenient to use dimensionless variables, but for the moment we shall denote the dimensional variables with a bar: $\bar{\rho}$, $\bar{u}_x$, $\bar{T}$. The dimensionless variables are normalized with respect to the values of the respective variables in the unperturbed flow (as $x \to -\infty$):

$$\rho = \bar{\rho}/\bar{\rho}_0 \quad , \quad u_x = \bar{u}_x/\bar{u}_{x0} \quad , \quad T = \bar{T}/\bar{T}_0 \quad . \tag{1.8.1}$$

The integral equations for the ideal gas ($\gamma = \text{const}$) in dimensional variables have the form

$$\bar{\rho}\bar{u}_x = \text{const}_1 \quad , \tag{1.8.2}$$

$$\bar{\rho}\bar{u}_x^2 + \bar{p} - \left(\frac{4}{3}\eta + \zeta\right)\frac{d\bar{u}_x}{dx} = \text{const}_2 \quad , \tag{1.8.3}$$

$$\bar{\rho}\bar{u}_x\left(\frac{\bar{u}_x^2}{2} + \frac{\gamma}{\gamma - 1}\frac{\bar{p}}{\bar{\rho}}\right) - \kappa\frac{d\bar{T}}{dx} - \left(\frac{4}{3}\eta + \zeta\right)\bar{u}_x\frac{d\bar{u}_x}{dx} = \text{const}_3 \quad , \tag{1.8.4}$$

where κ, η, ζ are the coefficients of viscosity and heat conductivity of the gas. The values of the constants on the right-hand sides of (1.8.2-4) are found by letting $x \to -\infty$, when all the spatial derivatives vanish, and

$\bar{\rho} = \bar{\rho}_0$, $\bar{u}_x = \bar{u}_{x0}$, $\bar{T} = \bar{T}_0$, or $\rho = u_x = T = 1$. The coordinate system can always be chosen so that $\bar{u}_{y0} = \bar{u}_{z0} = 0$ (then it can be easily proved that $\bar{u}_y = \bar{u}_z = 0$ throughout the front; in other words, the flow is unidimensional). Exactly such a coordinate system is used here.

From (1.8.2) we have

$$\rho u_x = 1 \quad . \tag{1.8.5}$$

Substituting ρ from (1.8.5) into (1.8.3,4) we obtain the shock-layer equations in dimensionless variables:

$$u_x - 1 + \frac{1}{\gamma M_0^2}\left(\frac{T}{u_x} - 1\right) - \Delta_V \frac{du_x}{dx} = 0 \quad , \tag{1.8.6}$$

$$\frac{u_x^2 - 1}{2} + \frac{1}{(\gamma - 1)M_0^2}(T - 1) - \Delta_V u_x \frac{du_x}{dx} - \Delta_T \frac{dT}{dx} = 0 \quad , \tag{1.8.7}$$

where

$$\Delta_V = \frac{\frac{4}{3}\eta + \zeta}{\bar{\rho}_0 \bar{u}_{x0}} \quad , \quad \Delta_T = \frac{\kappa \bar{T}_0}{\bar{\rho}_0 \bar{u}_{x0}^3}$$

are the characteristic scales of viscosity and heat conductivity, and the Mach number has been defined in Sect.1.6.3. Making use of the elementary gas-kinetic estimates of the coefficients of viscosity and heat conduction, we can demonstrate that for an ideal gas the values of Δ_V and Δ_T are functions of dimensionless temperature T and have the order of the mean free path of an atom or a molecule in the gas at a given temperature. The relative importance of the viscosity and heat conduction is defined by

$$\frac{\Delta_V}{T\Delta_T} = \gamma(\gamma - 1)\mathrm{Pr}\left(\frac{M_0^2}{T}\right) \quad ,$$

where the Prandtl number $\mathrm{Pr} = \nu/\chi$, (ν being the kinematic viscosity, χ being the temperature conductivity), which for gases is usually of the order of unity. The ratio (M_0^2/T) in the forefront region may be high with $M_0 \gg 1$, but reduces to unity as the temperature of the gas increases, since downstream $T_2 \sim M_0^2$. Consequently, the viscosity and the heat conduction play a more of less equal role in the shaping of the front.

The structure of the front of a gasdynamic shock wave is represented by the solution of the set of equations (1.8.6,7) with the boundary conditions (1.6.29,30):

$$\text{as} \quad x \to -\infty \quad u_x = 1 \quad , \quad T = 1 \quad ; \tag{1.8.8}$$

$$\text{as} \quad x \to +\infty \quad u_x \to u_{GD} = \frac{(\gamma - 1)M_0^2 + 2}{(\gamma + 1)M_0^2} \quad ,$$

$$T \to T_{GD} = \frac{[(\gamma - 1)M_0^2 + 2][2\gamma M_0^2 - \gamma + 1]}{(\gamma + 1)^2 M_0^2} \quad .$$

1.8.2 Shock Structure Shaped by Viscosity Alone

Now we consider an extreme case of low heat conductivity of the gas, when the Prandtl number is high: $Pr \gg 1$, $\Delta_v \gg \Delta_T$. Such a situation is not common with ordinary gases, but it may occur in a highly magnetized plasma (Chap.3). Assuming that the shaping of the front structure is then dominated by the viscous effects, i.e., the derivatives of the dimensionless velocity u_x have the order of Δ_v^{-1}, in the first approximation with respect to Δ_T/Δ_v we may disregard the effects of heat conduction and drop the last term in (1.8.7). Then (1.8.6,7) can be used to express du_x/dx and T via u_x:

$$\frac{du_x}{dx} = -\frac{\gamma + 1}{2u_x\Delta_v}(1 - u_x)(u_x - u_{GD}) \quad , \tag{1.8.9}$$

$$T = 1 + \gamma(\gamma - 1)(1 - u_x)\left(1 + \frac{2}{\gamma M_0^2} - \frac{2u_x}{\gamma M_0^2}\right)\frac{M_0^2}{2} \quad . \tag{1.8.10}$$

Then with the aid of (1.8.10) the temperature dependence $\Delta_v(T)$ can be reduced to the dependence on velocity, the value of $\Delta_v(u_x)$ in the interval $u_{GD} \leq u_x \leq 1$ always being positive and continuous. We see that the problem of the shock structure with only one dissipative mechanism taken into account reduces to a unidimensional phase space; in other words, its solution can be obtained in quadratures, since the corresponding differential equation is autonomous (the coordinate x enters the equation only as a derivative). If $\Delta_v \propto T^\sigma$, where σ is an integer or a semi-integer, the solution can be obtained in elementary functions; with $\sigma = 0$ it has the form

$$\frac{1 - u_x}{(u_x - u_{GD})^r} = \exp\left[(1 - M_0^{-2})\frac{(x + C)}{\tilde{\Delta}}\right] \quad , \tag{1.8.11}$$

where $r = u_{GD}$, $\tilde{\Delta} = \Delta_v$. In the notation used in (1.7.17), the coefficient at x in the exponent is $1/|\mathrm{Im}\{k_0\}|$, whereas the value of $1/|\mathrm{Im}\{k_2\}|$, according to

(1.8.11), is r times as small. Using (1.7.17), we estimate the width of the front of the viscous gasdynamic shock wave

$$\Delta_k = \frac{2(\gamma M_0^2 + 1)}{(\gamma + 1)(M_0^2 - 1)} \Delta_v \sim \begin{cases} \dfrac{2\Delta_v}{M_0 - 1} , & M_0 - 1 \ll 1 \\ \dfrac{2\gamma}{\gamma + 1} \Delta_v , & M_0 \gg 1 . \end{cases} \tag{1.8.12}$$

The estimate (1.8.12) can be compared to the Prandtl estimate of the front width:

$$\frac{\Delta_{Pr}}{\Delta_v} = \frac{1}{\Delta_k} \frac{1 - u_{GD}}{|du_x/dx|_{max}} = \frac{2(\gamma M_0^2 + 1)}{(\gamma + 1)(1 + u_{GD}^{1/2}) M_0^2} \sim \begin{cases} 1 , & M_0 - 1 \ll 1 \\ 1 + (1 - \gamma^{-2})^{1/2} , & M_0 \gg 1 . \end{cases} \tag{1.8.13}$$

We see that both estimates coincide for weak shock waves. For strong shock waves the Prandtl estimate, as expected, is the estimate from below (based on the maximum gradient), differing from (1.8.12) by a factor between 1 and 2, depending on γ. This difference arises from the difference between $1/|\mathrm{Im}\{k_1\}|$ and $1/|\mathrm{Im}\{k_2\}|$; that is, from the fact that r in (1.8.11) is not equal to unity. Here the difference between the two estimates is due only to the compression of the gas, which even with the strongest shocks has the order of unity. If we take into account the temperature dependence of Δ_v, the discrepancy between the estimates may become quite large. Generally speaking, it is the estimate of type (1.7.17), and not the Prandtl estimate, that corresponds to the experimentally measured width of the front.

For weak shock waves ($M_0 - 1 \ll 1$, $1 - r \ll 1$), disregarding the difference between r and unity, we can rewrite (1.8.11) in the form ($C = 0$)

$$u_x = 1 - \frac{1 - u_{GD}}{1 + \exp\left[(1 - M_0^{-2}) \frac{x}{\bar{\Delta}}\right]} . \tag{1.8.14}$$

This form of the solution is fairly universal for weak shock waves, regardless of the equation of state of the thermodynamically normal medium and the front-shaping dissipative processes. The solution (1.8.14) can be obtained by expanding the flow parameters in powers of small deviations from the unperturbed values, the deviation being proportional to $M_0 - 1$. As expected, this solution coincides with the solution (1.6.18) of Burgers' equation.

1.8.3 Shock-Front Structure in a Gas with High Heat Conductivity

Let us now consider another extreme case, when the heat conductivity dominates over the viscosity, i.e., $Pr \ll 1$. This case is also uncommon with gases, but is encountered with nonmagnetized plasmas (Chap.3). This time the zero-order approximation with respect to the Prandtl number as a small parameter means that we drop the viscous terms in (1.8.6,7) and consider only the heat conduction, getting from (1.8.6)

$$T = u_x(\gamma M_0^2 + 1) - \gamma M_0^2 u_x^2 \quad . \tag{1.8.15}$$

Expressing T via u_x with the aid of (1.8.15) and substituting in (1.8.7), we get

$$\frac{du_x}{dx} = \frac{(\gamma + 1)}{(\gamma - 1)\Delta_T} \frac{(1 - u_x)(u_x - u_{GD})}{\gamma M_0^2 + 1 - 2\gamma M_0^2 u_x} \quad . \tag{1.8.16}$$

Equation (1.8.16) is similar in form to (1.8.9). The difference between these two equations is that (1.8.9) always admits a continuous solution, representing the structure of the shock front, while (1.8.16) has a solution only when the denominator in its right-hand side is nonzero in the entire interval $u_{GD} \leq u_x \leq 1$. The formal solution of (1.8.16) with $\Delta_T = \text{const}$ has the form (1.8.11), where

$$\tilde{\Delta} = (\gamma - 1)(\gamma M_0^2 - 1)\Delta_T \quad ; \quad r = \frac{\gamma(3 - \gamma)(M_T^2 - M_0^2)}{(\gamma + 1)(M_0^2 - 1/\gamma)}$$

$$M_T^2 = \frac{3\gamma - 1}{\gamma(3 - \gamma)} > 1 \quad . \tag{1.8.17}$$

Such a solution satisfies the boundary conditions (1.8.8) only when $r > 0$, i.e., when

$$M_0^2 < M_T^2 \quad , \quad M_0^2 > 1/\gamma \quad . \tag{1.8.18}$$

The latter inequality is satisfied always ($M_0 > 1 > 1/\gamma$), whereas the former sets the upper limit for the velocity of the gasdynamic shock wave, front structure of which can be shaped by the heat conductivity alone. With $M_0^2 < M_T^2$ the heat-conduction-shaped front structure possesses all the qualities discussed above (Sect.1.8.2) with reference to the viscous structures. In particular, the Prandtl criterion tends to underestimate the width of the front with respect to the estimate given by (1.7.17), although the two estimates coincide for weak shock waves, when $M_0 - 1 \ll 1$ [then, according to (1.8.17), $1 - r \ll 1$], and the shock structure has the form (1.8.14).

The structure of a sufficiently strong shock wave cannot be shaped by the heat conductivity alone, however large it may be. Physically speaking, this is because the heat-conduction mechanism efficiently dissipates the inflow of the energy brought to the front by the moving gas. At the same time, the gas velocity can be changed only via the variations in the pressure, and therefore this dissipative mechanism (like other physical mechanisms incapable of directly dissipating the momentum of the incoming flow, e.g., diffusion, relaxation) is by itself incapable of arresting a sufficiently rapid flow of a gas. Accordingly, the viscous terms must be considered, no matter how small the Prandtl number.

Here we encounter a typical example of the multiple-scale asymptotic method, much used in analyzing the shock structures in more or less complicated cases. This method consists essentially of setting up a hierarchy of scales, pertaining to different physical processes and having highly different magnitudes. The analysis of a structure starts with the largest scale $\Delta^{(1)}_{max}$, on which all the derivatives have the order of $1/\Delta^{(1)}_{max}$, and all the terms pertaining to the physical processes whose scales are $\Delta_i \ll \Delta^{(1)}_{max}$ are small with respect to the parameters $\Delta_i/\Delta^{(1)}_{max}$. Disregarding these terms in the zero-order approximation, we get the equations of shock layer, which account only for the physical process with the scale $\Delta^{(1)}_{max}$. Such a set of equations can always be solved in quadratures, as demonstrated above.

If this procedure brings us to a solution satisfying the appropriate Rankine-Hugoniot boundary conditions, then the problem is settled: the structure of the front is shaped entirely by the physical mechanism in question, and the front width has the order of Δ_{max}. Such was the case considered in Sect.1.8.1, where the shock-layer equations contained the scales of viscosity Δ_v and heat conduction Δ_T. The front structure was found to be shaped by viscosity ($\Delta_v \equiv \Delta^{(1)}_{max} \gg \Delta_T$), and the account for the heat conduction results only in small corrections of the order Δ_T/Δ_v.

If, however, there is no representative solution on the length scale $\Delta^{(1)}_{max}$, then one has to consider the region of domination of another physical mechanism, whose characteristic scale is next in order to $\Delta^{(1)}_{max}$, $\Delta^{(2)}_{max} \ll \Delta^{(1)}_{max}$. In terms of the outer scale $\Delta^{(1)}_{max}$ in the zero-order approximation in the small parameter $\Delta^{(2)}_{max}/\Delta^{(1)}_{max}$, this region is the inner discontinuity in the solution with the characteristic scale $\Delta^{(2)}_{max}$, and it is the form of this solution that defines the location of the inner discontinuity in the structure. The necessary additional boundary condition on such a discontinuity usually reduces to requiring the continuity of certain flow parameters. This is because the derivatives describing the action of the mechanism with the scale $\Delta^{(1)}_{max}$ must have the order of smallness of $\Delta^{(2)}_{max}/\Delta^{(1)}_{max}$ within the limits of the discontinuity, so that the respective terms in the dimensionless equations should have the order of unity. If the inclusion of the physical mechanism with the scale $\Delta^{(2)}_{max}$ is sufficient for determining the structure of the internal discontinuity, the problem is solved. (Such is the problem of the structure of the gasdynamic shock wave when $Pr \ll 1$; then $\Delta^{(1)}_{max} = \Delta_T$, $\Delta^{(2)}_{max} = \Delta_v$, and the inclusion of viscosity is sufficient for setting up a solution). Otherwise, one has to consider yet another discontinuity on the scale $\Delta^{(3)}_{max} \ll \Delta^{(2)}_{max}$, nested within the internal discontinuity, and so on.

If the scales contained in the equations for the shock layer are arranged in a strict order ($\Delta^{(1)}_{max} \gg \Delta^{(2)}_{max} \gg \Delta^{(3)}_{max} \ldots$), then at each step of getting a solution the phase space is unidimensional, and the solution can be obtained in quadratures. Sometimes, however, several characteristic scales may have the same order of magnitude, in other words, several physical processes are equally important in shaping the front structure, then the dimensionality of the phase space m equals to number of such processes. The pertinent examples for $m = 2$ and $m = 3$ can be found below in this section and in Sect.3.5.3, respectively. In this case, obviously, it is much harder to determine the structure of the front, but the task is facilitated by the multiple-scale method, which allows one to single out the relevant characteristic scales.

In our present case the next characteristic scale after $\Delta^{(1)}_{max} = \Delta_T$ is $\Delta^{(2)}_{max} = \Delta_v$. Then from (1.8.7) we obtain

$$\frac{dT}{dx} = O(\Delta_v/\Delta_T) \quad . \tag{1.8.19}$$

Thus, at $M_0 > M_T$ the structure of the front contains a viscous discontinuity, which is isothermal in the zero-order approximation in Δ_v/Δ_T. The condition of continuity of the energy flux is then replaced by an additional condition $T = \text{const}$, while the continuity of the mass and momentum fluxes in the viscous discontinuity still takes place. According to the results in Sect.1.7, the three boundary conditions correspond to two outgoing waves. Therefore, the evolutionarity of this discontinuity requires satisfying the inequalities $M_0' > 1$, $M_2' < 1$, where $M_{0,2}'$ are the Mach numbers with respect to the isothermal speed of sound $a_T = (p/\rho)^{\frac{1}{2}}_T = a/\sqrt{\gamma}$ upstream and downstream of the viscous discontinuity. As the value $M_0' = 1$ can be surpassed only in a viscous isothermal discontinuity, these inequalities are equivalent to the conditions $M_0^2 > 1/\gamma$,

$M_2^2 < 1/\gamma$. Making use of (1.6.32), which relates M_2^2 to M_0^2, we find that these conditions are equivalent to (1.8.18), i.e., the viscous isothermal jump in the front structure of a gasdynamic shock wave for $Pr \ll 1$ appears exactly when it is evolutionary.

The boundary conditions on the isothermal jump at fixed T are found by solving (1.8.15), quadratic in u_x. Its structure is described by (1.8.6), where $T = const$ (and hence $\Delta_v = const$), and is similar to (1.8.11). For any $T < T_{max} = (\gamma M_0^2 + 1)^2/4\gamma M_0^2$ the isothermal jump is possible from one branch of the curve (1.8.15) to the other, from the higher value of velocity to the lower one.

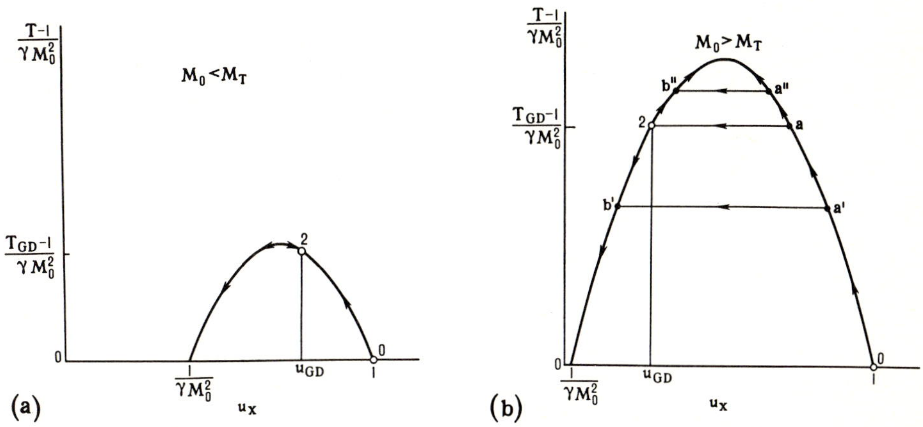

Fig.1.6a,b. Phase curves representing the structures of shock waves in the (u_x,T) plane for $Pr = 0$: (**a**) weak shock wave; (**b**) strong shock wave with viscous subshock

To determine the exact location of the viscous jump in the structure of the front, we plot the curve (1.8.15) in the (u_x,T) phase plane (Fig.1.6). The movement of the phase point, representing the state of the gas in accordance with (1.8.16), is indicated by arrows. For $M_0 < M_T$, moving along the curve (1.8.15) from point 0 we can get to point 2 (Fig.1.6a). If, however, $M_0 > M_T$, we cannot get from point 0 to point 2 along the same curve (Fig.1.6b): this would require an isothermal jump from one branch of the curve to the other. The boundary conditions admit the isothermal jump at any $T < T_{max}$ ($a' \to b'$, $a'' \to b''$, etc.). The arrows, however, do not enter point 2; therefore point 2 can be reached directly only from point a via the isothermal jump.

Consequently, the isothermal jump occurs at the rear of the front structure $(a \to 2)$.

We conclude that the structure of a gasdynamic shock wave for $M_0 > M_T$ commences with a thick ($\sim \Delta_T$) region of heating, shaped by heat conductivity, where the temperature rises from unity to a final value T_2, while the velocity decreases from unity to

$$u_a = \frac{2\gamma M_0^2 - \gamma + 1}{\gamma(\gamma + 1)M_0^2} \quad , \quad u_{GD} < u_a < 1 \quad .$$

The structure of this region is described by the solution (1.8.11), where $\tilde{\Delta}$ and r are given by (1.8.17). In the (u_x,T) phase plane this portion of the structure is depicted by the curve segment $0 \to a$ (Fig.1.6b). This portion is followed by the viscous isothermal jump, where at $T = T_2 = \text{const}$ the velocity goes down from u_a to u_{GD} (segment $a \to 2$). On the scale Δ_T this jump appears as a zero-thickness discontinuity, though actually it has a continuous viscosity-shaped structure. In the first-order approximation with respect to Pr it is described by the same solution (1.8.11), where $(1 - u_x)$ in the numerator in the right-hand side has to be replaced by $(u_a - u_x)$;

$$r = \gamma \frac{(\gamma - 1)M_0^2 + 2}{2\gamma M_0^2 - \gamma + 1} \quad ; \quad \tilde{\Delta} = \frac{(2\gamma M_0^2 - \gamma + 1)(M_0^2 - 1)}{\gamma(\gamma + 1)M_0^4} \, \nu \quad . \tag{1.8.20}$$

To get a clearer impression of why the structure of a gasdynamic shock wave $M_0 > M_T$ contains the isothermal jump in the zero-order approximation in Pr as a small parameter, it is expedient to plot the phase portrait of the integral curves of the complete set (1.8.6,7) for small Pr. By virtue of (1.8.6), (1.8.7) can be transformed into

$$2\Delta_T \frac{dT}{dx} = \frac{2}{\gamma(\gamma - 1)M_0^2}(T - 1) + (1 - u_x)\left(1 + \frac{2}{\gamma M_0^2} - u_x\right) \equiv F(u_x,T) \quad . \tag{1.8.21}$$

Hence, writing (1.8.6) in the form

$$\Delta_\nu \frac{du_x}{dx} = G(u_x,T) \quad , \tag{1.8.22}$$

we get

$$\frac{dT}{du_x} = \frac{\Delta_\nu}{2\Delta_T} \frac{F(u_x,T)}{G(u_x,T)} \quad . \tag{1.8.23}$$

We see that at $G(u_x,T) \neq 0$ the derivative $dT/du_x = O(Pr)$, and the integral curves of the set (1.8.21,22) on the phase plane (u_x,T) are arranged almost horizontally. The variations in T along each integral curve are mainly localized near the isocline $du_x/dx = 0$, whose equation (1.8.15) generates the phase curve plotted in Fig.1.6b. Knowing the isocline equation $dT/dx = 0$, or $F(u_x,T) = 0$, one can easily set up the phase portrait (Fig.1.7). Comparing the diagrams in Figs.1.6b,7 we see that the zero-order approximation in small Pr, in which the solution for $M_0 > M_T$ exhibits the zero-thickness isothermal jump, gives a fair approximation of the exact solution for small (finite) values of Pr.

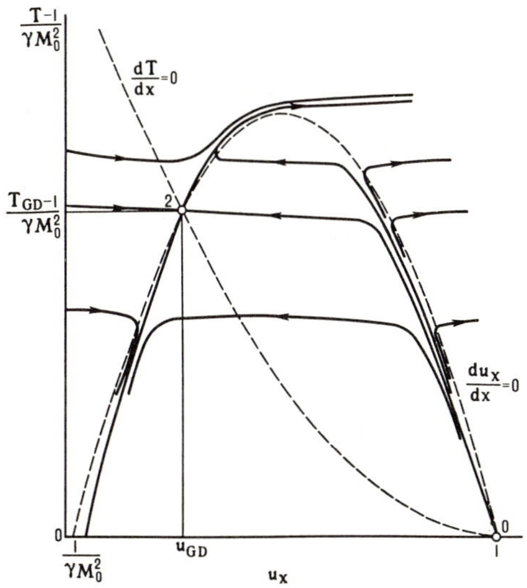

Fig.1.7. Integral curves of shock-layer equations in the (u_x,T) phase plane for finite-valued $Pr \ll 1$ (Fig.1.6b)

Figure 1.7 illustrates yet another property of the solutions for the set of shock-layer equations, representing the structures of the shock front. We see that the singular point 2 on the (u_x,T) plane is a saddle point, and, as $x \to \infty$, is entered by a single integral curve out of the number of the integral curves passing in the vicinity of point 0. It is this curve that gives the sought-for shock structure. Should we desire to integrate the set (1.8.21,22) numerically, we cannot start at point 0 because of its singularity: in this point the right-hand sides of both equations become zero. At the same time, having started with any point in the vicinity of point 0, we cannot get to point 2 as x is being increased (Fig.1.7). It is entirely impossible that the numeric (and hence approximative) integral curve should be that very unique curve which enters point 2 as $x \to \infty$. The proximity of the chosen starting point to point 0 does not even ensure that the calculated integral curve will pass anywhere near point 2 before going astray.

The situation is entirely different from that of the unidimensional problem. Then for an evolutionary shock wave $q_0 + q_2 = m + 1 = 2$, q_0 and q_2 being the dimensionalities of the manifolds of integral curves going out of point 0 and into point 2, respectively, as x was increased (Sect.1.7.3). Since in the unidimensional case $q_0 \leqslant 1$ and $q_2 \leqslant 1$, the only possibility is that $q_0 = q_2 = 1$. In other words, if the dissipative mechanism in question guarantees the existence of the front structure, then the only integral curve, emerging from point 0, inevitably tends to point 2, as $x \to \infty$, as long as x increases in the appropriate direction [according to the solution (1.8.11), in the sense of decreasing u_x].

For the two-dimensional phase space $q_0 + q_2 = m + 1 = 3$. Since $q_0 \leqslant 2$ and $q_2 \leqslant 2$, the existence of the structure is possible only with $q_0 = 2$, $q_2 = 1$ (or $q_0 = 1$, $q_2 = 2$), i.e., one of the singularities must be a node, and the other a saddle. In this case the direct numeric integration of the set of equations is possible; one has only to be careful in choosing the right starting point and proceeding in the right direction, from the saddle to the node. For instance, in the situation illustrated in Fig.1.7, the integration must start in the vicinity of point 2, to the right of the pair of separatrices that enter this singularity at $x \to -\infty$ (say, at $u_x(0) = u_{GD} - \varepsilon$, $T(0) = T_2 - \delta$, $\varepsilon, \delta > 0$), and proceed in the sense of decreasing x. Then, as follows from Fig.1.7, as $x \to -\infty$ we shall be sure to arrive at point 0, thus constructing an approximative numeric solution for the shock structure.

If, however, the phase space of the complete set of shock-layer equations is many-dimensional ($m \geqslant 3$), then there may be no nodal singularities at all: the singularities both upstream and downstream may be represented by saddles of varying orders ($q_0 < m$, $q_2 < m$). Consequently, despite that the stationary shock front structure is described by a self-contained set of ordinary differential equations—which poses no problem today to solve, however large the number of equations may be—the solution generally cannot be obtained by direct numerical integration. This is why the methods of approximation are needed, above all, the multiple-scale method, giving a qualitative result for the structure of the front. This approximate result can further be improved using iterative techniques.

1.9 Detonation and Deflagration

1.9.1 Propagation of an Exothermal Reaction. Equations of Structure of the Reaction Zone

The theory of detonation and combustion constitutes a vast area of chemical physics, dealt with in numerous specialized monographs [1.49-52], so here we need not to treat the fundamentals. We shall just use a few examples to demonstrate certain properties of discontinuities in media capable of maintaining exothermal chemical reactions.

We consider an explosive, an ideal gas A, in which at the temperature above $\bar{T} = \bar{T}^*$ an irreversible monomolecular reaction occurs

$$A \rightarrow B + Q \quad ; \tag{1.9.1}$$

the conversion of unit mass of gas A into the products of combustion (ideal gas B) is accompanied by the release of a certain amount of heat Q. An important parameter used in this theory is the dimensionless characteristic of the energy released in the reaction

$$q = Q/a_0^2 \quad , \tag{1.9.2}$$

where a_0 is the speed of sound in gas A at the initial temperature. Henceforth we assume that $q \gg 1$ (this means that the reaction, once initiated, strongly heats the gas). We are interested in the boundary conditions corresponding to the steady-state regimes of propagation of the reaction (1.9.1), and in the reaction front structures.

We shall immediately write the basic equations in dimensionless variables. The stationary kinetic equation of the reaction (1.9.1) has the following model form:

$$\Delta_r \frac{d\alpha}{dx} = \theta(T - T^*)(1 - \alpha) \quad , \tag{1.9.3}$$

where α is the fraction of the substance which reacted; Δ_r is the characteristic scale of the reaction (for simplicity, we further assume Δ_r = const, although accounting for the density and temperature dependence of Δ_r and other characteristic scales does not alter the results); T^* is the dimensionless critical temperature. The θ function in (1.9.3) ensures that the reaction at the initial temperature of the explosive does not occur, which allows us to view the set of equations as with dissipation included and so validates the results of Sect.1.7. The physical basis of this model is the fact that the unidimensional approximation becomes inadequate at low temperatures: the energy loss on the outer surface of the explosive is so high as to prevent the initiation of the reaction. However, sometimes it is necessary to consider the situation that the initial set of equations does not describe dissipation (Chap.4).

The first integrals of the equations of motion and energy balance in the dimensionless variables are similar to (1.8.2-4), the only difference being that now we assume $\gamma_A = \gamma_B = \gamma$ and take account of the heat released in the reaction [the last term in (1.9.5)]:

$$u_x - 1 + \frac{1}{\gamma M_0^2}\left(\frac{T}{u_x} - 1\right) - \Delta_v \frac{du_x}{dx} = 0 \quad , \tag{1.9.4}$$

$$\frac{1}{2}(u_x^2 - 1) + \frac{1}{(\gamma - 1)M_0^2}(T - 1) - \Delta_v u_x \frac{du_x}{dx} - \Delta_T \frac{dT}{dx} - \frac{\alpha q}{M_0^2} = 0 \quad . \tag{1.9.5}$$

In the uniform upstream and downstream flows the derivatives of velocity and temperature in (1.9.4,5) become zero. In the initial state $\alpha = 0$, and in the final state, according to (1.9.3), $\alpha = 1$. (A priori we cannot exclude the situation when the temperature at the edge of the reaction zone falls below T^*, and the reaction is prematurely quenched. Meanwhile we just postulate that this does not happen; later this will be proved in the analysis of the front structure). Eliminating the final temperature from (1.9.4,5), we get the equation for the velocity of the combustion products u_k, subscript $k = 1,2$ denoting the possible final states:

$$(1 - u_k)(u_k - u_{GD}) = \frac{2(\gamma - 1)}{(\gamma + 1)} \frac{q}{M_0^2} \quad , \tag{1.9.6}$$

where

$$u_{GD} = \frac{(\gamma - 1)M_0^2 + 2}{(\gamma + 1)M_0^2} \quad .$$

It is easily checked that (1.9.6) has two real roots u_1 and u_2, provided that one of the following inequalities is satisfied:

$$M_0 \geqslant M_{CJ}^{(1)} = [2(\gamma^2 - 1)q]^{\frac{1}{2}} + O(q^{-\frac{1}{2}}) \quad , \tag{1.9.7}$$

$$M_0 \leqslant M_{CJ}^{(2)} = [2(\gamma^2 - 1)q]^{-\frac{1}{2}} + O(q^{-3/2}) \quad . \tag{1.9.8}$$

These roots are confined to the following intervals:

$$\min(1, u_{GD}) < u_2 \leqslant u_{CJ} \leqslant u_1 < \max(1, u_{GD}) \quad , \tag{1.9.9}$$

where

$$u_{CJ} = \frac{\gamma M_0^2 + 1}{(\gamma + 1)M_0^2} \, .$$

Using the expression for the local Mach number

$$\frac{1}{M^2} = \frac{\gamma}{u_x}\left(1 - u_x + \frac{1}{\gamma M_0^2}\right) \, , \qquad (1.9.10)$$

we readily find that the flow with velocity u_1 is supersonic ($M_1 > 1$), and the flow with velocity u_2 is subsonic ($M_2 < 1$). They will coincide with one another only when $M_0 = M_{CJ}^{(1)}$ or $M_0 = M_{CJ}^{(2)}$. Then the downstream flow of the combustion products travels at the local speed of sound: $M_1 = M_2 = 1$, $u_1 = u_2 = u_{CJ}$. Such flow is called the Chapman-Jouguet flow, hence the subscript CJ.

From (1.9.7-9) we conclude that two regimes of propagation of the reaction exist, having essentially different speeds. The supersonic regimes are called the detonation ones: $M_0 \geqslant M_{CJ}^{(1)} \gg 1$. The detonation waves with the subsonic and the supersonic downstream velocities are called, respectively, the strong and weak detonation waves. Detonation, like a shock wave, is accompanied by compression of the gas. The subsonic propagation of the reaction is called combustion or deflagration; the deflagration waves travel at a speed much slower than the speed of sound ($M_0 \leqslant M_{CJ}^{(2)} \ll 1$) and are accompanied by the expansion of the gas ($1 < u_2 \leqslant u_1$), giving rise to a backward flow.

Let us determine the number of boundary conditions required for the above-named regimes to correspond to the evolutionary fronts of propagation of the reaction, considered as the discontinuity surfaces, separating the flows of gases A and B. Following the lines laid down in Sect.1.8 for the gasdynamic shock waves (the three basic boundary conditions here are the same) we get the results summarized in Table 1.1.

Table 1.1. Number of boundary conditions corresponding to evolutionary propagation regimes

Regime of propagation of reaction	Number of outgoing waves	Number of boundary conditions required for evolutionarity	Number of additional boundary conditions required for evolutionarity
Strong detonation	2	3	0
Weak detonation	3	4	1
Deflagration with subsonic downstream flow	3	4	1
Deflagration with supersonic downstream flow (this regime cannot be realized because the number of boundary conditions is too high)	4	5	2

We see that strong detonation waves are generally similar to shock waves: the incoming flow is supersonic, the outgoing flow is subsonic. There are no additional boundary conditions, i.e., for a given initial state the velocity of the strong detonation wave, depending on the circumstances, may take on any value within a certain range. The boundary conditions admit the existence of strong detonation waves in the entire interval $M_{CJ}^{(1)} \leqslant M_0 < \infty$. However, this interval may be narrowed by the condition of existence of a stationary structure (Sect.1.9.2). A weak detonation wave (as well as deflagration with subsonic downstream flow) requires one additional boundary condition; in other words, the velocity of propagation of the reaction in this regime depends on the initial parameters and the coefficients dissipation. Finally, deflagration with a supersonic downstream flow is impossible, since it requires five boundary conditions, which is too many. A dissipationless medium is described by $n = 3$ variables, and the only additional parameter available is the velocity of the front. Therefore, the flow in the deflagration wave can only be subsonic.

1.9.2 Structures of the Detonation and Deflagration Fronts

First we consider the structure of a detonation wave. For clarity we simplify the problem by disregarding the effect of heat conduction and taking account of viscosity alone, i.e., by setting $\Delta_T = 0$ in (1.9.5).

From (1.9.4,5), accounting for $M_0 > M_{CJ}^{(1)} \gg 1$, we easily find

$$\frac{2\Delta_v u_x}{\gamma - 1} \frac{du_x}{dx} = \frac{2q\alpha}{M_0^2} + \frac{(\gamma + 1)}{(\gamma - 1)} (u_x - 1)(u_x - u_{GD}) \quad , \tag{1.9.11}$$

$$\frac{2q\alpha}{M_0^2} + (u_x - 1)^2 - \frac{2}{\gamma(\gamma - 1)M_0^2} (T - 1) = 0 \quad . \tag{1.9.12}$$

The front structure is described by (1.9.3,11) in u_x and α, whereas the temperature T on the right-hand side of (1.9.3) is expressed via these variables by virtue of (1.9.12). The form of the structure depends essentially on the dimensionless parameter Δ_v/Δ_r, the ratio of the space scales of viscosity and chemical reaction. Figure 1.8 shows the phase picture of the given set of equations in the (u_x,α) plane for different values of Δ_v/Δ_r. The dashed lines in Fig.1.8 are the isocline $du_x/dx=0$ [bringing the right-hand side of (1.9.11) to zero] and the parabola $T(u_x,\alpha)=T^*$, which cuts out the region on the phase plane where the reaction occurs.

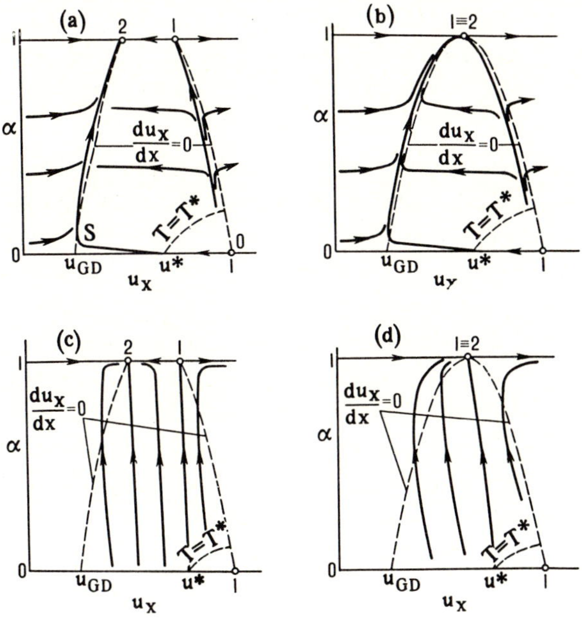

Fig.1.8a-d. Integral curves for the structure of the reaction zone in the (u_x,α) phase plane; (**a**) strong detonation wave, whose structure conforms with the ZND model for $M_0 > M_{CJ}^{(1)}$; (**b**) the same in the degenerate case $M_0 = M_{CJ}^{(1)}$; concerning the structure, the Chapman-Jouget detonation wave is strong; (**c**) weak detonation wave in the opposite extreme case of a narrow reaction zone for $M_0 > M_{CJ}^{(1)}$; (**d**) the same for the degenerate case $M_0 = M_{CJ}^{(1)}$; concerning the structure, the Chapman-Jouguet detonation wave is weak

Figure 1.8a corresponds to $\Delta_v/\Delta_r \ll 1$, $M_0 > M_{CJ}^{(1)}$. The integral curves are horizontal almost everywhere, except for the vicinity of the isocline $du_x/dx = 0$. The integral curves, tending to the singular point 2 at $x \to +\infty$, pass near this isocline (the singular point 2 corresponds to the strong detonation wave). The front structure is represented by the integral curve which connects points 0 and 2. It illustrates the Zel'dovich-von Neumann-Döring (ZND) detonation model, which assumes scale discrimination between the zones of shock compression and heating (the 0-S segment of the integral curve, corresponding to the ordinary gasdynamic shock wave with near-constant α) and the reaction (the portion S-2 close to the isocline corresponds to near-inviscid flow of the reacting gas). We see that a small perturbation of the phase picture will not affect the existence of the structure of the strong detonation wave: point 2 as $x \to +\infty$ is a stable node, entered by all integral curves close to $0 \to 2$. On the contrary, the saddle point 1 is entered by a single integral curve, the saddle separatrix, not passing via point 0, and therefore the weak detonation wave under these conditions has no structure and hence cannot exist.

Figure 1.8b is plotted under the same assumptions for $M_0 = M_{CJ}^{(1)}$. The singular points 1 and 2 then merge into a degenerate singularity $1 \equiv 2$, a saddle node, adjoining for $\alpha \leqslant 1$ with the nodal (on the left) and the saddle (on the right) sectors. The curve connecting the points 0 and $1 \equiv 2$ lies in the nodal sector, and the structure is qualitatively similar to that depicted in Fig. 1.8a and obviously corresponds to the strong detonation shock wave. Its peculiarity arises from the fact that the Chapman-Jouguet detonation wave is singular, and the linearization of (1.9.10) near point $1 \equiv 2$ yields

$$\frac{2\Delta_v u_{CJ}}{\gamma - 1} \frac{du_x}{dx} - \frac{\gamma + 1}{\gamma - 1} (u_x - u_{CJ})^2 - \frac{2q}{M_0^2} (1 - \alpha) \quad . \tag{1.9.13}$$

The second term on the right-hand side of (1.9.13) as $x \to +\infty$ decreases exponentially; therefore with large values of x the first term prevails, providing that it exhibits a slower decrease, and the asymptotic solution has the following form:

$$u_x \sim u_{CJ} + \frac{2\gamma\Delta_v}{(\gamma + 1)^2 x} \quad . \tag{1.9.14}$$

Such a slow approach to equilibrium, which does not even enable the width of the front to be estimated using (1.7.17), is typical of certain types of singular shock waves.

Figure 1.8c illustrates an extreme case of a very rapid reaction (which can hardly be realized), $\Delta_v/\Delta_r \gg 1$, $M_0 > M_{CJ}^{(1)}$. Here the integral curves, including the separatrix entering point 2, are almost vertical at $\alpha < 1$. For this reason, if the velocity u^*, which has to be attained to get the reaction started, is higher than u_1, neither the strong nor weak detonation waves are found to have any structure, and neither of them exists. At $u_1 = u^*$ (shown in Fig.1.8c) the front structure is represented by the separatrix entering point 1; in other words, a weak detonation wave exists. At $u_1 < u^*$ the reaction occurs in the domain of attraction of point 2, and the strong detonation wave is realized. Expressing the relations between u_1 and u^* via M_0 and T^* by dint of (1.9.12), we get after some straightforward transformations

$$M_0 = M^* \tag{1.9.15}$$

for weak detonation waves, and

$$\infty > M_0 > M^* > M_{CJ}^{(1)} \qquad (1.9.16)$$

for strong detonation waves.

For the extreme case in question

$$M^* = \frac{M_{CJ}^{(1)^2} + M'^2}{2M'} \quad , \qquad (1.9.17)$$

where

$$M' = (\gamma + 1) \left[\frac{2(T^* - 1)}{\gamma(\gamma - 1)}\right]^{\frac{1}{2}} \quad .$$

Thus, for weak detonation waves the condition of existence of a stationary structure gives us an additional boundary condition (1.9.15). It restricts the velocity of the waves. On the other hand, the velocities of the strong detonation waves may assume any value within the interval (1.9.16). Of course, this result fits in with the evolutionarity conditions listed in Table 1.1

From the arguments developed above it follows that a critical value μ of the parameter Δ_v/Δ_r must exist such that at $\Delta_v/\Delta_r < \mu$ the phase picture has the form shown in Fig.1.8a (strong detonation), while at $\Delta_v/\Delta_r > \mu$ the detonation wave may be strong or weak, depending on which of the conditions (1.9.15,16) is realized. Note that for finite values of Δ_v/Δ_r, (1.9.16) for M^*, which represents the limit of M as $\Delta_v/\Delta_r \to \infty$, is no longer valid. It is clear that the intermediate case $M^* = M_{CJ}^{(1)}$ corresponds to the front structure of the Chapman-Jouguet detonation wave, in which the integral curve, entering the point $1 \equiv 2$, is the separatrix of this degenerate singularity, which separates the nodal and saddle regions (Fig.1.8d). The boundary conditions here are the same as in Fig.1.8b, although the phase picture is similar to that in Fig.1.8c, and the Chapman-Jouguet detonation wave is weak. It is noteworthy that this characteristic of the wave is reflected in the asymptotic form of the structure. It is easily proved that for the integral curve, which enters point $1 \equiv 2$ along the separatrix, the power law (1.9.14) of decreasing v as $x \to \infty$ is not valid: all the variables close on their limits exponentially. In accordance with (1.7.17), the front has a finite thickness of the order of Δ_r. In other words, the Chapman-Jouguet detonation wave can alwys be attributed to a definite type of wave: it is strong for $\Delta_v/\Delta_r < \mu$ and weak for $\Delta_v/\Delta_r = \mu$. If $\Delta_v/\Delta_r > \mu$, this wave does not exist at all, as follows from (1.9.15,16).

The realization of different detonation regimes depends on the initial and boundary conditions. Of special practical importance is free, self-sustained detonation, when the movement of the combustion products in the direction of propagation of detonation is not maintained by the movement of, for instance, a piston, so therefore an expansion wave must exist in the downstream region, which separates the front from the stationary gas. Since in the strong detonation waves the downstream flow is subsonic, the expansion wave catches up with the strong detonation wave and weakens it for $M_0 \neq M_{CJ}^{(1)}$. If $\Delta_v/\Delta_r \ll 1$, which is the most physically real case, and the weak detonation wave cannot exist, then the only possible solution is

$$M_0 = M_{CJ}^{(1)} \quad . \tag{1.9.18}$$

We emphasize that (1.9.18) is *not* an additional boundary condition in the sense used in Sect.1.7.2. Its fulfillment is conditioned, on the one hand, by the structural restrictions (for $\Delta_v/\Delta_r \ll 1$ only the strong detonation wave has a stationary structure), and on the other hand, by the appropriate choice of the initial and boundary conditions, in accordance with which the velocity of the downstream flow cannot be subsonic. If, however, $\Delta_v/\Delta_r \geqslant \mu$, then the self-sustained detonation takes the form of the weak detonation wave, whose propagation speed is given by (1.9.15); the downstream flow then is supersonic and the expansion wave cannot catch up with the front. The strong detonation waves with velocities in the range (1.9.16) can be realized here only with the aid of a piston. Most experiments with detonation in gases confirm (1.9,18) in the first approximation. Some recent results [1.52] indicate that the self-sustained detonation is actually weak (its velocity is slightly above $M_{CJ}^{(1)}$, and compression slightly below $\gamma/(\gamma+1)$. This can hardly be attributed to the action of transfer effects, since the value of Δ_v/Δ_r is, as a rule, quite small. Evidently, treatment of this problem must involve complicated nonunidimensional structures, which certainly are beyond the scope of the present monograph.

Let us now consider the front structure of the deflagration wave. In contrast to the preceding case, we now neglect viscosity ($\Delta_v = 0$), and take into account the heat conduction. Then, in place of (1.9.12), we get from (1.9.4)

$$T = u_x(1 + \gamma M_0^2 - \gamma u_x M_0^2) \quad . \tag{1.9.19}$$

From (1.9.19) it follows that the deflagration wave is isobaric at the head of the front (since $M_0 \ll 1$), although on the whole the pressure in the wave falls owing to the large downstream expansion of the gas ($u_x \sim M_0^{-2} \gg 1$). Substituting (1.9.19) into (1.9.5), we get an equation which together with (1.9.3,19) describes the structure of the deflagration wave:

$$2\Delta_T \frac{du_x}{dx} = -\left[\frac{2q}{M_0^2} - \frac{\gamma + 1}{\gamma - 1}(u_x - 1)(u_{GD} - u_x)\right](1 + \gamma M_0^2 - 2\gamma u_x M_0^2)^{-1} \quad . \quad (1.9.20)$$

It is clear that a continuous, physically meaningful solution, passing via point 0, is possible only for $u_x < u'$, where $u' = (1 + \gamma M_0^2)/2\gamma M_0^2$ is the zero of the denominator of the right-hand side of (1.9.20). Noting that $u' < u_{CJ} < u_1$, we find that the deflagration wave with the supersonic downstream flow does not have a stationary structure and thus cannot be realized, which was predicted earlier on the basis of its nonevolutionarity. The singular point 2 in the (u_x, α) phase plane is a saddle (Fig.1.9), and therefore the deflagration wave can exist only if the separatrix, which enters the saddle point, passes via the point u^*. This is possible only with a certain value of the front velocity, so thus we get an additional boundary condition of the type (1.9.15), which determines the velocity of the deflagration wave, in full accordance with the conditions of evolutionarity. In the extreme case $\Delta_T/\Delta_r \gg 1$ we can write, similarly to (1.7.17), the explicit expression for M , which for $T^* \lesssim M_{CJ}^{(2)2}$ has the form

$$M^* = M_{CJ}^{(2)}\left[2(\gamma + 1)(T^* - 1)\left(1 - \frac{M_{CJ}^{(2)} T^*(\gamma + 1)}{2}\right)\right]^{\frac{1}{2}} M_{CJ}^{(2)} \quad . \qquad (1.9.21)$$

The rate of propagation of the deflagration wave in real physical conditions is given by the Zel'dovich-Frank-Kamenetski formula [1.51,53]. The initial and boundary conditions of deflagration are essentially different from those of detonation. For instance, in the piston problem, to sustain deflagration the piston must be moved not towards the front (as for strong detonation), but rather away from the front. The question regarding whether the self-sustained detonation of deflagration will take place in the absence of a moving piston is determined by the nonstationary (and non-self-similar)

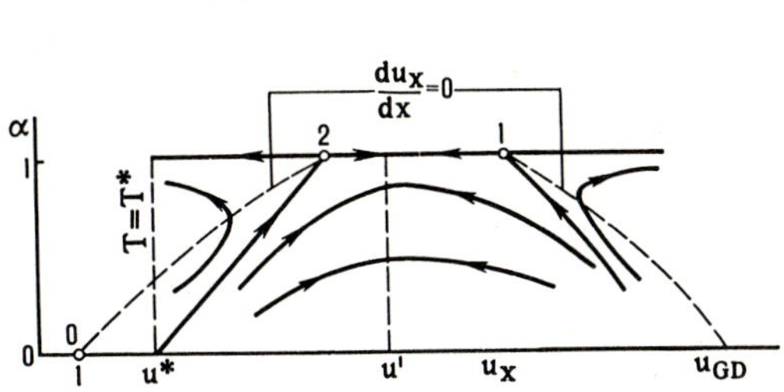

Fig.1.9. Integral curves for the reaction-zone structure in the (u_x, α) phase plane for a deflagration wave: $M_0 < M_{CJ}^{(2)}$

stage of the ignition. If the ignition is so slow that the pressure perturbations in the ignition region have time to be carried away by the linear sonic waves, then most probably the deflagration wave will arise. If the rapid release of energy creates the nonlinear compression waves, then the detonation wave is more likely to arise.

1.9.3 Realization of Different Propagation Regimes of the Reaction. The Piston Problem

We have seen that three different regimes of propagation of an exothermal reaction exist: two supersonic (the strong and the weak detonation) and one subsonic (deflagration). Each of them can, in principle, be realized with the appropriate initial and boundary conditions, which characterize the initiation and subsequent propagation of the reaction. Let us illustrate this point using the simplest example of detonation in the approximation $\Delta_v/\Delta_r \gg 1$ (Sect.1.9.2), when the Mach numbers for the weak and strong detonation waves are given by (1.9.5,16), respectively.

Consider the following problem. The explosive gas occupies the half-space $x>0$, limited from the left (at $x=0$) by an impenetrable piston. At $t=0$ the reaction is initiated near the piston, and the piston starts moving in the x direction with speed u_p, corresponding to the Mach number $M_p = u_p/a_0$. Presuming that the given conditions favor the supersonic (detonation) regime of propagation of the reaction, we analyze the nature of the flow ahead of the piston for different values of M_p.

The problem under consideration is self-similar; its initial and boundary conditions contain no length or time parameters, which enter only in the velocity-like combination $\eta = x/t$ [1.34]. This implies that the flow ahead of the piston is represented by a sequence of regions of constant flow (where $u_x = \mathrm{const}$, $p = \mathrm{const}$, etc.), separated by simple waves or discontinuities. Our problem admits the existence of the four types of simple waves and discontinuities: strong and weak detonation waves, shock waves and centered expansion waves. The following boundary condition must be satisfied on the surface of the piston: either the velocity of the flow equals the velocity of the piston ($M = M_p$), or the density of the gas turns to zero (cavitation).

The problem is most conveniently approached in the following manner: after having set up all the possible sequences of simple waves and discontinuities, one must find the piston velocities corresponding to each sequence. Since the reaction in the detonation waves of both kinds proceeds to the end, this sequence may include a detonation wave of only one kind. It is also clear, that it is the detonation wave that travels in the lead, in the unperturbed gas. Indeed, the front of the shock wave travels at a subsonic speed (with respect to the speed of sound in the shock-compressed substance), the velocity of the tail of the centered expansion wave equals the local speed of sound, whereas the detonation wave travels at a supersonic speed and would have overtaken the waves ahead of it (if any), which contradicts our presumption of the self-similarity of the problem. Further, the downstream flow in the strong detonation wave is subsonic, and thus it cannot be followed by either the supersonic shock wave, or by the centered expansion wave, which travels at the speed of sound. Consequently, the velocity of the combustion products in the homogeneous region downstream of the front of a strong detonation wave equals the velocity of the piston. On the contrary, the downstream flow in a weak detonation wave is supersonic, and such a wave can be followed either by a limited-strength shock wave, or by an expansion wave.

Having written the continuity equation for the discontinuity traveling in the resting gas with the density ρ_0 at the speed D in the laboratory system

$$\rho_0 D = \rho(D - U)$$

(U being the downstream velocity and ρ being the density of the gas), we find

$$U = D(1 - \rho_0/\rho) \quad ,$$

or in dimensionless variables

$$M = U/a_0 = M_0(1 - u) \quad , \tag{1.9.22}$$

where u is the dimensionless downstream velocity. With the aid of (1.9.6,22) we find the Mach number M_p, corresponding to the strong detonation wave, traveling ahead of the piston, the velocity of the wave being in the interval (1.9.16):

$$M_p = \frac{M_0}{\gamma + 1}\left[1 + \left(1 - \frac{M_{CJ}^{(1)^2}}{M_0^2}\right)^{\frac{1}{2}}\right] \quad . \tag{1.9.23}$$

The lower bound M_{p1} of the relevant range of values is given by (1.9.23), where we set $M_0 = M^*$. As it follows from (1.9.23), the Mach number M_0 of the strong detonation wave is related to M_p by

$$M_0 = \frac{(\gamma + 1)^2 M_p^2 + M_{CJ}^{(1)^2}}{2(\gamma + 1)M_p} \quad . \tag{1.9.24}$$

Similarly we find that the dimensionless velocity of the downstream flow of the combustion products in a weak detonation wave is

$$M_{p2} = \frac{M^*}{\gamma + 1}\left[1 - \left(1 - \frac{M_{CJ}^{(1)^2}}{M^{*2}}\right)^{\frac{1}{2}}\right] < M_{p1} \quad . \qquad (1.9.25)$$

The products of combustion downstream of the weak detonation wave may carry either a shock wave, which accelerates them in the direction of the front, or a centered expansion wave, in which the velocity, on the contrary, is reduced. It is clear that these two waves cannot occur simultaneously in the self-similar solution, since each of them would tend to catch up with the wave traveling ahead. Hence it follows that the range $M_{p2} < M_p < M_{p1}$ corresponds to the sequence weak detonation wave—shock wave, whereas the lower values of the piston velocity $M_p < M_{p2}$ correspond to the sequence weak detonation wave—centered expansion wave.

After some simple transformations, using (1.9.6,10) we find that the shock waves which travel in the wake of the weak detonation wave correspond to the following range of Mach numbers M_s (with respect to the speed of sound in the products of combustion):

$$1 < M_s \leqslant \left\{1 + (\gamma + 1)\left[\frac{\gamma M^{*2} + 1}{M^{*2} - 1}\left(1 - \frac{M_{CJ}^{(1)^2}}{M^{*2}}\right)^{-\frac{1}{2}} - \gamma\right]^{-1}\right\} \quad , \qquad (1.9.26)$$

while a given value of M_s in this range corresponds to the piston velocity

$$M_p = M_{p2} + \frac{2(M_s^2 - 1)}{(\gamma + 1)M_s}\,[\gamma\lambda(M^* - \lambda)]^{\frac{1}{2}} \quad , \qquad (1.9.27)$$

where

$$\lambda = \left(\frac{2(T^* - 1)}{\gamma(\gamma - 1)}\right)^{\frac{1}{2}} \quad . \qquad (1.9.28)$$

If the dimensionless piston velocity is below M_{p2}, then the weak detonation wave is followed by the centered expansion wave, in which the velocity of the combustion products slows down from M_{p2} to M_p, and their density is reduced. Since in the present case

$$u - \frac{2}{\gamma - 1}\,a = \text{const}$$

for the centered expansion wave (u being the flow velocity, and a the local speed of sound, Sect.1.4.5), the speed of sound is reduced to zero (together with the density, since $a \propto \rho^{(\gamma-1)/2}$) when the dimensionless flow velocity in the expansion wave is reduced to

$$M_{p3} = -\left\{\frac{2}{\gamma - 1}[\gamma\lambda(M^* - \lambda)]^{\frac{1}{2}} - \lambda\right\} < 0 \quad . \qquad (1.9.29)$$

Thus, when the dimensionless piston velocity is below M_{p3}, the centered expansion wave is followed by a vacuum region, which separates the wave's tail from the surface of the piston. The results are collected in Table 1.2.

Table 1.2. Solution of the piston problem for detonation waves

Dimensionless velocity of the piston	Flow pattern
$M_p > M_{p1}$	Strong detonation wave
$M_{p1} \geqslant M_p > M_{p2}$	Weak detonation wave + shock wave
$M_p = M_{p2}$	Weak detonation wave
$M_{p2} > M_p \geqslant M_{p3}$	Weak detonation wave + centered expansion wave
$M_{p3} > M_p$	Weak detonation wave + centered expansion wave, followed by vacuum space

We see that the realization of a particular type of detonation wave depends on the initial and boundary conditions. Within the context of the self-similar problem under consideration, the question is reduced to the relation between the dimensionless piston velocity M_p and the value of M_{p1}: for $M_p > M_{p1}$ the detonation is strong, for $M_p < M_{p1}$ the detonation is weak. In particular, the value $M_p = 0$, i.e., free propagation of the detonation in the half-space, corresponds to the weak detonation wave, followed by the centered expansion wave. Thus there is no room for uncertainty: either of the supersonic regimes can be realized under the appropriate conditions.

From Table 1.2 it is clear that any value of the piston velocity $-\infty < M_p < \infty$ corresponds to some self-similar solution of the piston problem with the detonation wave of some kind or other. Of course, this does not imply that the problem admits no self-similar solution in the subsonic regime (deflagration). Such solutions are also possible, and can be easily constructed. Indeed, observe that in a self-similar problem no other waves and discontinuities can travel behind the subsonic deflagration wave. Therefore, the piston velocity coincides with the velocity of the combustion products downstream of the deflagration wave front, while the unperturbed gas upstream may carry both the shock waves (although the downstream flow in the shock wave us subsonic, its Mach number is of the order of unity, while in a deflagration wave $M \ll 1$) and the centered expansion wave. The intensity of the shock wave is restricted from above (because the temperature jump in the shock wave for the assumed flow pattern should not exceed T^*, otherwise the reaction should start on the shock front, thus changing the pattern); restrictions are also imposed on the velocity variations in the expansion wave (the propagation of a deflagration wave requires nonzero density of the explosive). Consequently, the range of piston velocities is restricted: $M_{p6} < M_p < M_{p5}$. The propagation of the deflagration wave alone corresponds to a certain fixed value $M_p = M_{p4}$ in this range [as follows from (1.9.21), $M_{p4} < 0$, since in a deflagration wave $u > 1$]. With $M_{p6} < M_p < M_{p4}$ the deflagration wave is preceded by a centered expansion wave, and by a shock wave with $M_{p4} < M_p < M_{p5}$. The values of M_{p4}, M_{p5}, M_{p6} can be easily calculated on the basis of a concrete model of the deflagration front, but here we shall not go into these details. Note only that this range includes the piston velocities which correspond to the detonation regimes. Whichever regime is realized in a more realistic setup (when the process of ignition is taken into consideration, with due account of its space and time characteristics), can be determined only in the non-self-similar stage of development of the process, which, accordingly, has to be considered in order to remove the ambiguity.

For real experimental conditions the piston which drives the detonation wave ahead, can be represented by a current sheet (magnetic piston) in the electromagnetic shock tube [1.54]. Then, however, the analysis of the flow must account for the magnetohydrodynamic waves traveling ahead of the piston [1.54,55].

2. Gas Shock Ionization and Shock-Wave Structures in Plasmas

In this chapter we describe the structure of shock waves in plasma and the structures of strong ionizing shock waves. In the absence of an external magnetic field, when there are no electric currents across the shock, the conductive properties of plasma do not affect the shock structures. At the same time, the existence of the two plasma components, the light (electrons) and heavy (ions or atoms and ions) components, engaged in the Coulombian interaction, gives rise to the existence of several different length scales which characterize nonequilibrium processes in the plasma, resulting in rather complicated shock structures, essentially dissimilar from the structures of shocks in a simple one-component gas.

For the most simple problem of the shock structure in a fully ionized plasma (Sect.2.1) the Rankine-Hugoniot conditions are the same as for gasdynamics. Since the heat conduction in the plasma depends mainly on its electronic component, and the viscosity depends on the ionic component, the Prandtl number is as small as $(m_e/m_i)^{\frac{1}{2}}$, and the shock structures are generally similar to those described in Sect.1.8.3. The structures of weak shock waves (Sect. 2.1.2) are shaped by electronic heat conduction. The structures of strong waves (Sect.2.1.3) contain a "tongue" of heat-conductive heating, followed by a viscous subshock and the region of temperature relaxation. The two-fluid nature of the plasma, apart from causing nonisothermality in the strong shock waves ($T_e \neq T_i$), is manifest in that the electronic component diffuses upstream, and in the shock front the confining electric field arises aligned with the direction of propagation of the shock wave (Sect.2.1.4). The jump of electric potential across the shock thus produced, as opposed to the jumps of other variables across the ordinary gasdynamic shock front (Sect.1.6.3), is not determined by the boundary conditions, but rather reflects the structure of the shock wave in plasma.

Shock waves in plasma, which cause considerable change in the ionization, are considered in Sect.2.2. The endothermal reaction of ionization modifies both the Rankine-Hugoniot conditions and the shock structures (Sect.2.2.1). Section 2.2.2 is devoted to shock waves in a plasma with multiple ionization. In this case the viscous subshock is followed by the relaxation region, where the temperature and ionization relaxation take place simultaneously. In Sect. 2.2.3 we analyze the structure of a shock wave in partly ionized argon in the range of Mach numbers pertaining to single (first) ionization. The structure comprises four different successive layers, the scales of ionization and of heat conduction being different in this case.

Still more sophisticated is the morphology of strong ionizing shocks, discussed in Sect.2.3. As demonstrated in Sect.2.3.1, the structure includes the region of nonequilibrium precursor ionization and excitation, a viscous subshock, the region of ionization relaxation, followed usually by the region of radiative cooling. The precursor region, in turn, is subdivided into the near and far precursor regions (Sect.2.3.2). The near precursor region is shaped by direct photoionization of neutral gas ahead of the shock by radiation of shock-heated plasma, while the structure of the far precursor region is shaped by diffusion of resonant radiation. In electromagnetic shock tubes the source of nonequilibrium precursor ionization may also be the fast waves of electrostatic breakdown (Sect.2.3.4). The structures for the regions of ionization relaxation and radiative cooling are described in Sect.2.3.5.

Deviations from the idealized unidimensional flow pattern in experiments with ionizing shock waves are often rather considerable (Sect.2.4). The existence of wall boundary layers in shock tubes often affects the structure of ionizing shock waves (Sect.2.4.1). A related phenomenon, consisting of an increase in temperature in the region of ionization relaxation, i.e., in the shortening of this region, is fairly well explained by the quasi-unidimensional gasdynamic theory, whereas the phenomenon whereby the electron avalanche is displaced towards the shock front in the wall boundary layer seems to be unaccountable. In Sect.2.4.2 we describe the results of experiments aimed at exploring the instability of ionizing shock waves near the limits of stability, indicated in Sect.1.7.4. A manifestation of such an instability is the transition to the turbulent regime, observed in argon in the Mach-number range $M \sim 13$-16.

2.1 Shock Structures in a Completely Ionized Plasma

2.1.1 Equations for the Shock Layer and Boundary Conditions

We consider a shock wave in a simple plasma, i.e., a completely ionized plasma with one kind of singly charged ions ($Z=1$); $\gamma_e = \gamma_i = 5/3$. Although the experimental data on shock structures in such a plasma are lacking because the real shocks in plasmas are inseparably associated with factors such as the magnetic field, the ionization, and the radiative energy transfer, this problem is important from the methodological viewpoint, insofar as it illustrates many characteristic features of shocks in plasmas. Shock structures in plasmas were calculated in [2.1-13]; we shall rely on the results of [2.8].

Our description of the shock structure in a plasma is based on the two-fluid Navier-Stokes approximation, which uses 16 flow variables: the densities of electrons and ions n_e, n_i, the temperatures of electrons and ions T_e, T_i, the components of the velocities of electrons and ions $\mathbf{u}^e$, $\mathbf{u}^i$, and of the electric and the magnetic fields $\mathbf{E}$, $\mathbf{B}$ (three for each quantity). The equations of the shock layer evolve from Maxwell's equations, as well as from the equations of continuity of the fluxes of mass, momentum and energy of the plasma, and the equations of motion and heat balance for the electrons and ions.

We restrict our attention to the unidimensional stationary shock layer. Let the x axis be directed downstream, normal to the shock plane, and assume all the quantities to depend only on x, i.e.,

$$\nabla = \left\{\frac{d}{dx} \; , \; 0 \; , \; 0\right\} \quad , \quad \frac{\partial}{\partial t} = 0 \quad . \tag{2.1.1}$$

We seek a solution for the shock structure in which the motion of the plasma takes place only along the x axis, and does not give rise to transverse electric and magnetic fields. From Maxwell's equations

$$\operatorname{div} \mathbf{B} = 0 \quad , \quad \operatorname{curl} \mathbf{E} = -\frac{\partial \mathbf{B}}{\partial t} \quad . \tag{2.1.2}$$

Taking (2.1.1) into account, we get

$$B_x = \text{const}, \; E_y = \text{const}, \; E_z = \text{const} \quad .$$

Setting

$$B_x = B_y = B_z = 0 \quad , \quad E_y = E_z = 0 \quad ,$$

$$u_y^e = u_z^e = u_y^i = u_z^i = 0 \quad , \tag{2.1.3}$$

we find that (2.1.2) and the continuity equations for the flux of transverse momentum of the plasma and the transverse motion of electrons are satisfied identically. From

$$\text{curl}\,\mathbf{B} = \mu_0 \mathbf{j} + \varepsilon_0 \mu_0 \frac{\partial \mathbf{E}}{\partial t} \tag{2.1.4}$$

it follows that

$$j_x = e(n_i u_x^i - n_e u_x^e) = 0 \quad . \tag{2.1.5}$$

Thus, the problem is formulated in terms of seven variables: n_e, n_i, T_e, T_i, u_x^e, u_x^i, E_x. From the equations for the conservation of mass and number of electrons we obtain algebraic relations

$$n_e u_x^e = \text{const} \quad , \qquad n_i u_x^i = \text{const} \quad ,$$

whence, with due account taken of (2.1.5), we get

$$n_e u_x^e = n_i u_x^i = C_1 \quad , \tag{2.1.6}$$

the value of C_1 being determined by the boundary conditions.

The relation (2.1.6) does not imply that $n_e = n_i$. The shock wave produces plasma polarization, described by the one remaining electrodynamic equation, the Poisson equation

$$\frac{dE_x}{dx} = \frac{e}{\varepsilon_0}(n_i - n_e) \quad . \tag{2.1.7}$$

Now let us write the continuity equations for the fluxes of the x component of momentum and energy of the plasma as a whole:

$$m_e n_e (u_x^e)^2 + m_i n_i (u_x^i)^2 + n_e T_e + n_i T_i - \frac{\varepsilon_0 E_x^2}{2}$$
$$- \frac{4}{3}\eta_e \frac{du_x^e}{dx} - \frac{4}{3}\eta_i \frac{du_x^i}{dx} = C_2 \quad , \tag{2.1.8}$$

$$\frac{1}{2}\left[m_e n_e (u_x^e)^3 + m_i n_i (u_x^i)^3\right] + \frac{5}{2}(n_e T_e u_x^e + n_i T_i u_x^i)$$
$$- \frac{4}{3}\eta_e u_x^e \frac{du_x^e}{dx} - \frac{4}{3}\eta_i u_x^i \frac{du_x^i}{dx} - \kappa_e \frac{dT_e}{dx} - \kappa_i \frac{dT_i}{dx} + q_{ux} = C_3 \quad , \tag{2.1.9}$$

where η_e, κ_e and η_i,κ_i are the coefficients of viscosity and thermal conductivity of electrons and ions, respectively; q_u is the heat flux due to the relative motion of plasma components, C_2, C_3 are constants determined by the boundary conditions.

The equations of motion and heat balance of electrons and ions, respectively, have the form

$$m_e n_e u_x^e \frac{du_x^e}{dx} = -\frac{d}{dx}(n_e T_e) + \frac{d}{dx}\left(\frac{4}{3}\eta_e \frac{du_x^e}{dx}\right) - en_e E_x + R_{ux} + R_{Tx} \quad , \tag{2.1.10}$$

$$\frac{3}{2} n_i u_x^i \frac{dT_i}{dx} + n_i T_i \frac{du_x^i}{dx} = \frac{d}{dx} \kappa_i \frac{dT_i}{dx} + \frac{4}{3}\eta_i \left(\frac{du_x^i}{dx}\right)^2 - \frac{3m_e}{m_i}\frac{n_e}{\tau_e}(T_e - T_i) \quad , \tag{2.1.11}$$

where $\mathbf{R}_u$ is the force of friction acting upon electrons from the side of the ions; $\mathbf{R}_T$ is the thermal force.

We use the values of kinetic coefficients for a simple plasma, calculated in [2.14]:

$$\eta_e = 0.73 n_e T_e \tau_e \quad , \qquad \eta_i = 0.96 n_i T_i \tau_i \quad ,$$

$$\kappa_e = \frac{3.16 n_e T_e \tau_e}{m_e} \quad , \qquad \kappa_i = \frac{3.9 n_i T_i \tau_i}{m_i} \quad ,$$

$$R_{ux} = -\frac{m_e n_e}{\tau_e} 0.51(u_x^e - u_x^i) \quad , \tag{2.1.12}$$

$$R_{Tx} = -0.71 n_e \frac{dT_e}{dx} \quad ,$$

$$q_{ux} = 0.71 n_e T_e (u_x^e - u_x^i) \quad ,$$

where the periods between collisions for the electron and the ions are taken in the form

$$\tau_e = \frac{3(2\pi)^{3/2}\varepsilon_0^2 m_e^{1/2} T_e^{3/2}}{e^4 n_i \Lambda} \quad , \qquad \tau_i = \frac{12\pi^{3/2}\varepsilon_0^2 m_i^{1/2} T_i^{3/2}}{e^4 n_i \Lambda} \quad , \tag{2.1.13}$$

$\Lambda = \ln[12\pi\varepsilon_0(\varepsilon_0 T^3/n_e)^{1/2}/Ze^3]$ being the Coulomb logarithm (the dimensionless cutoff impact parameter).

Now we go over to dimensionless variables, relating each of the variables $n_{e,i}$, $T_{e,i}$, $u_x^{e,i}$, similarly to Sects.1.8,9, to their respective values upstream or downstream, and using the same alphabetic notation for the dimensionless variables. For the dimensionless electric field we use

$$E_x = \frac{e\bar{E}_x \, r_{Dk}}{\bar{T}_k} \quad , \qquad k = 0,2 \quad ,$$

which means work done by the electric field $\bar{E}_x$ on the scale of Debye's radius

$$r_{Dk} = \left(\frac{\varepsilon_0 \bar{T}_k}{e^2 \bar{n}_k}\right)^{\frac{1}{2}} \quad ,$$

related to the thermal energy of a plasma particle in either one of the equilibrium states: $k = 0$ upstream, or $k = 2$ downstream.

Let us rewrite the shock-layer equations, discarding the terms of the order ε^2 with $\varepsilon = (m_e/m_i)^{\frac{1}{2}}$. From (2.1.6) we find

$$n_e u_x^e = n_i u_x^i = 1 \quad . \tag{2.1.14}$$

Expressing n_e and n_i via u_x^e and u_x^i with the aid of (2.1.14), and using the dimensionless charge-separation variable

$$\delta n \equiv n_i - n_e = \frac{1}{u_x^i} - \frac{1}{u_x^e} \quad ,$$

we obtain a set of dimensionless shock-layer equations (for the sake of definiteness, we set $k = 0$):

$$\frac{dE_x}{dx} = \frac{\delta n}{r_{D0}} \quad , \tag{2.1.15}$$

$$u_x^i - 1 + \frac{3}{10M_0^2}\left(\frac{T_e + T_i}{u_x^i} - 2\right) - \Delta_{vi}\frac{du_x^i}{dx} - \Delta_{ve}\frac{du_x^e}{dx}$$

$$- \frac{3}{10M_0^2}\,\delta n T_e - \frac{3}{20M_0^2}E_x^2 = 0 \quad , \tag{2.1.16}$$

$$\frac{(u_x^i)^2 - 1}{2} + \frac{3}{4M_0^2}(T_e + T_i - 2) - \Delta_{vi}u_x^i\frac{du_x^i}{dx} - \Delta_{ve}u_x^e\frac{du_x^e}{dx}$$

$$- \Delta_{Te}\frac{dT_e}{dx} - \Delta_{Ti}\frac{dT_i}{dx} + \frac{3}{10M_0^2}\,0.71\delta n u_x^i T_e = 0 \quad , \tag{2.1.17}$$

$$\frac{d}{dx}\frac{T_e}{u_x^e} + \frac{E_x}{r_{D0}u_x^e} - \frac{10M_0^2}{3}\frac{d}{dx}\Delta_{ve}\frac{du_x^e}{dx} + \frac{0.71}{u_x^e}\frac{dT_e}{dx} - \frac{1}{\Delta_u}u_x^i\delta n = 0 \quad , \tag{2.1.18}$$

$$\frac{3}{2}\frac{dT_i}{dx} + \frac{T_i}{u_x^i}\frac{du_x^i}{dx} = \frac{10M_0^2}{3}\frac{d}{dx}\Delta_{T_i}\frac{dT_i}{dx} + \frac{10M_0^2}{3}\Delta_{vi}\left(\frac{du_x^i}{dx}\right)^2 + \frac{T_e - T_i}{\Delta_r} . \quad (2.1.19)$$

Here the characteristic scales of the different processes can be expressed as

$$\Delta_{ve} = \Delta_{ve}^{(0)}\frac{u_x^i}{u_x^e}T_e^{5/2} \quad , \quad \Delta_{ve}^{(0)} = \frac{0.206}{M_0}\varepsilon\ell_0 \quad ;$$

$$\Delta_{vi} = \Delta_{vi}^{(0)}T_i^{5/2} \quad , \quad \Delta_{vi}^{(0)} = \frac{0.384}{M_0}\ell_0 \quad ;$$

$$\Delta_{Te} = \Delta_{Te}^{(0)}\frac{u_x^i}{u_x^e}T_e^{5/2} \quad , \quad \Delta_{Te}^{(0)} = \frac{0.2}{M_0^3}\frac{1}{\varepsilon}\ell_0 \quad ; \quad (2.1.20)$$

$$\Delta_{Ti} = \Delta_{Ti}^{(0)}T_i^{5/2} \quad , \quad \Delta_{Ti}^{(0)} = \frac{0.35}{M_0^3}\ell_0 \quad ;$$

$$\Delta_r = \Delta_r^{(0)}u_x^i u_x^e T_e^{3/2} \quad , \quad \Delta_r^{(0)} = 0.236\,\frac{M_0}{\varepsilon}\ell_0 \quad ;$$

$$\Delta_u = \Delta_u^{(0)}\frac{u_x^i}{u_x^e}T_e^{3/2} \quad , \quad \Delta_u^{(0)} = \frac{0.416}{\varepsilon M_0}\ell_0 \quad ,$$

where $\ell_0 = a_0\tau_{i0}$ is the mean free path, and a_0 is the upstream speed of sound in the plasma. In (2.1.20) and henceforth we neglect variations of the Coulomb logarithm in the shock wave.

At equilibrium on either side of the front all the derivatives in (2.1.15-19) are zero, $u_x^e = u_x^i$, $T_e = T_i$, $E_x = 0$, and (2.1.15-19) represent the set of the boundary conditions for the common gasdynamic shock in a gas with $\gamma = 5/3$ (Sect.1.6.3). Hence,

$$u_x^e,\ u_x^i,\ T_e,\ T_i \to 1 \quad , \quad E_x \to 0 \quad \text{at} \quad x \to -\infty \quad , \quad (2.1.21)$$

and

$$u_x^e,\ u_x^i \to u_{GD} = \frac{M_0^2 + 3}{4M_0^2} \quad , \quad E_x \to 0 \ , \quad T_e,\ T_i \to T_{GD} = \frac{(M_0^2 + 3)(5M_0^2 - 1)}{16M_0^2}$$

$$\text{at} \quad x \to +\infty \quad , \quad (2.1.22)$$

further, the sonic Mach number tends to its downstream value

$$M^2 \to M_2^2 = \frac{M_0^2 + 3}{5M_0^2 - 1} \quad . \quad (2.1.23)$$

Should we choose to use dimensionless variables normalized to the final state $k = 2$, then (2.1.15-20,22,23) would require the transcription $0 \leftrightarrow 2$; the asymptotics (2.1.21) will then correspond to $x \to +\infty$, and (2.1.22) to $x \to -\infty$.

Insofar as in the limit $\omega \to 0$, $k \to 0$ only the simple sonic wave can propagate in the plasma, its speed being

$$a = \left(\frac{\gamma_e T_e + \gamma_i T_i}{m_i}\right)^{\frac{1}{2}} \quad . \tag{2.1.24}$$

The conditions of evolutionarity also coincide with those for a gasdynamic shock wave: $M_0 > 1$ and $M_2 < 1$. Due to the coincidence of boundary conditions, each of these inequalities is derivable from the other; see (2.1.23). On the grounds that this condition provides for the existence of a unique shock structure in the form (1.7.8), we can assert that if a unique solution of the equations for the shock layer (2.1.15-20) exists, satisfying the boundary conditions (2.1.21,22), then this solution is also the unique solution of the complete set of shock-layer equations. In this set the transverse components of fields and velocities, in compliance with the boundary conditions, turn to zero far ahead of and behind the front, although some of them may be nonzero in the shock region. In other words, the solution of the complex set of equations is obtained by supplementing the said solution by the conditions (2.1.3); these conditions do not restrict the generality of the solution, but rather require a careful choice of the coordinate system. In those cases when such simplifying assumptions are not justified by the evolutionarity conditions, they may lead to a loss of physically meaningful solutions (a relevant hydromagnetic example can be found in Sect.3.3.5).

2.1.2 Structure of a Weak Shock Wave

Now we consider shock waves with $M_0 \sim 1$, and, accordingly, $T_{GD} \sim 1$. Then all the characteristic scales in (2.1.20) are determined by the mean free path ℓ_0 and by the dimensionless small parameter $\varepsilon = (m_e/m_i)^{\frac{1}{2}}$. This allows us immediately to set up the hierarchy of the collision scales:

$$\Delta_{T_e}, \Delta_r, \Delta_u \sim \ell_0/\varepsilon \gg \Delta_{vi}, \Delta_{Ti} \sim \ell_0 \gg \Delta_{ve} \sim \varepsilon\ell_0 \quad . \tag{2.1.25}$$

The characteristic scale of charge separation is r_{D0}. If $\Delta \sim \ell_0$ (Δ being the shock width), then the separation of charges can be neglected as long as the ratio (with $n_0[\mathrm{cm}^{-3}]$ and $T_0[\mathrm{eV}]$)

$$r_{D0}/\ell_0 = 2\cdot 10^{-10}(\lambda_0/10)n_0^{1/2}T_0^{-3/2} \quad (2.1.26)$$

is small. The ratio (2.1.26) is, in fact, small for the typical plasma parameters in collisive shock waves: $r_{D0}/\ell_0 = 10^{-3}$ for $n_0 = 10^{16}\mathrm{cm}^{-3}$, $T_0 = 5$ eV. Consequently, in the zero approximation in r_{D0}/ℓ_0 we may set

$$u_x^e = u_x^i = u_x \ , \ E_x = 0 \ , \ \delta n = 0 \quad (2.1.27)$$

for calculating the shock structure. With this done, further approximations can be used for calculating the electric field strength, the charge separation and the potential jump on the shock front (Sect.2.1.4).

By virtue of (2.1.27), the term with the scale factor Δ_u drops out from (2.1.18), and the largest scales in the problem under investigation are those of the electronic heat conduction Δ_{Te} and of the electronic-ionic temperature relaxation Δ_r. As follows from (2.1.25), $Pr \sim \varepsilon \ll 1$. The shock structure therefore should be expected to bear qualitative resemblance with that considered in Sect.1.8.3: in weak shock waves it is shaped by heat conduction, and stronger shock waves exhibit a viscous jump. For the moment, we shall concentrate on the former case.

Neglecting ionic viscosity and heat conduction (ε being small on the scale of ℓ_0/ε), we obtain the equations for the shock layer

$$u_x - 1 + \frac{3}{10M_0^2}\left(\frac{T_e + T_i}{u_x} - 2\right) = 0 \ , \quad (2.1.28)$$

$$\frac{u_x^2 - 1}{2} + \frac{3}{4M_0^2}(T_e + T_i - 2) - \Delta_{Te}^{(0)} T_e^{5/2}\frac{dT_e}{dx} = 0 \ , \quad (2.1.29)$$

$$\frac{3}{2}\frac{dT_i}{dx} + \frac{T_i}{u_x}\frac{du_x}{dx} = \frac{1}{\Delta_r^{(0)}}\frac{T_e - T_i}{u_x^2 T_e^{3/2}} \ . \quad (2.1.30)$$

Note that for $\ell_0/\varepsilon \gg \ell_0$ the Navier-Stokes approximation can be justified, see criterion (1.2.16).

Using (2.1.28), we eliminate T_i and obtain

$$\frac{\Delta_{Te}^{(0)}}{2} T_e^{5/2}\frac{dT_e}{dx} = (1 - u_x)(u_x - u_{GD}) \ , \quad (2.1.31)$$

$$\frac{\Delta_r^{(0)}}{2} u_x^2 T_e^{5/2}\frac{du_x}{dx} = -\left\{(1 - u_x)\left[1.76M_0^4 u_x^2 T_e^{-1}(u_x - u_{GD}) - \frac{5M_0^2}{3}u_x + T_e\right]\right\}$$
$$\times\left[\frac{T_e}{u_x} + \frac{40}{3}M_0^2 u_x - 5\left(\frac{5M_0^2}{3} + 1\right)\right]^{-1} \ . \quad (2.1.32)$$

As indicated in Sect.1.8.3, the set of (2.1.31,32) has a solution representing the evolutionary shock-wave structure, if in the (u_x, T_e) phase plane one of the singular points is a node, and the other a saddle. Linearizing these equations, we find the product of eigenvalues at the singular point $k = 0, 2$:

$$q_1(k)q_2(k) \propto \frac{1 - M_k^2}{5M_k^2 - 4} \quad . \tag{2.1.33}$$

At point 0 the right-hand side of (2.1.33) is always negative, and so this point is a saddle point. Accordingly, (2.1.31,32) describe the shock structure as soon as the singular point 2 is a stable node, which requires

$$M_2^2 > 4/5 \quad , \tag{2.1.34a}$$

or, by virtue of (2.1.23),

$$M_0^2 < 19/15 \quad . \tag{2.1.34b}$$

As follows from (2.1.34b), the electronic heat conduction can only shape the structure of a weak shock wave. Provided condition (2.1.34) is satisfied, (2.31,32) can be numerically integrated, starting at point 0. As $x \to \infty$, the integral curve in the (u_x, T_e) plane tends to the stable nodal point 2. It can be demonstrated that this curve does not cross the parabola

$$\frac{T_e}{u_x} + \frac{40}{3} M_0^2 u_x - 5\left(\frac{5M_0^2}{3} + 1\right) = 0 \quad , \tag{2.1.35}$$

on which the right-hand side of (2.1.32) displays a singularity (Fig.2.1). For the condition (2.1.34b) which implies that $(M_0 - 1) \ll 1$, we can also find an approximative analytic solution. Indeed, to within $O((M_0 - 1)^3)$ the compression of plasma in weak shock waves is adiabatic, i.e.,

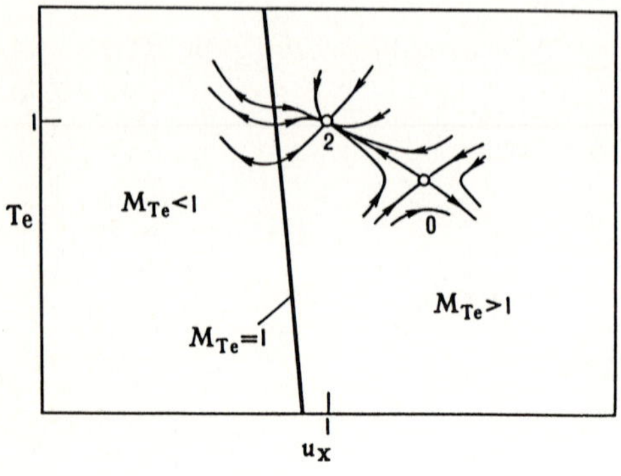

Fig.2.1. Integral curves of (2.1.31,32) in the (u_x, T_e) phase plane for a weak shock wave in plasma. Line $M_{Te} = 1$ is a segment of parabola (2.1.35)

$$T_e = T_i = u_x^{-2/3} \quad . \tag{2.1.36}$$

Substituting (2.1.36) into the left-hand side of (2.1.31) and retaining the terms up to second order in $(M_0 - 1)$, we get a solution of the form of (1.8.11) with

$$r = \frac{7}{2} - \frac{5}{2M_0^2} \quad , \quad \tilde{\Delta} = \frac{4}{9}\Delta_{T_e}^{(0)} \quad . \tag{2.1.37}$$

We see that in weak shock waves the plasma behaves as a one-fluid one-temperature medium with $Pr \ll 1$ (in the zero approximation in r_{D0}/ℓ_0 and ε).

2.1.3 Structure of a Strong Shock Wave

Let us now consider the strong shock waves, for which the inequalities (2.1.34) are not valid ($M_0 \gg 1$). It must be acknowledged that in this case the head portion of the shock front is dominated by electronic heat conduction. Since the rate of propagation of perturbations upstream in the electronic component is by a factor of $1/\varepsilon$ greater than in the ionic component, downstream heating of the plasma results when a heat-conductive "tongue" appears at the head of the shock, in which the electrons are heated while the ambient plasma is motionless and the ions cold.

For $M_0 \gg 1$ we find from (2.1.16,17,19) for the vicinity of the initial point ($T_e = T_i = u_x = 1$)

$$|u_x - 1| = O(1/M_0^2) \quad , \quad dT_i/dx = O(1/M_0^2) \quad . \tag{2.1.38}$$

This allows us to construct an approximate solution, which neglects compression and heating of ions against the background of strong heating of electrons. Expressing $M_0^2(1 - u_x)$ via T_e with the aid of (2.1.28) (assuming $T_i = 1$), we find from (2.1.29)

$$\Delta_{T_e}^{(0)} T_e^{5/2} \frac{dT_e}{dx} = \frac{9}{20M_0^2}(T_e - 1) \quad . \tag{2.1.39}$$

The solution of this equation has the form

$$\frac{9}{40M_0^2} \frac{x - x_0}{\Delta_{Te}^{(0)}} = \frac{T_e^{5/2}}{5} + \frac{T_e^{3/2}}{3} + T_e^{1/2} + \frac{1}{2}\ln\frac{T_e^{1/2} - 1}{T_e^{1/2} + 1} \quad , \tag{2.1.40}$$

where x_0 is the integration constant.

As $T_e \to 1$, on the right-hand side of (2.1.40) the last term dominates, which takes care of the exponential asymptotic of the solution as $x \to -\infty$, see (1.7.15). For $T_e \gg 1$ the first term dominates, and the profile is typical of a heat wave dominated by nonlinear heat conduction [2.15]

$$T_e = 2.0\left(\frac{m_e}{m_i}\right)^{1/5}\left(M_0\,\frac{x - x_0}{\ell_0}\right)^{2/5} . \tag{2.1.41}$$

Thus, for $x \sim x_0$ the exponential solution of the form (1.7.15) goes over into the profile of the form (2.1.41). As follows from (2.1.41), the length of the heating zone exhibits a rapid increase with M_0. Indeed, if the electrons there are heated to the near-final temperature ($T_{GD} \sim M_0^2$), then, according to (2.1.41), $x - x_0 \sim M_0^4 \ell_0$.

We see that compression of the plasma for $M_0 \gg 1$ takes place mostly when the temperature of the electrons approaches its final value. For this reason it is more convenient to normalize the shock-layer equations with respect to state 2. Then in the region of interest $T_e \sim 1$, and, since $M_2 \sim 1$ (as a matter of fact, $M_2 \to 1/\sqrt{5} \simeq 0.45$ as $M_0 \to \infty$), the hierarchy of scales (2.1.25) remains unchanged after the transcription $0 \to 2$. When compression of the plasma and heating of ions become so strong that the departure from (2.1.40) becomes significant, the electronic temperature is already high enough to justify the use of (2.1.31,32) to describe the shock structure.

However, as we have already indicated, these equations do not have a solution which would describe the entire shock structure, since condition (2.1.34) is not valid for strong shock waves. Point 0 is then outside, and point 2 within the region enclosed by the parabola (2.1.35). Therefore, point 2 cannot be reached along the integral curve, going out of point 0, in the sense of increasing x: at the intercept with the parabola the direction of variation of x along this integral curve is reversed, and the second boundary condition cannot be satisfied (Fig.2.2).

To illustrate the physical sense of the restriction (2.1.34), let us consider the propagation of density perturbations in the plasma, whose wavelengths are small on the scale of heat conduction, within the limits of the shock region described by (2.1.31,32). With respect to such perturbations the electrons of the plasma are isothermal, and hence their rate of propagation is given by (2.1.24), where we must set $\gamma_e = 1$. Introducing the local Much number

$$M_{Te} = u_x/a_{Te} = u_x\left(\frac{\frac{5}{3}T_i + T_e}{m_i}\right)^{-\frac{1}{2}} , \tag{2.1.42}$$

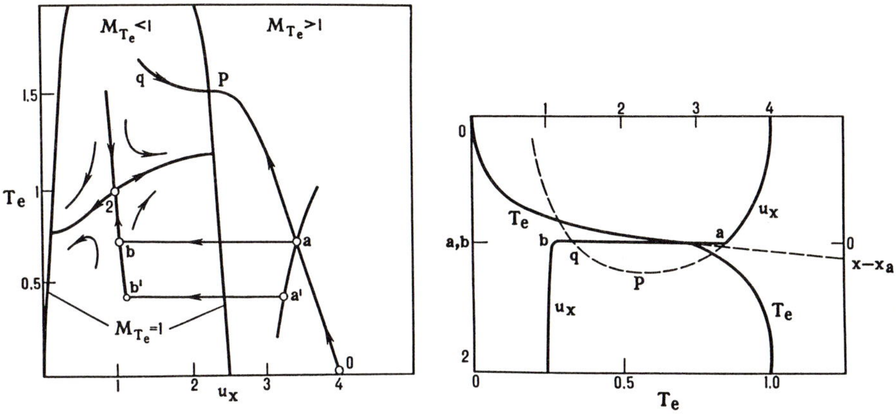

Fig.2.2. Integral curves of (2.1.31,32) in the (u_x, T_e) phase plane (*left*) and profiles of velocity and electronic temperature (*right*) for a strong shock wave in plasma. (---) in the right part of the diagram carry on the integral curve $0 \to a$ into the region $a \to p$ instead of transition $a \to b$. Ion shock $a \to b$ in this approximation is represented by an infinitely thin discontinuity $(x_a = x_b)$

we apprehend that the second term in brackets on the right-hand side of (2.1.32) can be rewritten in the form

$$\frac{T_e}{u_x} + \frac{40}{3} M_0^2 u_x - 5\left(\frac{5M_0^2}{3} + 1\right) = 5M_0^2 u_x\left(1 - \frac{1}{M_{Te}^2}\right) \quad . \tag{2.1.43}$$

If the condition (2.1.36) is satisfied, the upstream and downstream flows are supersonic with respect to the electron-isothermal speed of sound a_{Te}, i.e., $M_{Te}(0) > 1$, $M_{Te}(2) > 1$, or, in terms of equilibrium values, $M_0 > \sqrt{4/5}$, $M_2 > \sqrt{4/5}$, and therefore the transition from one state to the other may be accomplished by adiabatic compression of the ion plasma component. The violation of (2.1.36) implies that the downstream flow becomes subsonic. As an adiabatic supersonic-subsonic transition is impossible, an internal discontinuity must arise. Its evolutionarity, and hence the uniqueness of its structure, is ensured by the conditions $M_{Te}(0) > 1$, $M_{Te}(2) < 1$.

The internal discontinuity is shaped by ion viscosity and heat conduction ("ion shock"). Since the ions there are heated to temperatures close to the final one, it would be natural to carry out the normalization with respect to state 2: $\Delta_{vi} \sim \Delta_{vi}^{(2)}$, $\Delta_{Ti} \sim \Delta_{Ti}^{(2)}$. As follows from (2.1.17), $dT_e/dx \sim \varepsilon$ if $d/dx \sim 1/\ell_2$, and thus in zero approximation in ε the ion shock can be considered isothermal in terms of electron temperature. In this approximation the structure of the ion shock is described by

$$\Delta_{vi}\frac{du_x}{dx} + u_x - 1 + \frac{3}{10M_2^2}\left(\frac{T_e + T_i}{u_x} - 2\right) = 0 \quad , \tag{2.1.44}$$

$$\frac{3}{2}\frac{dT_i}{dx} + 2\left(\frac{T_i}{2u_x} - \frac{5M_2^2}{3}\Delta_{vi}\frac{du_x}{dx}\right)\frac{du_x}{dx} - \frac{10M_2^2}{3}\frac{d}{dx}\left(\Delta_{Ti}\frac{dT_i}{dx}\right) = 0 \quad . \tag{2.1.45}$$

For $T_e = \text{const.}$ (2.1.45) is integrable with the aid of (2.1.44):

$$\frac{10M_2^2}{3}\Delta_{Ti}^{(2)}T_i^{5/2}\frac{dT_i}{dx} = \frac{3}{2}T_i - T_e \ln u_x + 2\left(\frac{5M_2^2}{3} + 1\right)u_x - \frac{5M_2^2}{3}u_x^2 + F \quad , \tag{2.1.46}$$

F being the integration constant. Equations (2.1.44,46) describe the ion shock structure. On the scale of ℓ_0 the condition of applicability of the gasdynamic equations (1.2.16) is no longer satisfied, so (2.1.44-46) should rather be considered. A more correct calculation of the ion shock structure would require employing the kinetic equations. This problem was solved by the Mott-Smith technique in [2.10]; it was found that the gasdynamic estimate of the shock width constitutes from one-half to one-fourth of the kinetic estimate.

The boundary conditions for the ion shock are obtained by nullifying the right-hand sides of (2.1.44,46). If u_b is the downstream velocity of the ion shock, then the upstream velocity u_a can be found as a root of

$$f(u_x) \equiv 5\left(\frac{5M_2^2}{3} + 1\right)(u_x - u_b) - \frac{20}{3}M_2^2(u_x^2 - u_b^2) - T_{ea}\ln\frac{u_x}{u_b} = 0 \quad , \tag{2.1.47}$$

which exists always and is unique for $u_x > u_b$ if $M_0^2 > 19/15$. Note that the boundary conditions are obtained on the order ℓ_0/ε, when the gasdynamic approximation is valid, and hence they are exact in zero approximation in ε, where the ion shock is considered a discontinuity.

On the whole, the shock structure is constructed in the following way. According to (2.1.33), for $M_0^2 > 19/15$ the singular point 2 in the (u_x, T_e) phase plane is a saddle point; therefore, as $x \to +\infty$, this point is entered (in the sense of increasing T_e) by a single integral curve (Fig.2.2). For each point b' of this integral curve, setting $u_x = u_b$ and using (2.1.47), we can find point a' on the $(u_x = u_{a'}, T_e = T_{b'})$ phase plane, from which a transition is allowed across the ion shock into the given point b' $(u_x = u_{b'}, T_e = T_{b'})$. The locus of points a' is a curve in that region of the (u_x, T_e) phase plane where $M_{Te} > 1$. Its intercept a with the unique integral curve, going out of point 0 in the direction of increasing T_e, corresponds to the origin of the ion shock wave.

Insofar as points 0 and 2 are saddle points, the numerical integration can be performed easily: (2.1.31,32) are integrated from the initial point in the vicinity of singular point 0 (x increasing) and from the terminal point near singular point 2 (x decreasing). The resulting integral curves $0 \to a$ and $b \to 2$ are close to the true separatrices of the saddle points. The curve $b \to 2$ is then used for constructing the locus of the points a', and the origin of the ion shock is defined as the intercept of the latter with the curve $0 \to a$.

Note that at $M_0^2 = 19/15$ $(M_2^2 = 4/5)$ for $u_b = 1$ we have

$$\left.\frac{\partial f(u_x)}{\partial u_x}\right|_{u_x=1} = 0 \quad ,$$

i.e., in this case the points 2, a, b coincide. If the flow parameters allow the emergence of the ion shock, it will be located at the rear of the front, and the local Mach number at point a is $M(a) = 1$. In stronger shock waves the ion shock is also enhanced, although not unrestrictedly: $M(a) \to (5/3)^{\frac{1}{2}}$ as $M_0 \to \infty$ [2.9]. The finiteness of the maximum strength of the ion shock is due to ion heating by heat exchange. The ion shock divides the structure into two parts. The region $0 \to a$, as we have seen, is the zone of heat-conduction heating. Then follows the ion shock. It can be readily demonstrated with the aid of (2.1.28,31) that near the singular point $2 dT_i/dx < 0$ already for $M_0 > (9/5)^{\frac{1}{2}}$. Accordingly, the region $b \to 2$ of the strong shock structure can be termed the zone of temperature relaxation, where the ions, heated in the ion shock wave above the equilibrium temperature, give their thermal energy to electrons in elastic collisions. The mechanism of electronic heat conduction carries the thermal energy upstream, thus causing a slight compression of the plasma in the relaxation zone. With $M_0 = 3 \div 10$ the width of the relaxation layer is about one-fifth of the width of the heating layer [2.9].

Figure 2.3 shows the structure of a strong shock in a plasma [2.8]. With $\Delta_{Te} = \ell_2/\varepsilon$ the ion shock wave is represented by a zero-thickness discontinuity, whose structure can be made visible only by locally extending the scale. We see that plasma in the shock wave is an essentially two-temperature medium. The two-fluid effects, which are manifest in the polarization of the plasma, will be discussed below.

It would be interesting to investigate whether a shock wave is possible in a nonequilibrium, completely ionized plasma, retaining its initial nonequilibrium. If possible, this would imply the existence of new classes of

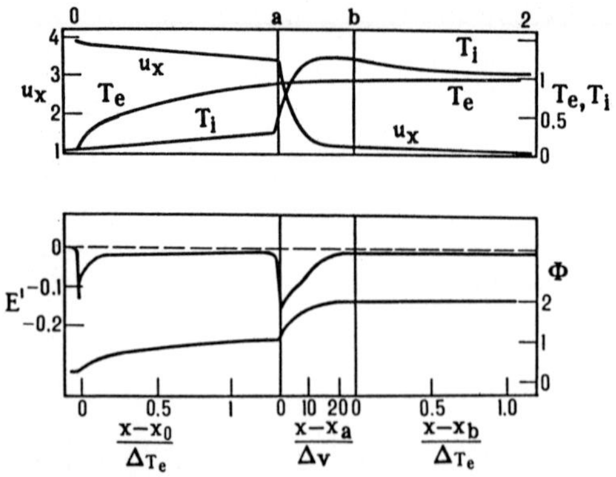

Fig.2.3. Structure of a strong shock wave in plasma [2.8]: the profiles of velocity and temperature (*top*), electric field and potential (*bottom*) for $M_0 = 10$. The scales Δ_{Te}, Δ_v pertain to the equilibrium downstream state

shock waves, not encountered in conventional gasdynamics. The slowest relaxation process in a completely ionized plasma is the heat exchange between electrons and ions. In the gasdynamic approximation of a plasma with cold ions and hot isothermal electrons the characteristic rate of propagation of perturbations is the speed of ion sound [2.16]

$$a_i = (T_e/m_i)^{1/2} \quad .$$

The shock structure in the initially nonisothermal plasma ($T_e > T_i$) was explored in the gasdynamic approximation in [2.11]. It was found that even a strong nonisothermality ($T_{e0}/T_{i0} \simeq 30$) has almost no effect on the shock structure because of the electronic heat-conduction mechanism, which preheats plasma ahead of the ion shock. As the scales of electronic heat conduction and temperature relaxation are of the same order of magnitude, in the downstream region the nonequilibrium state of plasma is not preserved: as $x \to +\infty$ $T_i \to T_e$. Accordingly, the only distinction of the shock waves in question is that the initial Mach number is defined in terms of the nonisothermal speed of sound: in (2.1.24) we must set $T_{e0} + T_{i0}$. Evidently, to make the nonequilibrium effects manifest, the electronic-heat conduction should be suppressed. This can be accomplished, for instance, by applying a strong transverse magnetic field (Sect.3.4.2).

2.1.4 Polarization of Plasma in Shock Waves

Now let us use subsequent approximations in r_{D0}/ℓ_0 (or in r_{D2}/ℓ_2) to calculate the effects of charge separation, i.e., polarization of plasma in the shock wave.

Assuming that the characteristic scale of gasdynamic variables is $\Delta \gg r_D$, from (2.1.15,18) we find

$$E_x = O(r_D/\Delta) \quad , \quad \delta n = O((r_D/\Delta)^2) \quad . \tag{2.1.48}$$

In accordance with (2.1.48), we can first derive the electric polarization field E_x, and then, knowing the profile of E_x, proceed to find the charge separation δn with the aid of (2.1.15). Introducing the dimensionless potential Φ, defined by

$$E_x = - r_{D0} \frac{d\Phi}{dx} \quad , \tag{2.1.49}$$

we note that Φ, as follows from (2.1.48), is not small in r_D/Δ. Even assuming the quasi-neutrality of the plasma in the shock front, the jump of the dimensionless electric potential across the front $\Delta\Phi = \Phi_2 - \Phi_0$ [which corresponds to the dimensional potential difference $\Delta U = (T_0/e)\Delta\Phi$] has a finite value. It is noteworthy that this value cannot be calculated from the known velocity of the front and the initial parameters of plasma. This comes from the fact that the shock structure is described by a larger number of variables than is the nondissipative flow on either side of the front: within the limits of the shock the plasma may suffer polarization, while being quasi-neutral at the infinities. This does not contradict the results of Sect.1.7.2, since the local electric potential Φ is not an observable quantity and is not included in a number of gasdynamic variables: the downstream state of the plasma does not depend on Φ_2. At the same time, the potential jump $\Delta\Phi$ can be directly measured in an experiment; in a sense such measurements are even more informative than the measurements of the jumps of gasdynamic variables, the latter being dependent only on the Hugoniot conditions, whereas the values of $\Delta\Phi$ allow conclusions to be drawn regarding the structure of the shock.

If the effects of electron viscosity can be neglected on the scale Δ, we get from (2.1.18) with due account taken of (2.1.48)

$$E_x = - r_{D0}\left(u_x \frac{d}{dx}\frac{T_e}{u_x} + 0.71 \frac{dT_e}{dx}\right) \quad . \tag{2.1.50}$$

The first term in parentheses is due to the contribution of the electronic pressure, the second to that of thermal force.

Especially easy is the calculation of $\Delta\Phi$ in weak shocks ($M_0 < \sqrt{19/15}$). With an accuracy of 10^{-2} the electronic temperature T_e and velocity u_x in such shocks are linked by the adiabaticity condition (2.1.36), which allows us to rewrite (2.1.50) in the form

$$E_x = - r_{DO} \cdot 3.21 \frac{d}{dx} u_x^{-2/3} \quad , \tag{2.1.51}$$

whence

$$\Phi - \Phi_0 = 3.21(u_x^{-2/3} - 1) \quad ; \quad \text{and} \quad \Delta\Phi = 3.21(u_{GD}^{-2/3} - 1) \quad . \tag{2.1.52}$$

A ready explanation exists for the direction of the electric polarization field E_x (in the upstream direction) and the positive sign of $\Delta\Phi$. The plasma being a mixture of a light gas (electrons) and a heavy gas (ions), the lightweight component, with the higher coefficient of diffusion, migrates upstream, creating an excess of negative charge. This process is counteracted by the emergent polarization field. The profiles of E_x and Φ are constructed on the basis of the analytic solutions (1.8.11) and (2.1.37). For the strongest shock wave, still described by these solutions ($M_0 = \sqrt{19/15} = 1.13$), the dimensional potential difference across the front is $\Delta U = 0.4\ T_0$, T_0 being the initial temperature of the plasma in electronvolts.

In a strong shock wave the temperature profile in the heating zone is given by the solution (2.1.40). The profiles of the electric field and the potential can be constructed with the aid of (2.1.40,50):

$$E_x = - 3.8 \frac{r_{DO}}{\ell_0} M_0 \left(\frac{m_e}{m_i}\right)^{\frac{1}{2}} T_e^{-5/2}(T_e - 1) \quad , \tag{2.1.53}$$

$$\Phi - \Phi_0 = 1.71(T_e - 1) \quad . \tag{2.1.54}$$

Expression (2.1.53) describes, in particular, a sharp peak of the electric field at the head of the shock (the precursor electric layer), the high value of E_x being due to the large temperature gradient at the head of the nonlinear heat-wave profile (2.1.41) for $x \sim x_0$, and a rapid decline of the field as the electronic temperatures goes up [the factor $T_e^{-5/2}$ in (2.1.53)].

In the ion shock the gradient of electron temperature can be neglected compared with the velocity gradient; accordingly,

$$E_x = r_{DO} \frac{T_{ea}}{u_x} \frac{du_x}{dx} \quad , \tag{2.1.55}$$

$$\Phi = \Phi_a - T_{ea} \ln \frac{u_x}{u_a} \quad . \tag{2.1.56}$$

The potential jump in the ion shock wave is defined by

$$\Phi_b - \Phi_a = T_{ea} \ln \frac{u_a}{u_b} \quad . \tag{2.1.57}$$

In the zero approximation in $(m_e/m_i)^{1/2}$ the derivation dT_e/dx is zero within the ion shock and nonzero outside; the discontinuity of this derivative would result, following (2.1.50), in a jump in the electric field. Actually, the electric field at the head of the ion shock rapidly increases over a small distance, where the effects of charge separation and electron viscosity may be of equal importance. A detailed analysis of these effects [2.8] reveals the existence of polarization fine structure in the ion shocks—damped oscillations around the solution (2.1.55,56). These strongly damped oscillations are the plasma oscillations of ions. It is due exactly to this phenomenon that the charge separation is capable of producing the oscillatory shock structure in a plasma. In our present case, however, this effect is of little prominence; besides, the plausibility of the results obtained by employing the Navier-Stokes approximation on a scale much smaller than the free path length, is questionable. This effect can be important in shocks in cold plasma in the presence of strong magnetic field, (Sect.3.3.4).

The potential jump at the front of a strong shock is defined by (2.1.49, 50):

$$\Delta\Phi = 1.71(T_2 - 1) - \int_{-\infty}^{+\infty} T_e(x') \frac{1}{u_x(x')} \frac{du_x(x')}{dx'} dx' \quad . \tag{2.1.58}$$

The integral on the right-hand side of (2.1.58) is obviously positive. To obtain an upper estimate, we observe that

$$\int_{-\infty}^{\infty} \left(\frac{3}{2} \frac{dT_e}{dx} + \frac{T_e}{u_x} \frac{du_x}{dx}\right) dx \propto \int_{-\infty}^{\infty} T_e \frac{dS_e}{dx} dx > 0 \quad , \tag{2.1.59}$$

where S_e is the dimensionless entropy per one electron. Substituting (2.1.59) in (2.1.58) we get

$$1.71 < \frac{\Delta\Phi}{T_2 - 1} < 3.21 \quad . \tag{2.1.60}$$

From (2.1.57,58) we easily infer that the potential jump $\Delta\Phi$ across the shock cannot be calculated in a general way. Indeed, the integrand in (2.1.58), as we know from thermodynamics [2.17], is not a total differential, and the integral therefore depends not only on the terminal points 0 and 2, but also on the path of integration from 0 to 2, i.e., on the shock structure. The results of calculations carried out in [2.8] for strong shock waves indicate that $\Delta\Phi$ tends to be closer to the upper bound in (2.1.60).

The account of finiteness of the Debye radius on the scale of ℓ_0 does not bring about any qualitatively new results [2.8]. The interaction of the polarization field with the gasdynamic variables, as ought to be expected, reinforces the plasma's tendency to neutralization: according to (2.1.48), the increase in r_{D0} is accompanied by the increase in the absolute values of E_x and δn, although the relative values of the same quantities, reduced to the relevant scales, $E_x/(r_D/\Delta)$ and $\delta n/(r_D/\Delta)^2$, tend to decrease. This results in a certain smoothing out of the polarization-field profile. The profiles of gasdynamic variables are little affected by assuming the Debye radius to be finite.

2.2 Shock Structure in a Plasma with Ionization

2.2.1 Shock-Layer Equations and Boundary Conditions

Inclusion of plasma ionization in the shock complicates the investigation of the shock structure. Even if we restrict our consideration to not too high temperatures, when only single ionization is essential, the description of a plasma composed of electrons, ions and neutral atoms would call for the use of a three-fluid approximation. The transfer coefficients of a three-component plasma show a strong dependence on the type of gas; calculation of them, accounting for inelastic processes of ionization and recombination, is a task so formidable that it has not yet been accomplished for any gas with such a degree of consistency as achieved in [2.14] for a completely ionized plasma. It must be admitted, therefore, that our present discussion will have to be more speculative than the treatment of Sect.2.1.

We begin with making some simplifying assumptions. First, the masses of ions and atoms are practically the same, and the collision cross section between ions and atoms is large due to the process of resonant charge exchange. The time to establish thermal equilibrium between the heavy components is of the same order as the time of "Maxwellization" of each of these components (or the one with the higher concentration, if $n_a \gg n_i$ or $n_i \gg n_a$). For this reason we make no distinction between the temperatures of the heavy components, $T = T_a = T_i$. Additionally, the large collision cross section prevents the slippage of ions with respect to atoms. The existence of external electromagnetic fields, which act only on the charged components of the plasma, may change this situation, assisting the ion slip (Sect.4.4). However, the electric field of plasma polarization is estimated to be insufficiently strong to accomplish this. Accordingly, we may set $\mathbf{u} = \mathbf{u}_a = \mathbf{u}_i$, which amounts to assuming that the heavy components of the plasma are, in zero approximation in ℓ_{ia}/Δ (Δ being the characteristic scale of the shock layer), effectively represented by one fluid, described by the gasdynamic velocity $\mathbf{u}$ and temperature T. (This approximation may be inadequate for a gasdynamic viscous jump, where the characteristic scale Δ is of the same order as the mean free path of atoms or ions, and the electric polarization field is at its maximum.) Viscosity, heat conductivity and heat exchange with the electrons for this fluid are directly composed of the respective values for the ion and atom components.

Secondly, both the heavy and light components are assumed to have nearly equilibrium Maxwellian distributions with temperatures T and T_e, respectively. The inelastic processes of ionization and recombination may distort the distribution function of electrons, especially near the value of the ionization energy. The assumption of rapid Maxwellization of electrons is justified as long as their density is sufficiently high. It is only then that the electron component may be described by the gasdynamic equations.

Thirdly, at this point we neglect the radiative effects, on grounds that the radiation path length, which determines the spatial scale Δ_{ph} of the processes of radiative energy transfer, is much larger than the scales of electron heat conduction and shock ionization pertinent to this problem. The radiation coming from the downstream region at equilibrium slightly affects the state of plasma in the far upstream region, due to the radiative heating, ionization and excitation of atoms. Since the strong shock structures in plasmas depend little on the parameters of the initial state, these effects may be disregarded. Photorecombination can also be neglected as long as the electron density downstream is high enough.

Fourthly, we assume the Debye radius to be small, and thus the plasma to be quasi-neutral in the first approximation in r_D/Δ. As (2.1.5) remains valid, $u_x^i = u_x^e$, i.e.,

$$u_x^a = u_x^i = u_x^e \equiv u_x \quad .$$

After implementing all these simplifying assumptions, we are left with only one additional variable, the dimensionless degree of ionization

$$\alpha = \frac{n_e}{n_i + n_a} \quad . \tag{2.2.1}$$

Denoting the heavy-particles density $n_i + n_a$ by N, we may write $n_e = n_i = \alpha N$, $n_a = (1-\alpha)N$. At thermodynamic equilibrium the equilibrium value $\alpha = \alpha_{eq}(T)$ at temperature T is found from the Saha formula

$$\frac{\alpha^2}{1-\alpha} = \frac{2g_+}{g_0 N}\left(\frac{m_e T}{2\pi\hbar^2}\right)^{3/2} \exp\left(-\frac{J}{T}\right) \quad , \tag{2.2.2}$$

where g_+, g_0 are the statistical weights of the ion and atom, respectively; J is the atom's ionization energy.

The impact ionization of atoms takes place chiefly in either of two ways: directly from the ground state by collision with another atom or electron

$$A + A \to A^+ + e + A \quad ; \quad e + A \to A^+ + e + e \quad , \tag{2.2.3}$$

or through a two-step process, in which the atom is first excited and then ionized:

$$A + A \to A^* + A \quad , \quad A^* + A \to A^+ + e + A \quad ;$$
$$e + A \to A^* + e \quad , \quad e + A^* \to A^+ + e + e \quad . \tag{2.2.4}$$

In inert gases with long-lived excited states, the processes of two-step ionization (2.2.4) at not too high temperatures (the region of first ionization) are much faster than ionization from the ground state. The rate of these processes is determined by the rate of excitation: the excited atoms become almost instantaneously ionized. Accordingly, disregarding the processes (2.2.3), we may write the kinetic equation of ionization in the form

$$\dot{n}_e = k_{*a} n_a^2 - k_{ra} n_a n_e^2 + k_{*e} n_e n_a - k_{re} n_e^3 \quad , \tag{2.2.5}$$

where k_{*a}, k_{*e} are the kinetic coefficients of impact excitation by atoms and electrons, respectively; k_{ra}, k_{re} are the coefficients of three-particle recombination.

With due account of (2.1.1) we get

$$\frac{d}{dx} n_e u_x = \dot{n}_e \quad .$$

Now we turn once again to dimensionless variables, denoting $u_x = \bar{u}_x/\bar{u}_{x0}$, etc. Making use of the principle of detailed balancing, we can rewrite (2.2.5) in the form of a kinetic equation of ionization in the dimensionless variable

$$\frac{d\alpha}{dx} = \frac{1-\alpha}{\Delta_{ion}^{(a)}}\left(1 - \frac{\alpha^2}{1-\alpha}\,\frac{1-\alpha_{eq}(T)}{\alpha_{eq}^2(T)}\right) + \frac{\alpha(1-\alpha)}{\Delta_{ion}^{(e)}}\left(1 - \frac{\alpha^2}{1-\alpha}\,\frac{1-\alpha_{eq}(T_e)}{\alpha_{eq}^2(T_e)}\right) , \tag{2.2.6}$$

where

$$\Delta_{ion}^{(a)} = \frac{\bar{u}_{x0}u_x}{k_{*a}} , \qquad \Delta_{ion}^{(e)} = \frac{\bar{u}_{x0}u_x}{k_{*e}} , \tag{2.2.7}$$

and the values of $\alpha_{eq}(T)$, $\alpha_{eq}(T_e)$ are calculated with the aid of (2.2.2).

The coefficients k_{*s} (s = a,e) are defined by [2.15]

$$k_{*s} = \int_{(2E/\mu_s)^{1/2}}^{\infty} \sigma_s^*(v) v f_s(v) dv , \tag{2.2.8}$$

where E is the excitation energy, μ_s is the reduced mass:

$$\mu_s = \frac{m_s m_a}{m_s + m_a} = \begin{cases} m_e , & s = e \\ m_a/2 , & s = a \end{cases} .$$

To calculate k_{*s} by (2.2.8) with the known (Maxwellian) distribution function f_s, we need to know the excitation cross section σ_s^* as a function of the relative velocity of colliding particles v. If the temperature is not too high (the region of single ionization), the relevant energy is close to the excitation threshold, where the cross section can be presented as a linear function of the energy

$$\sigma_s^*(v) = \begin{cases} C_s^*\left(\frac{\mu_s v^2}{2} - E\right) , & \frac{\mu_s v^2}{2} \geq E , \\ 0 , & \frac{\mu_s v^2}{2} < E . \end{cases} \tag{2.2.9}$$

Then

$$k_{*s} = C_s^* T_s \left(\frac{8T_s}{\pi\mu_s}\right)^{1/2} \left(\frac{E}{T_s} + 2\right) \exp\left(-\frac{E}{T_s}\right) . \tag{2.2.10}$$

The ratio of excitation coefficients at the same temperature is

$$\frac{k_{*e}}{k_{*a}} = \frac{\Delta_{ion}^{(a)}}{\Delta_{ion}^{(e)}} = \frac{C_e^*}{C_a^*} \left(\frac{m_a}{2m_e}\right)^{1/2} . \tag{2.2.11}$$

For example, this ratio for argon is equal to or greater than 10^4, and seems to be even larger for heavy inert gases (Kr, Xe). Consequently, the ionization due to collisions of heavy particles at the shock front may be important either when the density of electrons is very low, or when $T \gg T_e$.

The remaining shock-layer equations will be written directly in dimensionless form. Each plasma component is considered as an ideal gas with $\gamma = 5/3$. The continuity equations for the fluxes of energy and momentum of the plasma, disregarding the inertia of the electrons, the pressure of the electric polarization field and the electron viscosity, have the form

$$u_x - 1 + \frac{3}{5M_0^2(1 + \alpha_0)}\left(\frac{T + \alpha T_e}{u_x} - 1 - \alpha_0\right) - \Delta_v \frac{du_x}{dx} = 0 \quad , \tag{2.2.12}$$

$$\frac{u_x^2 - 1}{2} + \frac{3}{2M_0^2(1 + \alpha_0)}(T + \alpha T_e - 1 - \alpha_0) + \frac{3I}{5M_0^2(1 + \alpha_0)}(\alpha - \alpha_0)$$
$$- \Delta_v u_x \frac{du_x}{dx} - \Delta_T \frac{dT}{dx} - \Delta_{T_e} \frac{dT_e}{dx} = 0 \quad . \tag{2.2.13}$$

The equation for the electrons' motion is

$$\frac{d}{dx}\frac{T_e}{u_x} + \frac{E_x}{r_{D0}u_x} + \frac{q}{u_x}\frac{dT_e}{dx} = 0 \quad , \tag{2.2.14}$$

and the equations of heat balance for the electrons and the heavy particles are

$$\frac{3}{2}\frac{dT_e}{dx} + \frac{T_e}{u_x}\frac{du_x}{dx} = \frac{5M_0^2(1 + \alpha_0)}{3}\frac{d}{dx}\Delta_{Te}\frac{dT_e}{dx} + \frac{T_i - T_e}{\Delta_r}$$
$$- \frac{3}{2}T_e\left(\frac{d\alpha}{dx}\right)_a - \left(\frac{3}{2}T_e + I\right)\left(\frac{d\alpha}{dx}\right)_e \quad , \tag{2.2.15}$$

$$\frac{3}{2}\frac{dT}{dx} + \frac{T}{u_x}\frac{du_x}{dx} = \frac{5M_0^2(1 + \alpha_0)}{3}\frac{d}{dx}\Delta_T\frac{dT}{dx} + \frac{5M_0^2(1 + \alpha_0)}{3}\Delta_v\left(\frac{du_x}{dx}\right)^2$$
$$+ \alpha\frac{T_e - T}{\Delta_r} - I\left(\frac{d\alpha}{dx}\right)_a \quad . \tag{2.2.16}$$

The characteristic scales of viscosity, heat conduction of heavy particles and electrons, and temperature relaxation in (2.2.12-16) are

$$\Delta_v = \frac{\frac{4}{3}\eta}{m_a\bar{N}_0\bar{u}_{x0}} \quad , \quad \Delta_T = \frac{\kappa\bar{T}_0}{m_a\bar{N}_0\bar{u}_{x0}^3} \quad ,$$
$$\Delta_{Te} = \frac{\kappa_e\bar{T}_0}{m_a\bar{N}_0\bar{u}_{x0}^3} \quad , \quad \Delta_r = \frac{m_a\bar{u}_{x0}}{2m_e(\nu_{ea} + \nu_{ei})} \quad . \tag{2.2.17}$$

Note that the expression for thermal conductivity does not account for the inelastic contribution responsible for a sharp peak of the heat conductivity of neutral atoms at $T \sim 0.1$ J [2.18]. This effect is commonly neglected, as the behavior of κ_a in a relatively narrow temperature range does not much affect the shock structure.

Here the viscosity and heat conductivity of particles of kinds s = a, i, e are given by

$$\eta_s = \frac{5}{4}\frac{n_s T_s}{\nu_s} \quad , \quad \kappa_s = \frac{75}{16}\frac{n_s T_s}{m_s \nu_s} \quad ; \quad \eta = \eta_a + \eta_i \quad , \quad \kappa = \kappa_a + \kappa_i \quad .$$

The collision rates are

$$\nu_a = n_a \sigma_{aa}\left(\frac{16T}{\pi m_a}\right)^{\frac{1}{2}} + n_e \sigma_{ia}\left(\frac{16T}{\pi m_a}\right)^{\frac{1}{2}} + \frac{2m_e}{m_a} n_e \sigma_{ea}\left(\frac{8T_e}{\pi m_e}\right)^{\frac{1}{2}} \quad ,$$

$$\nu_i = n_a \sigma_{ia}\left(\frac{16T}{\pi m_a}\right)^{\frac{1}{2}} + n_e \sigma_{ii}\left(\frac{16T}{\pi m_a}\right)^{\frac{1}{2}} + \frac{2m_e}{m_a} n_e \sigma_{ei}\left(\frac{8T_e}{\pi m_e}\right)^{\frac{1}{2}} \quad , \qquad (2.2.18)$$

$$\nu_e = 2n_a \sigma_{ea}\left(\frac{8T_e}{\pi m_e}\right)^{\frac{1}{2}} + 2n_e \sigma_{ei}\left(\frac{8T_e}{\pi m_e}\right)^{\frac{1}{2}} + n_e \sigma_{ee}\left(\frac{16T_e}{\pi m_e}\right)^{\frac{1}{2}} \equiv \nu_{ea} + \nu_{ei} + \nu_{ee} \quad .$$

The cross sections of atom-atom σ_{aa}, ion-atom σ_{ia}, electron atom σ_{ea}, ion-ion σ_{ii}, electron-ion σ_{ei} and electron-electron σ_{ee} collisions, here are assumed averaged over the Maxwellian distributions of the colliding particles. Using products of averages instead of average products introduces a slight uncontrollable error into the values of the kinetic coefficients, which thus can be taken for known with accuracy only up to a certain factor of the order of unity. (Such a factor in the expression for the thermal force—the last term in (2.2.14)—is designated by q; for a completely ionized plasma $q = 0.71$ [2.14].) The temperature dependence of the cross sections σ_{aa}, σ_{ia}, σ_{ea} is established for each gas separately; for the Coulomb cross sections we have

$$\sigma_{ss} = \frac{e^4 \Lambda}{32\pi\varepsilon_0^2 T_s^2} \quad , \quad (s = e, i) \quad , \quad \sigma_{ei} = \sigma_{ee} \quad , \qquad (2.2.19)$$

Λ being the Coulomb logarithm. Note that unlike the case of a completely ionized plasma we cannot directly state the explicit temperature and density dependences of characteristic scales.

In (2.2.13-16) we denoted the dimensionless ionization energy by $I = J/T_0$. The downstream Mach number M_0 is found with the aid of (2.1.24):

$$M_0 = \frac{u_{x0}}{a_0} = u_{x0}\left(\frac{5}{3}\frac{(1+\alpha_0)T_0}{m_a}\right)^{-\frac{1}{2}} .$$

The boundary conditions for (2.2.6,12,13) are obtained by equating to zero the derivative-containing terms. The resulting set of three equations can be rewritten in the form

$$(u_2 - 1)(u_2 - u_{GD}) = \frac{3I}{10M_0^2(1+\alpha_0)}(\alpha_2 - \alpha_0) , \quad (2.2.20)$$

$$T_2 = \frac{5M_0^2(1+\alpha_0)}{3(1+\alpha_2)} u_2\left(1 + \frac{3}{5M_0^2} - u_2\right) , \quad (2.2.21)$$

$$\frac{\alpha_2^2}{1-\alpha_2} = \frac{2g_+}{g_0\bar{N}_0}\left(\frac{m_e\bar{T}_0}{2\pi\hbar^2}\right)^{3/2} T_2^{3/2}u_2 \exp\left(-\frac{I}{T_2}\right) . \quad (2.2.22)$$

The solutions of this set depend explicitly on the ionization potential of the gas, I, and the parameters of the initial state, N_0, T_0 [$\alpha_0 = \alpha_{eq}(T_0)$]. At $u_2 = u_{GD}$ the left-hand side of (2.2.20) is zero, and the right-hand side is positive; as $u_x \to 0$ its left-hand side tends to $u_{GD} > 0$. At the same time, according to (2.2.21,22), $T_2 \to 0$, $\alpha_2 \to 0$, and the right-hand side of (2.2.20) tends to zero or becomes negative. Consequently, there always exists a (unique) solution of the set (2.2.20-22), and $u_2^{-1} > u_{GD}^{-1}$; in other words, the compression in an ionizable plasma is always higher than in a gasdynamic shock wave at the same Mach number. In principle, the shock compression of ionizable plasma can be arbitrarily high, if only the right-hand side of (2.2.20) can be made large enough. Equation (2.2.21) indicates that the large amount of compression is favored by low densities, when the term $\alpha_2 - \alpha_0$ on the right-hand side of (2.2.20) has the order of unity, while the temperature (and hence the denominator of the right-hand side) is not yet too high. Quite conceivable is, for instance, the 20-fold shock compression of hydrogen plasma. Similarly extended is the range of variation of M_2: with a sufficiently low initial density the values of M_2 can become arbitrarily small, $0 < M_2 < 1/\sqrt{5}$.

Now let us explore the effects of multiple ionization. Let $\alpha_0, \alpha_1, \ldots, \alpha_Z$ be the dimensionless concentrations of neutral atoms and ions with the charges 1,2,...,Z, respectively, Z being the atomic number of the pertinent element:

$$\alpha_0 + \alpha_1 + \ldots + \alpha_Z = 1 \quad ; \tag{2.2.23}$$

(the dimensional concentration of ions with charge m is $N_m = \alpha_m N$). The condition of quasineutrality of the plasma can be written as

$$\sum_{m=0}^{Z} m\alpha_m = \alpha \quad , \tag{2.2.24}$$

where α is the degree of ionization, $\alpha = n_e/N$. The Saha equations for $1 \leqslant m \leqslant Z$ have the form

$$\frac{\alpha\alpha_m}{\alpha_{m-1}} = \frac{2g_m}{g_{m-1}N}\left(\frac{m_eT}{2\pi\hbar^2}\right)^{3/2} \exp\left(-\frac{J_m}{T}\right) \quad , \tag{2.2.25}$$

where J_m is the mth ionization energy. Multiplying both sides of (2.2.25), written for $m=1$, by the respective left and right sides of the same equation for $m=2,3,\ldots n$, we obtain

$$\frac{\alpha_n}{\alpha_0} = \frac{R_n(N,T)}{\alpha^n} \quad , \quad \text{where} \tag{2.2.26}$$

$$R_n(N,T) = \frac{g_n}{g_0}\left[\frac{2}{N}\left(\frac{m_eT}{2\pi\hbar^2}\right)^{3/2}\right]^n \exp\left(-\frac{1}{T}\sum_{m=0}^{n} J_m\right)$$

(we assume $J_0 = 0$).

Taking into account (2.2.23,24), we get

$$\sum_{m=1}^{Z} \alpha_m(m - \alpha) = \sum_{m=1}^{Z} m\alpha_m - \alpha\left(\sum_{m=0}^{Z} \alpha_m - \alpha_0\right) = \alpha\alpha_0 \quad ,$$

which allows us to write the Saha equation immediately, which connects the values of α, N, T:

$$\alpha = \sum_{m=1}^{Z} \frac{R_m(N,T)}{\alpha^m}(m - \alpha) \quad . \tag{2.2.27}$$

Equation (2.2.27) should be solved numerically. Having found $\alpha(N,T)$ we can determine the concentrations of n-ly charged ions by

$$\alpha_n = \frac{R_n(N,T)}{\alpha^n \sum_{m=0}^{Z} [R_m(N,T)/\alpha^m]} \quad , \quad n = 0,1,\ldots,Z \quad . \tag{2.2.28}$$

In the expression for the energy flux (2.2.13) the difference $I(\alpha - \alpha_0)$ must be replaced by

$$\sum_{n=1}^{Z} Q_n(\alpha_n - \alpha_{0n}) \quad , \tag{2.2.29}$$

where $Q_n = \sum_{m=1}^{n} I_m$; $I_m = J_m/T_0$ is the dimensionless m^{th} ionization energy.

The expressions (2.2.27-29) can be used for calculating the shock adiabat with due account taken of multiple ionization. If necessary, they can also be corrected for the decrease of ionization energy in a plasma; such corrections, however, are typically small.

From (2.2.27) it follows that any given value of temperature corresponds to the predominance of ions of a definite charge. In other words, in the temperature region of the mth ionization the $(m-1)^{st}$ ionization is mostly completed, and the (m+1)st ionization has not yet started. The definition of the region of nth ionization is based on the circumstance that the ratio of the nth summand in (2.2.27) to the preceding one is of the order of unity, while the subsequent terms are still small. Hence, we have the estimate

$$n \sim \frac{2}{N}\left(\frac{m_e T}{2\pi\hbar^2}\right)^{3/2} \exp\left(-\frac{J_n}{T}\right) \quad . \tag{2.2.30}$$

This is the basis of a useful approximation method of accounting for multiple ionization, the so-called method of equivalent ions [2.15]. The main idea of the method consists of replacing the mixture of differently charged ions by the homogeneous collection of ions, all having the same effective nonintegral charge, corresponding to the given temperature. From (2.2.23,24) it is clear that this average charge of the ions is precisely equal to the degree of ionization α. The degree of ionization is estimated by a simple formula of type (2.2.30), with

$$J(\alpha) = T \ln \frac{2}{\alpha N}\left(\frac{m_e T}{2\pi\hbar^2}\right)^{3/2} \quad , \tag{2.2.31}$$

where the ionization energy $J(\alpha)$ of the ion with nonintegral charge is obtained by interpolating between the adjacent discrete values, e.g., by the Lagrange technique. Likewise, interpolating between the values of Q_n, we get instead of (2.2.29)

$$Q(\alpha)\alpha - Q(\alpha_0)\alpha_0 \quad . \tag{2.2.32}$$

In particular, in the region of first ionization $I(\alpha) = Q(\alpha) = I$.

In the calculations of the shock adiabat (2.2.31) replaces the Saha formula, and (2.2.32) is substituted into (2.2.13) in place of $I(\alpha - \alpha_0)$. The method of equivalent ions proves useful for describing the kinetics of ionization.

Indeed, this approximation reduces the set of kinetic equations of ionization of varying order to a single equation. We assume that the ionization occurs by the two-step mechanism (via the excited state), and the degree of ionization is large enough to justify the neglect of ionization by impact of heavy particles. We also presume that the recombination takes place via the capture of electrons on the excited level. Then the kinetic equation of ionization can be written in the form

$$\dot{n}_e = k_* n_e N - k_d n_e N^* \quad , \tag{2.2.33}$$

where k_* and k_d are the coefficients of impact excitation and deactivation (we disregard the effects of radiative deactivation, which at small distances are compensated by excitation due to resonant absorption). Seeing that the distribution of ions over the excited levels is established much sooner than the ionization equilibrium (and thus the excited ions are at thermodynamic equilibrium with the actual electron temperature), and making use of the principle of detailed balancing, we obtain from (2.2.33) an equation similar to (2.2.6) [2.19]:

$$\frac{d\alpha}{dx} = \frac{\alpha}{\Delta_{ion}} \left(1 - \frac{\alpha}{\alpha_{eq}(T_e)}\right) \quad , \tag{2.2.34}$$

where the value of $\alpha_{eq}(T_e)$ is calculated from (2.2.31), and

$$\Delta_{ion} = \frac{\bar{u}_{x0} u_x}{k_*} \quad . \tag{2.2.35}$$

The ion excitation rate constant is calculated by (2.2.8), where we set $\sigma_i^* = \text{const}$, in compliance with the actual behavior of the excitation cross section near the threshold. Integration brings us to an expression of type (2.2.10), save that the term $(E/T_s + 2)$ now becomes $(E/T_s + 1)$. Reliable data on the excitation cross sections for multiply charged ions are lacking; calculations based on the Born approximation [2.20] give an estimate of

$$\sigma_i^* = \frac{9}{25} \pi a_0^2 \left(\frac{J_H}{J}\right)^2 \quad , \quad E = \frac{3}{4} J \quad , \tag{2.2.36}$$

where a_0 is the Bohr radius, and J_H is the ionization energy of hydrogen.

The ionization kinetic equation (2.2.34) gives a correct qualitative description of the two-step ionization process of multiply charged ions. It is not quite adequate in the region of single ionization, where the process of recombination is different, see (2.2.6); however, the recombination in ionizing shock waves usually becomes important only near the equilibrium state downstream. If this state corresponds to single ionization, then (2.2.6) must be used, while multiple ionization requires (2.2.34).

The transfer coefficients in a plasma in the approximation of equivalent ions are calculated by the conventional formulas for a simple plasma with ion charge $Z = \alpha$. Since these coefficients are known for any integral Z [2.14], the intermediate values can easily be obtained by interpolation.

2.2.2 Shock Structure Associated with Multiple Ionization

The question of shock structure associated with multiple ionization becomes immensely complicated if we choose to use the rigorous description of the kinetics of ionization. The approximation of equivalent ions, however, is quite adequate for a qualitative analysis of this problem (Sect.2.2.1).

Making use of this approximation we find that for temperatures $T_e \gtrsim 5$ eV, when multiple ionization becomes considerable, and with the degree of ionization not too small ($\alpha_0 \gtrsim 0.05$), the hierarchy of characteristic scales is

$$\Delta_{Te} \sim \Delta_{ion} \gg \Delta_v \ , \quad \Delta_T \ . \tag{2.2.37}$$

Now we write the shock-layer equations with regard to only those phenomena whose scales are the largest, and neglecting viscosity and heat conduction of heavy particles:

$$\Delta_{ion} \frac{d\alpha}{dx} = \alpha\left(1 - \frac{\alpha}{\alpha_{eq}(T_e)}\right) \ , \tag{2.2.38}$$

$$u_x - 1 + \frac{3}{5M_0^2(1 + \alpha_0)}\left(\frac{T + \alpha T_e}{u_x} - 1 - \alpha_0\right) = 0 \ , \tag{2.2.39}$$

$$\frac{u_x^2 - 1}{2} + \frac{3}{2M_0^2(1 + \alpha_0)} (T + \alpha T_e - 1 - \alpha_0)$$
$$+ \frac{3}{5M_0^2(1 + \alpha_0)} [Q(\alpha)\alpha - Q(\alpha_0)\alpha_0] - \Delta_{Te} \frac{dT_e}{dx} = 0 \ , \tag{2.2.40}$$

$$\frac{3}{2}\frac{dT}{dx} + \frac{T}{u_x}\frac{du_x}{dx} = \alpha \frac{T_e - T}{\Delta_r} \ . \tag{2.2.41}$$

As before (in Sect.2.1.2), (2.2.39) can be used to eliminate T, which brings us to equations similar to (2.1.31,32):

$$\frac{1}{2}\Delta_{Te}\frac{dT_e}{dx} = (1 - u_x)(u_x - u_{GD}) + \frac{3}{10M_0^2(1 + \alpha_0)} [Q(\alpha)\alpha - Q(\alpha_0)\alpha_0] \ , \tag{2.2.42}$$

$$\Delta_r \frac{du_x}{dx} = - \Big\{(1 - u_x) \left[\frac{\Delta_r}{\Delta_{Te}} (u_x - u_{GD}) + (1 + \alpha_0)\left(\frac{5M_0^2}{3} u_x - 1\right)\right]$$

$$+ (1 + \alpha)(T_e - 1) + (\alpha - \alpha_0) + \frac{9\Delta_r}{10M_0^2(1 + \alpha_0)} [Q(\alpha)\alpha - Q(\alpha_0)\alpha_0]$$

$$+ \frac{3\Delta_r}{\Delta_{ion}} \alpha \left[1 - \frac{\alpha}{\alpha_{eq}(T_e)}\right]\Big\} \times \Big\{\frac{1 + \alpha_0}{2} \Big[\frac{2\alpha T_e}{(1 + \alpha_0)u_x} - 5\left(\frac{5M_0^2}{3} + 1\right)$$

$$+ \frac{40}{3} M_0^2 u_x\Big]\Big\}^{-1} \quad . \tag{2.2.43}$$

The set (2.2.38,42,43) describing the shock structure on the scales Δ_{Te}, Δ_{ion} is obviously analogous to the set (2.1.31,32): in both cases the accepted approximation corresponds to the Prandtl number zero, and the physical processes considered (which now include ionization) do not dissipate the momentum of the incident flow directly. The structure therefore should be expected to contain a viscous jump, due to the collision processes of transport of heavy particles, the ion-atom shock. This term, strictly speaking, is justified only in the region of first ionization. At higher temperatures the internal shock is obviously due to ions. We use this term to emphasize the common language with the theory of ionizing shock waves.

Indeed, the denominator of (2.2.43) (the last term in square brackets) can be presented in a form similar to (2.1.43). It can be expressed via the electron-isothermal Mach number as

$$\frac{5M_0^2(1 + \alpha_0)u_x}{3} \left(1 - \frac{1}{M_{Te}^2}\right) \quad . \tag{2.2.44}$$

Point 0 is always "supersonic" with respect to the electron-isothermal speed of sound. Point 2 is subsonic as long as

$$M_2^2 < 1 - \frac{2}{5} \frac{\alpha_2}{1 + \alpha_2} \quad . \tag{2.2.45}$$

For weak shock waves with

$$M_0 - 1 \ll \frac{2}{5} \frac{\alpha_0}{1 + \alpha_0} \, ,$$

the inequality (2.2.45) will not be satisfied. This case can easily be approximated by an analytic solution. As yet, however, we refrain from doing this, since we are interested in the strong shock waves, which directly affect the state of ionization. As M_0 increases above unity, the left-hand side of (2.2.45) quickly falls below 1/5 and does not rise above this value. The right-hand side varies within the following limits:

$$\text{from } 1 - \frac{2}{5}\frac{\alpha_0}{1+\alpha_0} \qquad \text{as } M_0 \to 1$$

$$\text{to} \quad 1 - \frac{2}{5}\frac{Z}{1+Z} > \frac{3}{5} \quad \text{as } M_0 \to \infty \quad ,$$

(Z being the atomic number of the gas under study). It is clear that the inequality (2.2.45) is satisfied beginning with a certain value $M_0 = M_{0cr}$. No definite expression for M_{0cr} exists that is similar to (2.1.34), since the value of M_{0cr} depends on the initial temperature, the density of the gas and its ionization energy. In all cases of practical importance, however, $M_{0cr} \sim 1$.

For $M_0 > M_{0cr}$ the shock structure contains an ion-atom shock, whose evolutionarity is ensured by the inequality (2.2.45). By virtue of (2.2.37) the shock is isothermal with respect to the electron temperature, and the ionization there is frozen:

$$T_e = T(a) = \text{const} \quad ; \quad \alpha = \alpha(a) = \text{const} \quad , \tag{2.2.46}$$

where point a of the (u_x, T_e, α) phase space corresponds to the onset of the ion-atom shock.

The equations describing the structure of the ion-atom shock in the gasdynamic approximation are derived similarly to (2.1.44,46) and have the form

$$\Delta_V \frac{du_x}{dx} = u_x - 1 + \frac{3}{5M_0^2(1+\alpha_0)}\left(\frac{T + \alpha(a)T_e(a)}{u_x} - 1 - \alpha_0\right) \tag{2.2.47}$$

$$\Delta_T \frac{dT}{dx} = \frac{3}{2}T + (1+\alpha_0)\left[\left(\frac{5M_0^2}{3} + 1\right)u_x - \frac{5M_0^2}{6}u_x^2\right] - \alpha(a)T_e(a)\ln u_x - F \quad , \tag{2.2.48}$$

where F is the integration constant. The boundary conditions are obtained by equating the right-hand sides of these equations to zero.

The location of point a in the phase space is found by solving

$$f(u_x) \equiv 5\left(\frac{5M_0^2}{3} + 1\right)[u_x - u_x(b)] - \frac{20M_0^2}{3}[u_x^2 - u_x^2(b)] - \frac{2\alpha(a)T_e(a)}{1+\alpha_0}\ln\frac{u_x}{u_x(b)} = 0 \quad , \tag{2.2.49}$$

where point b corresponds to the end of the ion-atom shock, and in accordance with (2.2.46) $T_e(a) = T_e(b)$, $\alpha(a) = \alpha(b)$; for every $u_x(b)$ (2.2.49) has a unique root $u_x(a) > u_x(b)$. Setting $u_x(b) = 1$ and normalizing to state 2, we find that at $M_0 = M_{0cr}$

$$\left(\frac{df}{du_x}\right)_{u_x=1} = 0 \quad ,$$

i.e., the weak ion-atom shock, like the ion shock in a plasma, originates at the tail of the front. For $M_0 > M_{0cr}$ the strong ion-atom shock wave, dissipating the main portion of the kinetic energy of the incident flow, divides the front into the region of heat-conduction heating and the region of temperature and ionization relaxation, where an equilibrium of ionization downstream is established.

Joining together separate portions of the shock structure here is a more complicated task than in Sect.2.1.3, because the phase space has more dimensions. Linearization of (2.2.38,42,43) near the singular points for $M_0 \gg M_{0cr}$ demonstrates [2.19] that both points 0 and 2 are saddle points. As x is increased from point 0 in the direction of increasing α and T_e only one integral curve originates, which is part of the unidimensional attraction manifold of the singular point 0 as $x \to -\infty$, i.e., the locus of points of the phase space such that the integral curves passing through them tend to the singular point 0 as $x \to -\infty$. We denote this manifold by $S_1^-(0)$. Point 2 as $x \to +\infty$ is entered by the integral curves which fill up the two-dimensional attraction manifold of the singular point 2 as $x \to +\infty$; we denote it by $S_2^+(2)$.

The solution of (2.2.49) allows us to establish one-to-one correspondence between the point of the manifold $S_2^+(2)$ with coordinates $(u_x(b), T_e(b), \alpha(b))$ and a point with coordinates $(u_x(a), T_e(a), \alpha(a))$. The transition from these into the given point is allowed by the boundary conditions for the ion-atom shock. Such points form a two-dimensional surface $\tilde{S}_2^+(2)$ in the (u_x, T_e, α) phase space. The intercept of this surface with the curve $S_1^-(0)$ determines point a, the origin of the ion-atom shock (Fig.2.4).

This way of constructing the shock structure is associated with laborious calculations, necessary for covering the two-dimensional surface by integral curves. In this particular case, however, the task can be made easier by directly pinpointing the integral curve on the surface $S_2^+(2)$, which enters point 2. Indeed, with a typical temperature and a typical density of the plasma near point 2 we have

$$\Delta_r \ll \Delta_{ion} \quad . \qquad (2.2.50)$$

If (2.2.50) is true, then near point 2 the thermal equilibrium is established much faster than the equilibrium ionization, and the plasma may be considered

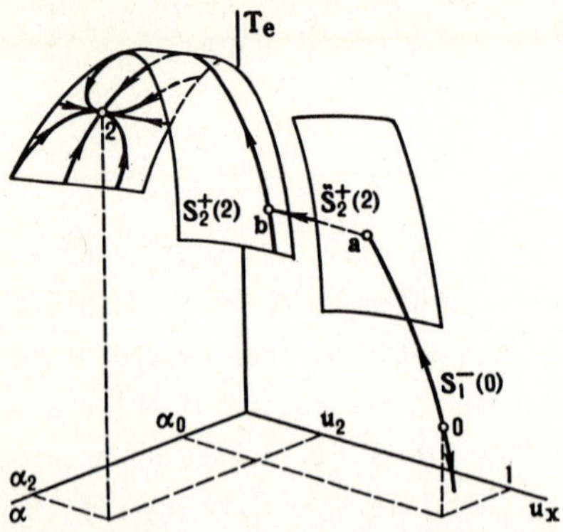

Fig.2.4. Integral curves of (2.2.38,42,43) in the (u_x, T_e, α) phase space for a strong shock in plasma with multiple ionization. Segment a→b pertains to the ion-atom subshock with T_e = const, α = const

isothermal. With the aid of (2.2.39) the condition $T = T_e$ can be transformed into the equation of a surface in the phase space, whose intersection with $S_2^+(2)$ gives us the direction in which the integral curve, going out of point b, enters point 2. This trick reduces the problem to the task already done in Sect.2.1.3. From (2.2.39), under the assumption of plasma isothermality, we obtain for the vicinity of point 2

$$(1 + \alpha_2) \frac{dT}{dx} = - T_2 \frac{d\alpha}{dx} + \frac{5M_2^2(1 + \alpha_2)}{3} \left(\frac{3}{5M_2^2} - 1\right) \frac{du_x}{dx} < 0 \quad , \qquad (2.2.51)$$

as the compression of plasma and the degree of ionization exhibit a monotonic increase, while $M_2^2 < 1/5$ for $M_0 \gg 1$. Consequently, the temperatures of both the electrons and heavy particles in the ionizing shock pass through a maximum, if (2.2.50) is satisfied.

A numerical solution of the set equivalent to (2.38,42,43) was obtained in [2.19] for air. The calculated profiles of shock waves are shown in Fig. 2.5. In these calculations the initial ionization was assumed low: $\alpha_0 = 0.01$. However, the heat-conduction thermal flux from behind the ion-atom shock, in this case, turns out to be so strong that the atoms, coming to the front, are already singly ionized. As the energy transfer from the heavy plasma component to the electrons ensures exponential growth of the degree of ionization, the shock structure should exhibit no dependence on the choice of α_0, whose only prerequisite is not too small. This point was confirmed by appropriate calculations [2.19].

We see that the strong shock wave in a plasma—with the processes of multiple ionization taken into account—is generally similar in kind to an ordinary shock in a plasma. The finer details of the shock structure are considered below.

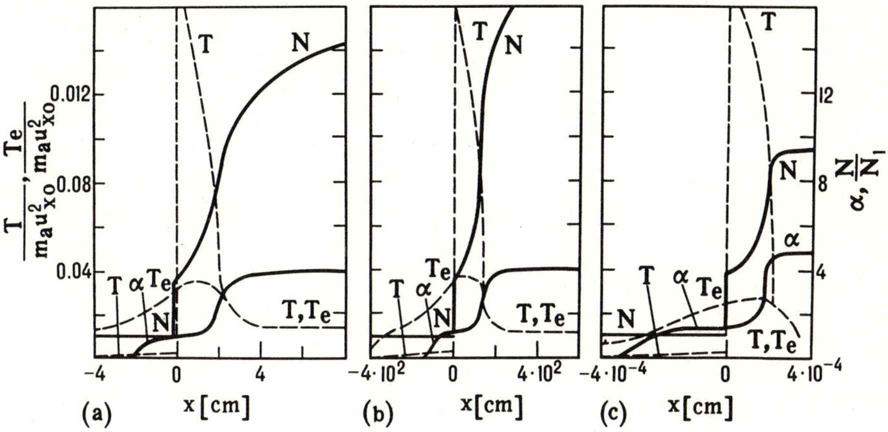

Fig.2.5a-c. Structures of strong shock waves in air plasma [2.19].
(**a**) $N_0 = 3.5 \cdot 10^{15} cm^{-3}$, $T_0 = 9000$ K , $M_0 = 28$;
(**b**) $N_0 = 4.2 \cdot 10^{13} cm^{-3}$, $T_0 = 7500$ K , $M_0 = 29$;
(**c**) $N_0 = 5.4 \cdot 10^{20} cm^{-3}$, $T_0 = 46783$ K , $M_0 = 26$

2.2.3 Shock Structure in Partially Ionized Argon

The peculiar features of the shock profile in a plasma in the region of single ionization can become clear only if the calculations are done for a specific gas. The experiments on ionizing shock waves are mostly performed with the heavy inert gases (argon, krypton, xenon), thus theoreticians also concerned themselves with these gases (mostly argon).

The formulas for collision cross sections, averaged over a Maxwellian distribution, for the temperature region of first ionization in argon can be found in [2.21-23]:

$$\sigma_{ia} = 1.4 \cdot 10^{-14} cm^2 = const \quad , \tag{2.2.52}$$

$$\sigma_{aa} = 1.7 \cdot 10^{-14} T^{-1/4} \quad [cm^2] \quad , \tag{2.2.53}$$

$$\sigma_{ea} = \begin{cases} (0.713 - 4.5 \cdot 10^{-4} T_e + 1.5 \cdot 10^{-7} T_e^2) \cdot 10^{-16} \ [cm^2] \quad , \\ \qquad\qquad T_e < 3000 \text{ K} \quad ; \\ (-0.488 + 3.96 \cdot 10^{-4} T_e) \cdot 10^{-16} \ [cm^2] \quad , \\ \qquad\qquad 3000 \text{ K} \leqslant T_e \leqslant 20000 \text{ K} \quad , \end{cases} \tag{2.2.54}$$

(the values of T and T_e are expressed in kelvins). Some other pertinent figures are: the energy of ionization $J = 15.759$ eV [2.24]; the energy of excitation $E = 11.5$ eV [2.25]; the constants of effective cross sections of impact excitation $C_a^* = 1.2 \cdot 10^{-19}\,cm^2/eV$ [2.26], $C_e^* = 7 \cdot 10^{-18}\,cm^2/eV$ [2.25]; the statistical weights in the Saha formula (2.2.2) $g_+ = 5.6$, $g_0 = 1.0$ [2.27].

The shock structure in argon was thoroughly investigated in [2.28] on the basis of the unidimensional gas-dynamic equations of a stationary shock layer (Sect.2.1) and an interpolation of effective cross sections (2.2.52-54). The degree of ionization of an initial plasma α_0 was assumed to be high enough to take care of the inequality

$$\alpha_0 \sigma_{ei}(0) > \sigma_{ea}(0) \quad , \tag{2.2.55}$$

and therefore the electron path length upstream is determined by Coulombian collisions. Then the relation between the characteristic scales of viscosity of heavy particles and the electron heat conduction in the initial state is

$$\frac{\Delta_v(0)}{\Delta_{Te}(0)} \sim \left[\left(\frac{m_e}{m_a}\right)^{\frac{1}{2}} \frac{\ell_{aa}(0)}{\ell_{ee}(0)}\right] \frac{M_0^2}{\alpha_0} \equiv \delta \, \frac{M_0^2}{\alpha_0} \quad , \tag{2.2.56}$$

where $\ell_{aa}(0)$, $\ell_{ee}(0)$ are the mean free paths (upstream), corresponding to the collision rates (2.2.18); $\delta = 7 \cdot 10^{-3}$ for $T_0 = 7500$ K, $\alpha_0 = 6 \cdot 10^{-3}$. As the heat-conduction heating of electrons quickly raises the value of Δ_{Te} by $\sim M_0^5$ ($T_e \sim T_2 \sim M_0^2$; $\kappa_e \sim T_e^{5/2}$), the ratio (2.2.56) for $\alpha_0 \gtrsim 10^{-2}$ can be considered a small parameter. This implies that the Prandtl number is small, and hence the shock structure in it is similar in kind to those discussed in Sects. 2.1.3 and 2.2. By expanding the shock-layer equations in the small parameter δ down to the second order of smallness, *Shanmugasundaram* and *Murty* [2.28] calculated the shock structure in partly ionized argon with a high degree of accuracy.

Similarly to the shock wave with multiple ionization, the flows far upstream and downstream are determined chiefly by the action of electron heat conduction, and the integral curve entering point 2, as $x \to +\infty$, is fixed by the condition $T = T_e$. The new circumstance compared with the results of Sect. 2.2 is that the transition from the integral curve going from point 0 to the integral curve entering point 2 is impossible via the ion-atom shock with $\alpha = \text{const}$, $T_e = \text{const}$, (Fig.2.6) [the inequality (2.2.37) here remains valid, like the boundary condition (2.2.49) for the ion-atom shock]. Here the characteristic scale of ionization Δ_{ion} is somewhat smaller than the scale of electron heat conduction, and therefore joining the solutions for

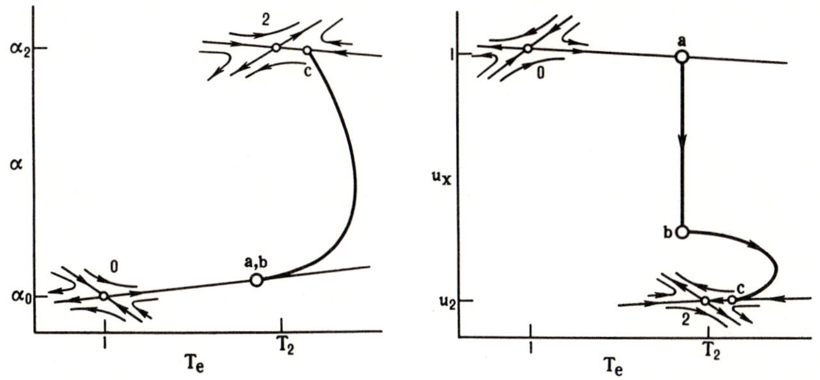

Fig.2.6. Phase curves of shock-layer equations for a strong shock wave in argon in the (T_e,α) planes (*left*) and (T_e,u_x) (*right*) [2.28]. Here a→b is the ion-atom subshock

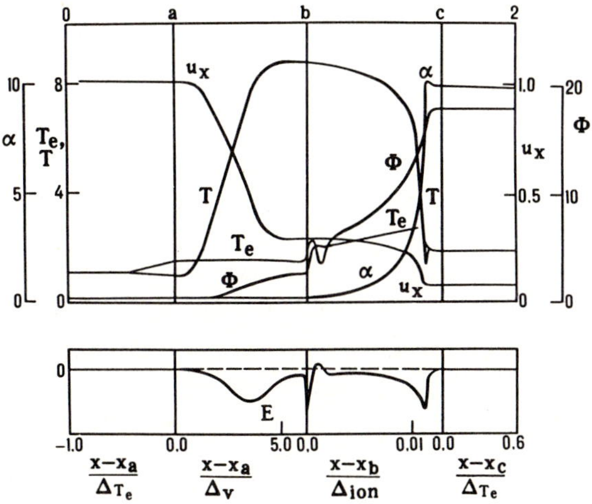

Fig.2.7. The structure of a strong shock wave in partly ionized argon [2.28]. The scales Δ_{Te}, Δ_v pertain to the equilibrium downstream state; the scale Δ_{ion} corresponds to a combination of parameters for a far-off downstream state (2) and directly behind the ion-atom shock (b) (by courtesy of S.S.R. Murty)

the different portions of the shock requires an additional layer (section b-c in Fig.2.6), corresponding to the domain of ionization relaxation [2.28]. In the next heat-conduction layer the variables change negligibly, tending to their downstream values. Figure 2.7 shows a typical shock profile in partially ionized argon, as calculated in [2.28] for $T_0 = 7500$ K, $N_0 = 10^{16}\,cm^{-3}$, $\alpha_0 = 6.5 \cdot 10^{-3}$, $M_0 = 5$ (for clarity the different layers are not shown to scale). In such a shock the plasma is heated to $T_2 = 13000$ K, compressed tenfold, and its degree of ionization is raised to $\alpha_2 = 0.62$. Owing to the smallness of α_0, the variables at the head of the front show little change; the profile of T_e does not display the shape typical of the nonlinear thermal wave (Fig.2.3). All the variables in the ion-atom shock wave change monotonically. In general, this portion of the front differs little from the ordinary gasdynamic shock. If we account for the fact that ionization occurs also in the ion-atom shock wave, the results will be slightly different: the shock width Δ_{ia} is found to increase a little with the rise in M_0 [2.28], whereas under the assumption of frozen ionization the value of Δ_{ia} decreases with a rise in M_0 [2.22].

Especially interesting is the functional behavior in the region of ionization relaxation. Oscillations of the electron temperature at the front are due to the simultaneous action of the heat conduction, the large amount of heat transferred to the electrons from the heavy component of the plasma, heated in the ion-atom shock, and the large energy loss by ionization. This peculiar profile of the electron temperature complicates numerical integration and necessitates special precautions [2.28]. Following (2.2.14), in the region of the abrupt drop in temperature at the near-constant plasma velocity the sign of the electric field of polarization is reversed, and a region of decreasing potential appears.

The profile of the electric field exhibits three peaks: in the ion-atom shock wave, and at the fore and rear of the region of ionization relaxation. For high Mach numbers ($M_0 \gtrsim 10$), yet another peak arises at the head of the heating zone [2.28], which is entirely similar to the precursor electric layer in the shock structure in a completely ionized plasma. Note that the calculations performed in [2.28] disregard the thermal force acting upon the electrons. Accordingly, the profiles of the electric field and potential are determined with a low degree of precision and must rather be considered as qualitative estimate.

The minimum of T at the rear part of the region of ionization relaxation is not easily interpreted. The heavy component cools mainly at the expense of elastic heat transfer to the electrons, and therefore the decrease in T at $T_e \geqslant T$ can be explained by the prevailing action of the heat conduction in the heavy component (provided that the value of $|\kappa\, dT/dx|$ in the structure in question reaches its maximum near the point where T is minimum, already for $T \leqslant T_e$). The minimum can also be attributed to the so-called expansion cooling of atoms as $\alpha \to 1$, since in the atom energy equation

$$\frac{dT_a}{dx} \sim \frac{2T_a}{3(1-\alpha)} \frac{d\alpha}{dx} \quad , \tag{2.2.57}$$

so that the negative value of the right-hand side of (2.2.57) can be quite large [2.29].

2.3 Structure of an Ionizing Shock Wave

2.3.1 Morphology

An ionizing shock wave in which the initially nonconducting gas converts into a plasma represents an intermediate case between the gasdynamic shock wave and the shock wave in a plasma. If the initial state of the gas is characterized by the room temperature, then the specific features of the ionizing shock are evident in a restricted range of Mach numbers, e.g., $10 \lesssim M_0 \lesssim 50$ for argon at $p_0 = 1$ Torr. For $M_0 \lesssim 10$ we are dealing with an ordinary gasdynamic shock; for $M_0 \gtrsim 50$ the shock wave is associated with such a strong ionizing action (the gas ahead of the shock is ionized by the radiation of the shock-heated plasma from behind it) that it actually propagates in the plasma (Sect.2.2.3). The importance of the ionizing shock waves arises from the fact that the strong shock waves, produced in the shock tubes or associated with meteorites and spacecraft traveling through atmosphere, usually belong exactly to this class.

Theoretical analysis of the ionizing shock waves is much more complicated than investigating shock waves in collision-dominated plasma. This is due to the much larger number of physical mechanisms shaping the ionizing shock-wave structure, and the actual experimental conditions which can influence the shock structure. In particular, account must be taken of the finiteness of the shock-tube diameter and the duration of the experiment; it is also important whether the ionizing shock wave is observed in the ordinary or in the electromagnetic shock tube (Sect.2.3.3). Shock-structure calculations must account for the kinetics of numerous collisive and radiative processes which take place in the shock layer. Even with the simple model of an argon atom [2.30], in which the resonant and metastable levels are considered as one low-lying excited level with the energy $E^* = 11.65$ eV (denoted A^*), and all the remaining levels are unified into one high-lying level with the energy $E^{**} = 14.72$ eV (denoted A^{**}), in the region of primary ionization one has to consider the kinetics of the following 24 reactions:

i) photoexcitation and deexcitation

$$A + h\nu \rightleftharpoons A^* \quad , \qquad (2.3.1)$$

$$A + h\nu \rightleftharpoons A^{**} \quad , \tag{2.3.2}$$

$$A^{*} + h\nu \rightleftharpoons A^{**} \quad ; \tag{2.3.3}$$

ii) photoionization from the ground and excited levels, and radiative capture of electrons

$$A + h\nu \rightleftharpoons A^{+} + e \quad , \tag{2.3.4}$$

$$A^{*} + h\nu \rightleftharpoons A^{+} + e \quad , \tag{2.3.5}$$

$$A^{**} + h\nu \rightleftharpoons A^{+} + e \quad ; \tag{2.3.6}$$

iii) production of molecules in the metastable excited state A_2^* with subsequent ionization, and the corresponding inverse processes of recombination

$$A^{*} + 2A \rightleftharpoons A_2^{*} + A \quad , \quad A_2^{*} + h\nu \rightleftharpoons A_2^{+} + e \quad , \tag{2.3.7}$$

$$A^{**} + A \rightleftharpoons A_2^{+} + e \quad ; \tag{2.3.8}$$

iv) ionization, excitation and deactivation due to collisions of atoms in the ground and excited states; recombination on capture to the ground and excited levels involving a neutral atom:

$$A + A \rightleftharpoons A^{*} + A \quad , \tag{2.3.9}$$

$$A + A \rightleftharpoons A^{**} + A \quad , \tag{2.3.10}$$

$$A^{*} + A \rightleftharpoons A^{**} + A \quad , \tag{2.3.11}$$

$$A^{*} + A^{*} \rightleftharpoons A^{**} + A \quad , \tag{2.3.12}$$

$$A + A \rightleftharpoons A^{+} + A + e \quad , \tag{2.3.13}$$

$$A + A^{*} \rightleftharpoons A^{+} + A + e \quad , \tag{2.3.14}$$

$$A + A^{**} \rightleftharpoons A^{+} + A + e \quad , \tag{2.3.15}$$

$$A^{*} + A^{*} \rightleftharpoons A^{+} + A + e \quad , \tag{2.3.16}$$

$$A^{*} + A^{**} \rightleftharpoons A^{+} + A + e \quad , \tag{2.3.17}$$

$$A^{**} + A^{**} \rightleftharpoons A^{+} + A + e \quad ; \tag{2.3.18}$$

v) ionization, excitation and deactivation of atoms in the ground and excited states by electron impact, and triple recombination involving an electron:

$$A + e \rightleftharpoons A^{*} + e \quad , \tag{2.3.19}$$

$$A + e \rightleftharpoons A^{**} + e \quad , \tag{2.3.20}$$

$$A^* + e \rightleftharpoons A^{**} + e \quad , \tag{2.3.21}$$

$$A + e \rightleftharpoons A^+ + e + e \quad , \tag{2.3.22}$$

$$A^* + e \rightleftharpoons A^+ + e + e \quad , \tag{2.3.23}$$

$$A^{**} + e \rightleftharpoons A^+ + e + e \quad . \tag{2.3.24}$$

Of course, we do not list here all the possible reactions, but rather only those which can noticeably affect the shock structure. It must be observed that the excitation kinetics is interesting not only in itself. To a large extent it also determines the kinetics of ionization, namely reactions (2.3.14-18,23,24), i.e., it shapes the profile of the electron density, which strongly affects the values of transfer coefficients and thus the shock structure on the whole.

The shock-layer equations must account for the processes of radiative energy transfer, essential for the reactions (2.3.1-6). In contrast to the gasdynamic transfer processes, these processes in a plasma are nonlocal: the nonequilibrium density of radiation at a given point is determined by the integral over the entire volume of plasma. This brings us to a set of integral-differential equations, the unknown profiles of density and temperature being contained in the integral. The search for the solution of this set, which would connect the initial and final singular points in the many-dimensional phase space, is associated with immense difficulties.

It is not difficult but rather time-consuming to write out the complete set of gasdynamic equations of the shock layer for the ionizing shock wave for the adopted model. However, this is hardly necessary, since the complexity of this system defies all attempts of either analytical or numerical treatment. Numerous calculations of the structures of ionizing shock waves [2.23,31-41] are based on the simplified equations, and are therefore more or less reliable. Being unable to solve the problem from first principles, we shall resort to considering the shock structure of the ionizing shock wave qualitatively, relying on the available theoretical and experimental results.

Let us consider the change in the shock profile, as the velocity of the shock wave increases. With Mach numbers below the ionizing shock threshold, the structure of a strong shock wave is determined by viscosity and heat conductivity of the gas, and the shock width is of the order of several mean free path lengths of atoms ℓ_a. An increase in M_0 leads to an ionization the of the gas in the equilibrium state downstream; at high temperature the neutral gas downstream of the gasdynamic shock does not assume the state of ionization equilibrium. The process of attaining ionization equilibrium is characterized by the scale Δ_{ion}, see (2.2.7). As the characteristic scale of, say, two-step ionization at low temperatures grows exponentially, $\Delta_{ion} \sim \exp(E/T)$, we may reach the inequality

$$\Delta_{ion} \gg \ell_a \quad , \tag{2.3.25}$$

which in the ionizing shock waves is satisfied practically always. By virtue of (2.3.25) the processes of ionization are negligible on the scale ℓ_a, and the gasdynamic transfer processes on the scale Δ_{ion}. We are clearly dealing with a typical shock wave in a medium with delayed excitation of some of its degrees of freedom [2.15]. For this reason the structures of relatively weak ionizing shock waves ($M_0 < 15$ for Ar) are reasonably well described by the Zel'dovich-von Neumann-Döring model (Sect.1.9.2). First comes the strong gasdynamic shock wave, viewed as a discontinuity on the scale Δ_{ion}, and then a wide layer of ionization, taking place in an ideal medium. In the present case the reaction (ionization) is endothermal, and, in contrast to the detonation wave, the gas in the reaction region cools and compresses, see (2.2.20). Such shock waves have been thoroughly investigated during the 1950-1960s; the results are exhaustively treated in [2.15].

Further increase in M_0 creates a situation where the hot downstream plasma abuts the cold neutral gas across the thin gasdynamic shock. The radiation emitted by plasma obviously affects the state of the upstream gas, although at temperatures typical of the ionizing shock waves this radiation is not yet sufficient to heat the gas substantially by the reactions (2.3.1-6). This implies that the gasdynamic jump is preceded by a region with nonequilibrium density of excited and ionized atoms, maintained by the radiation coming from downstream (the so-called precursor).

If the strength of precursor radiation is not too high, the equilibrium in the precursor region is not too disturbed, and the structure of this region can still be described on the basis of the Zel'dovich-von Neumann-Döring model for the shock structure, assuming its independence of the precursor phenomenona. Generally, the problem must be approached self-consistently. The profile of electron density in the precursor region is determined by the structure of the region of ionization relaxation, which in turn depends on the levels of excitation and ionization, maintained in the precursor region directly ahead of the gasdynamic shock front. This shock appears as an internal discontinuity in the structure of an ionizing shock wave. The boundary conditions for this shock—at the levels of precursor ionization typical of ionizing shock waves—do not differ from the gasdynamic Rankine-Hugoniot conditions. As M_0 increases further, the precursor region becomes the heating region, where the main transfer process is the electron heat conduction; the gasdynamic shock wave becomes the ion-atom shock (Sect.2.2).

The precursor phenomena are characterized by two length scales: the path length of ionizing quanta ℓ_{ph}, which are responsible for the reaction (2.3.4), and the diffusion length Δ_{ph} of radiation at wavelengths which comply with the reactions (2.3.1-3). Assuming the absorption line shape in the cold gas to be Lorentzian with the width $\Delta\omega$ (collisional broadening), we get

$$\frac{\ell_{ph}}{\Delta_{ph}} \sim \frac{\sigma^*_{ph}}{\sigma_{ph}} \cdot \frac{1}{1 + \left(\frac{\omega - \omega_0}{\Delta\omega/2}\right)^2} \quad , \qquad (2.3.26)$$

where σ_{ph} is the ionization cross section from the ground level, σ^*_{ph} is the excitation cross section for the middle of the absorption line; for the resonant argon line ($\lambda = 1048$ Å) $\sigma^*_{ph}/\sigma_{ph} \sim 10^6$. Accordingly, the radiation near the middle of the line is trapped, although for the far-off wings of the line ($\omega - \omega_0 > 10^3 \Delta\omega$) the path length of the resonant radiation is large compared with ℓ_{ph}. On this ground the precursor region can be subdivided into the near precursor region (with the width ℓ_{ph}) and the far precursor region (with the width $\Delta_{ph} \gg \ell_{ph}$), the former being dominated by photoionization from the ground level, and the latter by the diffusion of radiation in the lines and photoionization from the excited levels.

Note finally that the radiative processes extend the nonequilibrium region not only upstream, but also downstream. In the idealized unidimensional problem in which the stream of plasma is assumed endless on either side, the equilibrium between radiation and matter is attained at a distance of the order of the largest radiation path length from the end of the relaxation region. In a real finite-length shock tube equilibrium may be not established at all, and then the relaxation region is followed by the region of radiative cooling. In any case, the radiation absorbed in the precursor region is released in the relaxation region and the next nonequilibrium region.

Thus, the morphology of the strong ionizing shock wave may be pictured in the following way. At the head of the shock is the far precursor region, shaped by the diffusion of resonant radiation and photoionization of excited atoms. Then follows the near precursor region, dominated by photoionization directly from the ground level. This region abuts on the gasdynamic (ion-atom) shock wave, where the density, presure and temperature all sharply increase in the relatively narrow front, shaped mainly by the atom viscosity. Next come the relaxation region and the region of radiative cooling.

For the ionizing shock wave in argon at $p_0 = 1$ Torr, $M_0 = 20$, the width of the far precursor region is about 20 cm, the near precursor region is ~3 cm wide, the gasdynamic shock 0.1 cm, the relaxation region 3 cm, and the following nonequilibrium region 10 cm wide. If equilibrium between radiation and matter is not established, as commonly is the case [2.42-45], the relaxation region is followed by the region of radiative cooling, whose width is about 20 cm.

2.3.2 Structure of the Precursor Region

In experiments with ionizing shock waves the precursor concentrations of the excited and ionized atoms are usually not high. This allows one to study the structure of the precursor region, disregarding the compression and heating of the gas ahead of the gasdynamic shock and assuming $u_x = u_{x0}$, $T = T_0$.

We begin by considering the structure of the near precursor region, dominated by the photoionization of nonexcited atoms by the radiation coming from behind the shock front, located at $x = 0$. (The precursor region corresponds to negative x.) For a rough estimate we neglect the nonequilibrium of radiation downstream of the shock, assuming that plasma at $\omega > J/\hbar$ emits as a black body. Our "experimental set-up" will be most simple, corresponding to the ordinary cylindrical shock tube.

Let us find the number of ionizing quanta absorbed in unit time by unit volume near point M on the axis of the shock tube of radius R, at the distance $|x|$ from the surface of radiant plasma (Fig.2.8). The element $d\Sigma$ of the radiating surface, viewed from point M at an angle θ ($d\Sigma = 2\pi\rho\, d\rho = 2\pi x^2 \sin\theta\, d\theta/\cos^3\theta$), emits in unit time into the unit solid angle $d\Omega_M$ in

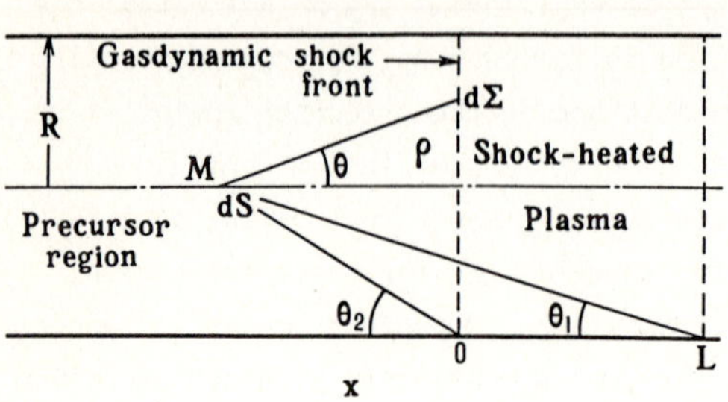

Fig.2.8. Scheme to calculate precursor ionization and excitation in a cylindric shock tube

the direction of point M the following number of quanta in the frequency range $(\omega,\omega+d\omega)$:

$$\frac{B_\omega^0}{\hbar\omega}\, d\omega\, \cos\theta\, d\Omega_M d\Sigma \quad . \tag{2.3.27}$$

Here B_ω^0 is the density of black-body radiation at the equilibrium temperature downstream:

$$d\Omega_M = \frac{dS}{x^2}\cos^3\theta \quad , \tag{2.3.28}$$

where dS is the surface element normal to the x axis near point M. Assuming the degree of ionization to be low, we find that the part of these quanta reaching point M constitutes

$$\exp\left(-\frac{\kappa_\omega |x|}{\cos\theta}\right) \quad , \tag{2.3.29}$$

where κ_ω (the absorption coefficient at frequency ω) is assumed to be constant. The fraction of quanta absorbed in the interval dx near point x is $\kappa_\omega\, dx/\cos\theta$. From (2.3.27-29), integrating over frequencies and angles with the substitution $v=1/\cos\theta$, we find

$$u_{x0}\frac{d\alpha}{dx} = \int_{J/\hbar}^{\infty} \frac{2\pi B_\omega^0}{\hbar\omega}\sigma_\omega\, d\omega \int_1^{\frac{(R^2+x^2)^{1/2}}{|x|}} \exp\left(-\kappa_\omega |x| v\right)\frac{dv}{v^3} \quad , \tag{2.3.30}$$

where σ_ω is the effective cross section of photoionization and $\kappa_\omega = 1/N_0\sigma_\omega$.

If the temperature of the radiating plasma is $T \ll J$, then, providing that $\omega \geqslant J/\hbar$, we may set $\sigma_\omega = \sigma_{ph}$, since the number of quanta with larger energy is exponentially small. Substituting in (2.3.30)

$$B_\omega^0 = \frac{\hbar\omega^3}{4\pi^3c^2[\exp(\hbar\omega/T_2) - 1]} \quad ,$$

and dropping the one in the denominator as small compared with the exponential, we integrate (2.3.30), obtaining

$$\frac{d\alpha}{dx} = j(x) \quad , \qquad \text{where} \tag{2.3.31}$$

$$j(x) = \frac{\sigma_{ph}J^2T_e\exp(-J/T_2)}{2\pi^2c^2\hbar^3u_{x0}}\left(1+\frac{2T_2}{J}+\frac{2T_2^2}{J^2}\right)$$

$$\times\left[E_2\left(-\frac{x}{\ell_{ph}}\right) - \frac{|x|}{(x^2+R^2)^{1/2}}E_2\left(\frac{(x^2+R^2)^{1/2}}{\ell_{ph}}\right)\right] \quad ,$$

$\ell_{ph} = 1/N_0\sigma_{ph}$,

and

$$E_m(z) = \int_1^\infty e^{-zt} \frac{dt}{t^m}$$

is the exponential integral of mth order.

The solution of (2.3.31) with the boundary condition $\alpha(-\infty) = 0$ is given by

$$\alpha(x) = \frac{J^2 T_2 \exp(-J/T_2)}{2\pi^2 c^2 \hbar^3 u_{x0} N_0} \left(1 + \frac{2T_2}{J} + \frac{2T_2^2}{J^2}\right)\left[E_3\left(-\frac{x}{\ell_{ph}}\right) - E_3\left(\frac{(x^2 + R^2)^{1/2}}{\ell_{ph}}\right)\right] . \tag{2.3.32}$$

The second term in square brackets is a correction for finiteness of the angular dimension of the radiating surface. For $|x| \ll R$ this correction is negligible; for $|x| \gg R$ the correction reduces the result by the factor of $R^2/2|x|\ell_{ph}$. If $R \gg \ell_{ph}$, then the correction is notable only in the region of very low ionization, beyond the limits of the near precursor region. Then the structure of the latter may generally be considered unidimensional and described by the first term in (2.3.32).

Direct measurements of the precursor density can be made only some distance off the shock front, usually in the rear part of the near precursor region. Let us use (2.3.32) to find the degree of ionization at the distance $|x| = 4$ cm from the shock front in argon with $M_0 = 23$ ($T_0 = 300$ K, $p_0 = 1$ Torr, $T_2 = 14{,}000$ K): $\alpha(-4\text{ cm}) \cong 4 \cdot 10^{-5}$. This value fits experimental results well [2.45,46]. As $|x|$ increases further, the ionization is predicted by (2.3.32) to drop according to the law

$$\alpha(x) \propto \frac{\ell_{ph}}{|x|} \exp\left(-\frac{|x|}{\ell_{ph}}\right) ,$$

the decline being much faster than observed experimentally. This implies that we enter the far precursor region.

First we consider the excitation kinetics in the far precursor region [2.47]. The atoms are excited from the ground state by the radiation coming from downstream or emitted by some of the surrounding atoms due to their deexcitation. The stationary equation of excitation kinetics has the form

$$\frac{d\beta}{dx} = j^*(x) + D(\beta,x) - \frac{\beta}{\Delta_\tau} , \tag{2.3.33}$$

where β is the dimensionless degree of excitation, $\beta = n^*/N$. Here the first term describes excitation by the radiation coming from downstream, the second accounts for the diffusion of resonant radiation, and the third term describes radiative deexcitation with the characteristic time τ ($\Delta_\tau = u_{x0}\tau$).

Assuming the absorption line shape to be Lorentzian, i.e.,

$$\kappa_\omega = \frac{\kappa_0}{1+y^2} \quad , \tag{2.3.34}$$

where $y = (\omega - \omega_0)/(\Delta\omega/2)$, and the radiating plasma to be optically dense and in thermodynamic equilibrium, we obtain an expression similar to (2.3.30):

$$j^*(x) = \frac{\beta_{eq}(T_2)}{2\pi\Delta_\tau} \int_1^{\frac{(R^2+x^2)^{1/2}}{|x|}} \frac{dv}{v^2} \int_{-\infty}^{\infty} \exp\left(-\frac{\kappa_0|x|v}{1+y^2}\right) \frac{dy}{1+y^2} \quad , \tag{2.3.35}$$

where $\beta_{eq}(T_2)$ is the equilibrium degree of excitation corresponding to the downstream temperature T_2; extending the integration limit to $y = -\infty$ does not introduce any considerable error. The second term in (2.3.33), which accounts for photoexcitation of atoms due to the diffusion of the resonant radiation, has the form

$$D(\beta,x) = \frac{1}{2\pi\Delta_\tau} \int_{-\infty}^{0} d\xi \int_{-\infty}^{\infty} dy \int_1^{\frac{(R^2+|x-\xi|^2)^{1/2}}{|x-\xi|}} \frac{dv}{v} \frac{\kappa_0}{(1+y^2)^2} \beta(\xi)$$

$$\times \exp\left(-\frac{\kappa_0|\xi - x|v}{1+y^2}\right) \quad . \tag{2.3.36}$$

Assuming that the resonant excitation of atoms depends mainly on the radiative deexcitation of excited atoms in a close neighborhood, we use the diffusion approximation, expanding $\beta(\xi)$ in powers of $\xi - x$ and retaining only the first terms of the expansion:

$$\beta(\xi) \cong \beta(x) + (\xi - x)\frac{d\beta}{dx} \quad . \tag{2.3.37}$$

Now we substitute (2.3.37) into (2.3.36) and take advantage of the fact that the transfer of resonant radiation depends mainly on the far wings of the lines ($y \gg 1$), which allows us to disregard unity in (2.3.35,36) since it is then small compared to y. Integration transforms (2.3.33) to the form

$$\left(1 - \frac{(R^2 + x^2)^{\frac{1}{4}} - |x|^{\frac{1}{2}}}{a}\right) \frac{d\beta}{dx} = - \frac{\beta}{a} \left\{ |x|^{-\frac{1}{2}} + \frac{1}{(2R)^{\frac{1}{2}}} F\left[\arc\cos\left(\frac{x^2}{R^2} + 1\right)^{-\frac{1}{4}} \quad , \quad 2^{-\frac{1}{2}}\right]\right.$$

$$\left. + \frac{1}{4(2\pi)^{\frac{1}{2}} R^{\frac{1}{2}}} [\Gamma(\tfrac{1}{4})]^2 \right\} + \frac{\beta_{eq}(T_2)}{a|x|^{\frac{1}{2}}} \left[1 - |x|^{3/2}(x^2 + R^2)^{-3/4}\right] \quad , \tag{2.3.38}$$

where $a = 3(\pi\kappa_0)^{\frac{1}{2}}\Delta_\tau$, $F(\varphi,k)$ is the elliptic integral of the first kind, $\beta_{eq}(T_2)$ being the equilibrium degree of excitation at $T = T_2$.

Note, that the diffusion approximation can be used only with the finite values of R; in the limit $R \to \infty$, corresponding to the strictly unidimensional problem, the second term in brackets on the left-hand side of (2.3.38) becomes divergent. With the finite values of R this term can be maximized by the value $(R/\pi\kappa_0)^{\frac{1}{2}}/3u_{x0}\tau$, which at typical experimental conditions ($p_0 = 1$ Torr, $\kappa_0 = 10^5 cm^{-1}$, $R = 10$ cm, $u_{x0} = 10^6$ cm/s) constitutes about $2 \cdot 10^{-9}/\tau$, which usually allows one to disregard this term as small compared with unity.

If photoionization of excited atoms is not significant, (2.3.38) determines $\beta(x)$ for each level individually. The asymptotic solution for $x \gg R$

$$\beta(x) = \frac{3(2\pi)}{2[\Gamma(\frac{1}{4})]^2} \left(\frac{R}{x}\right)^{5/2} \beta_{eq}(T_2) \tag{2.3.39}$$

depicts the comparatively slow (power-law) decrease of the density of non-equilibrium excited atoms with the increasing distance from the shock. Deviations from the Lorentzian line shapes do not essentially modify the results obtained.

The kinetics of photoionization of excited atoms in the far precursor region is described by

$$\frac{d\alpha}{dx} = \sum_i \beta_i 4\pi \int_{\omega_i^*}^{\infty} \frac{J_\omega(T_2)}{\hbar\omega} \sigma_i^*(\omega) d\omega \quad , \tag{2.3.40}$$

where $J_\omega(T_2)$ is the bulk luminosity of radiating plasma at the frequency ω, $\omega_i^* = E_i/\hbar$ is the threshold frequency of photoionization of the excited i^{th} level, $\sigma_i^*(\omega)$ is the effective photoionization cross section. Equation (2.3.40) is solved together with the set (2.3.33) for all the releavant excited levels [a corresponding term from the summation in (2.3.40) must be subtracted from the right-hand side of each of the equations in (2.3.33)]. It can be demonstrated that the contribution from the discrete emission spectrum to photoionization of excited atoms in the far precursor region is small, and hence one may be content with taking into account only the recombination radiation due to electron capture in the downstream plasma on the excited levels.

Strictly speaking, the problem of the stationary structure of the precursor region is physically correct only when the radiating plasma is optically dense, $J_\omega(T_2) = B^0_\omega(T_2)$. Otherwise, J_ω will be time-dependent, insofar as the thickness of the layer of radiant plasma builds up as the shock advances. However, since the time of establishment of the stationary profile of excitation and ionization is roughly equal to the lifetime of the excited states τ, it can be assumed that as long as the thickness of the radiating layer $L \gg \Delta_\tau$, the profiles of ionization and excitation follow the slowly changing luminosity of the plasma, so that at any given instant we are dealing with a quasi-stationary structure. Assuming the radiating plasma to be optically thin, i.e., setting

$$J_\omega(\theta) = \kappa_\omega \ell(\theta) B^0_\omega \quad ,$$

where $\ell(\theta)$ is the thickness of the layer of radiating plasma in the direction θ (Fig.2.8), we get [2.47,48]

$$\frac{J_\omega(T_2)}{\hbar\omega} = \frac{e^6}{4\sqrt{3}\pi^5\varepsilon_0^3\hbar^4c^3}\frac{T_2}{\omega}N_0(T_2)\xi(\omega,T_2)\exp(-E_i/T_2)$$

$$\times [R(\theta_2-\theta_1) + x\ln(\cos\theta_2/\cos\theta_1) - L\ln(\cos\theta_1)] \quad , \qquad (2.3.41)$$

where

$$\theta_1 = \arc\cos\frac{|x|+L}{[R^2+(|x|+L)^2]} \quad ; \quad \theta_2 = \arc\cos\frac{|x|}{(R^2+x^2)} \quad .$$

Here $N_0(T_2)$ is the equilibrium density of atoms at the temperature T_2, $\xi(\omega, T_2)$ is a known function, tabulated in [2.49]. The effective photoionization cross section of the excited atoms is calculated according to the semiclassical Kramers formula

$$\sigma_i^*(\omega) = \frac{m_e e^{10} G}{384\sqrt{3}\pi^5\varepsilon_0^5\hbar^6 c\omega^3(n_i^*)^5} \quad , \qquad (2.3.42)$$

where n_i^* is the effective principal quantum number, characterizing the given excited level, and G is the Gaunt factor.

The results of solving this set of equations for the ionizing shock wave in argon ($p_0 = 1$ Torr, $T_0 = 300$ K, $M_0 = 23$) are presented graphically in Fig. 2.9 [2.47]. The calculations accounted for the multiple reflections of radiation from the walls of the shock tube, which effectively increase the

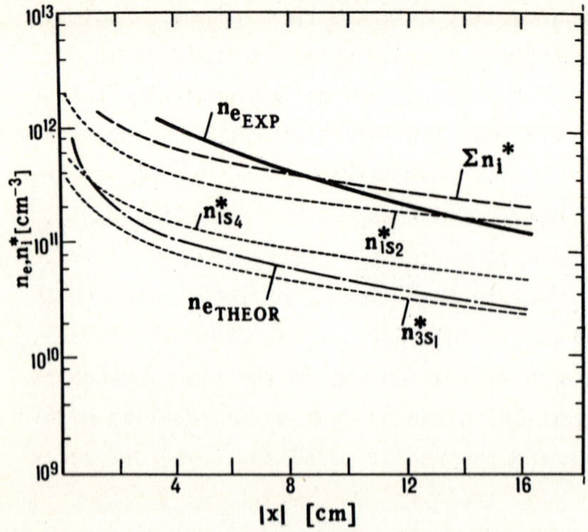

Fig.2.9. The structure of the far precursor region according to [2.47]. Profiles n_i^*, $n_{e_{THEOR}}$ are calculated, $n_{e_{EXP}}$ is experimental (by courtesy of M. Pinègre, P. Valentin)

radius of the radiating surface R with large values of x; $R_{eff} = const(|x| - R)$, the value of the constant depends on the reflection coefficient.

A comparison of the calculated profile of electron density with the experimental data [2.46] reveals that the above theory more or less correctly describes the trends in the far precursor region, although the calculated values are by an order of magnitude lower than the experimental ones (the discrepancy could have been even larger if multiple reflections were overlooked). On the other hand, the experiments [2.46] failed to discover the predicted high density of atoms in the resonant excited states $1s_2$ and $1s_4$; the experimental accuracy allows one to assert that this density does not exceed $10^{13} cm^{-3}$, one order of magnitude below the theoretical estimate.

Thus we come to the conclusion that the nonequilibrium ionization in the far precursor region seems to replace the nonequilibrium excitation. In other words, the experimental results point to the existence of an efficient mechanism, which acts to convert the energy of excited atoms into ionization. Since this effect cannot be attributed solely to the action of radiative mechanisms, an important role must be ascribed to collisions.

In the far precursor region the atoms are cold, and therefore the reactions (2.3.9-11,13) do not take place. The energy of photoelectrons also is not sufficient for excitation and ionization of atoms (under the conditions corresponding to Fig.2.9 the energy of photoelectrons does not exceed 0.5 eV). Actually, the elastic collisions quickly cool down the electrons to about room temperature. This was clearly demonstrated in [2.46] by directly measuring the collision rate of electrons, which turned out to be rather high, $\nu_e \sim 3 \cdot 10^{10} s^{-1}$, corresponding to $\langle\sigma v\rangle \sim 10^{-6} cm^3/s$. This high rate cannot arise from collisions with neutral atoms ($\sigma \sim 10^{-15} cm^2$), since this would imply an incredibly high electron energy. It must be assumed that the electrons collide mainly with ions, despite the latter's low concentration ($\alpha \lesssim 10^{-4}$), because with low-energy electrons the cross section for Coulomb interactions is high, being inversely proportional to the square of electron energy. The temperature of electrons, as estimated from the observed collision rate, does not exceed 0.04 eV.

High-energy electrons, capable of ionizing the excited atoms, can be produced in the reactions (2.3.16-18), which give rise to electrons with energies not less than $(2 \cdot 11.5 - 15.5)$ eV $= 7.5$ eV, which may take part in the reactions (2.3.23,24). The sequence of these reactions represents the possible path of conversion of the excitation energy into the ionization energy. However, the kinetics of reactions (2.3.16-18) seems too slow to explain the experimental results [2.46]: the number of electrons produced in these reactions constitutes no more than 10^{-2} of their actual density. The experimental results can be interpreted fairly well by assuming simply that a certain proportion of all available electrons $\varepsilon \sim 10^{-2}$ takes part in the reactions (2.3.23,24) ("electron avalanche" [2.29]) and carefully choosing the adjustment parameter ε. Then, however, it is not quite clear whence comes the energy required for impact ionization with the electrons of the second and subsequent generations. It is equally hard to interpret the need to use lower values of ε for describing the experiments with higher values of M_0. No fully satisfactory explanation has yet been given of these facts.

Although it is still impossible to pinpoint the exact mechanism of nonequilibrium ionization in the far precursor region, many options can safely be rejected. For instance, free diffusion of electrons from downstream cannot maintain the high precursor electron density, because the electric polarization field prevents migration of electrons away from ions to distances exceeding the Debye radius r_D. As $r_D \ll 1$ cm already at $n_e = 10^6 cm^{-3}$, the diffusion of charged particles to the precursor region can be ambipolar only, the diffusion coefficient being

$$D_A = D_i\left(1 + \frac{T_e}{T}\right) \quad \text{with} \quad D_i = \frac{2T}{m_a \nu_{ia}}$$

being the coefficient of diffusion of ions. Consequently, the characteristic length of diffusion upstream is

$$\Delta_D = \frac{D_A}{u_{x0}} = \frac{2(T + T_e)}{m_a \nu_{ia} u_{x0}} \cong \frac{2\ell_{ia}}{\gamma_0 M_0} \left(1 + \frac{T_e}{T_0}\right) \quad ,$$

where γ_0 is the initial adiabatic exponent, which means that Δ_D is of the order of the mean free path of ions and much smaller than any of the characteristic scales of the precursor region. This allows diffusion to be disregarded in the description of the precursor phenomena.

The exact values of the effective cross sections which govern the kinetics of associative ionization [reactions (2.3.3,8)] for heavy inert gases are not known. However, the calculations [2.48], in which those values were varied within reasonable limits, indicate that the contribution of these processes to the observed level of precursor ionization does not exceed 1%.

A certain role in the creation of precursor ionization may belong to foreign atoms, contained in the test gas. These impurity atoms may be ionized directly by radiation from downstream, or through collisions with excited atoms (Penning effect). The contribution of photoionization is calculated in the same way as for the original gas and depends on the photoionization threshold, the cross section and the concentration of the impurity atoms. The relevant estimates [2.46] indicate that direct photoionization of the impurity oxygen atoms (below 0.1% in concentration) has almost no effect on the precursor ionization.

The Penning effect also seems incapable of affecting the precursor's electron density in heavy inert gases. As demonstrated in [2.50], even with high impurity concentration (0.5%), the precursor region of the shock wave in xenon is dominated by the xenon ions rather than by the impurity ions.

A certain amount of precursor electrons can be produced by photoemission, e.g., the knocking-out of electrons from the walls of the shock tube by uv radiation of the shock wave. The efficiency of this mechanism is restricted by the limited luminosity of the shock and by the electric polarization field, produced by the spatial charge in the precursor region. The contribution of this mechanism under typical experimental conditions is insignificant.

2.3.3 Precursor Ionization in Electromagnetic Shock Tubes

All the above-listed mechanisms of precursor ionization have a certain common feature: the source of nonequilibrium excitation and ionization in the precursor region is the ionizing shock itself, or, more precisely, the hot downstream gas. Accordingly, the precursor region may be viewed as the frontal region of the shock, described as a whole by the unified set of shock-layer equations with the conventional Rankine-Hugoniot boundary conditions for the ionizing shock wave. This situation is precisely that realized in ordinary shock tubes, in which the hot gas region is constrained on one side by the gasdynamic shock front, and by the contact discontinuity which separates the hot shock-compressed gas from the cold driver gas on the other. In ionizing shock waves of moderate strength the perturbations of density, pressure, etc., cannot overtake the gasdynamic shock, and the influence of the shock-heated gas on the unperturbed upstream gas is restricted to the radiative effects, which shape the region of nonequilibrium excitation and ionization. As M_0 increases, the energy transfer from the hot region by radiation, electron heat conduction, gains increasing importance.

Entirely different mechanisms of precursor ionization are possible in the electromagnetic shock tubes. In the first place, the role of the piston, which drives the shock wave, is played here not by the cold driver gas, but rather by the discharge current sheet, which is propelled forward by the pressure of the magnetic field of the discharge current (Sect.5.2). The large amount of energy dissipated in the current sheet causes strong heating of plasma, and the current sheet itself becomes the source of ionizing uv radiation. Secondly, at the initial instant of operating the electromagnetic shock tube, when the discharge current has not yet reached its steady-state level, the high tension of the discharge bank is applied to the shock tube, and for a short time the unperturbed neutral gas is in a strong electric field. This transition period, however, may be long enough for electric breakdown to occur, which results in nonequilibrium ionization of the gas. These two effects do not depend on the parameters of the shock wave, but rather are determined by the design and operating conditions of experimental hardware; they slightly modify the boundary conditions of the shock wave.

The precursor photoionization by the uv radiation of the current sheet is calculated similarly to the photoionization by the radiation of the hot gas (Sect.2.3.2). At the initial moment, while the layer of hot gas is optically thin for the ionizing quanta, the radiation emitted by the current sheet shapes the profile of precursor ionization similarly to (2.3.32). The degree of ionization depends on the discharge current and on the structure of the current sheet; typically, $\alpha = 10^{-5}$-10^{-6} over the distance equal to several path lengths of ionizing quanta. Then the radiation emitted by the current sheet is trapped in the heated layer and partly reradiated forward, partly converted into thermal energy, which also reinforces the precursor photoionization. Consequently, the ionizing shock wave propagates not in a completely neutral gas, but rather in a nonuniform nonequilibrium region of precursor ionization.

The strong electric field in a neutral gas produces ionization waves which travel both parallel and antiparallel to the lines of force acting upon the electrons from the side of the electric field (proforce and antiforce ionization waves). Several explanations of the nature of these waves have been advanced, based on different concepts of the ionization mechanism [2.51,52]. The most plausible theory assumes that the ionization in the electric breakdown waves is due mainly to the electron impact [2.51,53,54]. The electron density n, produced in the neutral gas by the uniform electric field E_0 as a result of the passage of an ionization wave, can be estimated as

$$nJ \sim \frac{\varepsilon_0 E_0^2}{2} \quad , \tag{2.3.43}$$

and rigorous consideration confirms this estimate [2.52]. The degree of ionization

$$\alpha \sim \frac{\varepsilon_0 U_0^2}{2JN_0R^2} \tag{2.3.44}$$

(where U_0 is the initial voltage of the discharge bank, R is the characteristic transverse dimension of the electromagnetic shock tube) is established over period

$$t_i \sim \frac{R}{2v_i} \quad , \tag{2.3.45}$$

where v_i is the speed of the ionization wave. Typically, with electromagnetic shock tubes, $U_0 = 20$ kV, $R = 10$ cm, $N_0 = 3 \cdot 10^{15}\,\mathrm{cm}^{-3}$. Under these conditions the speed of the ionization wave in argon is $v_i \sim 10^9$ cm/s. Substituting these reasonable parameter values into (2.3.44,45), we find that the degree of ionization of $2 \cdot 10^{-5}$ is attained in about 5 ns; this period usually does not exceed the settling time of the discharge current. This implies that the profile of precursor ionization is superimposed on the ionization profile (2.3.44). In a coaxial shock tube (where R is the gap between the inner and outer tubes) the electric field is uniform over the length of the tube, and the breakdown precursor ionization is about the same in the entire volume of the gas. In cylindrical shock tubes, owing to the nonuniformity of the electric field, the degree of ionization is higher near the central electrode, gradually declining into the unperturbed gas.

In addition, the nonequilibrium ionization ahead of the ionizing shock waves is also created by mechanisms discussed in Sect.2.3.2. Accordingly, other conditions being equal, the precursor ionization in the electromagnetic shock tube is higher. This circumstance may be responsible for the different shock structures observed in installations of different types. In particular, this phenomenon is important in the theory of ionizing shock waves in a magnetic field (Chap.4).

2.3.4 Structure of the Ionization-Relaxation and Radiative Cooling Regions

The ionization-relaxation region is separated from the precursor region by a gasdynamic jump. The temperature of the heavy plasma component and the compression of plasma in the head of the relaxation region are determined by the boundary conditions for the gasdynamic shock, (1.6.29,30). The degrees of ionization and excitation, as well as the electron temperature, are determined by the profiles of these variables in the precursor region, i.e., ultimately they depend on the ionizing shock structure as a whole.

Calculations of such structures based on different model assumptions regarding the kinetics of collision and radiative nonequilibrium processes in plasmas were attempted in [2.23,25,31-41,55-58]. The most consistent and detailed calculations were carried out in [2.38,40,41] for ionizing shock waves in argon in the traditional formulation of the problem of a shock wave in a spatially unrestricted plasma, with regard to the excitation of only one low-lying level (A^*) and the kinetics of reactions (2.3.1,4,5,9,14,19, 22).

These works, based on quite realistic values for the kinetic coefficients in a plasma, demonstrated that the results of the shock structure calculations are sensitive to the basic assumptions of the shock structure models varied within the reasonable limits. Even for the most elaborate models [2.38,40] the results are noticeably different. One of the most important features of the precursor region is the degree of ionization near the gasdynamic shock front α_a, whose magnitude determines whether the gasdynamic shock is immediately followed by the electron avalanche of ionization, or the required initial number of electrons must be produced by atom-atom collisions, reactions (2.3.9,14). The low rate of these reactions, as compared with the rate of ionization and excitation by electron impact, is responsible for the crucial dependence of the relaxation region's width on the magnitude of α_a. So far, no reliable theoretical methods for calculating this parameter are available. The expression (2.3.32) for the near precursor region is based on the assumption of black-body radiation of shock-heated plasma, and gives only order-of-magnitude estimation of α_a. For a shock wave in argon, the magnitude of α_a was estimated as $3 \cdot 10^{-2}$ ($p_0 = 76$ mTorr, $M_0 = 29$) in [2.38], and as 10^{-5} ($p_0 = 1$ Torr, $M_0 = 24$) in [2.40]; the former is far above, and the latter far below the measured values [2.46,47]. The general appearance of the nonequilibrium ionization and excitation profiles in the precursor region, calculated in [2.39], as $x \to -\infty$, is consistent with the experimental findings [2.46,47], although the predicted degree of nonequilibrium ionization α is much higher than the degree of nonequilibrium excitation β. The predictions of [2.40] are exactly the opposite: $\beta \gg \alpha$. Thus, it must be acknowledged that so far there is no reliable theory for describing strong ionizing waves in an unrestricted medium; the results obtained so far must be considered preliminary.

The progress made in this direction should have a great impact on the theory of ionizing shock waves in astrophysics, which imposes no real restrictions on the stretch of the shock-heated plasma, which cannot be said of the laboratory conditions. As the flow of plasma is restricted by the walls of the shock tube, the energy radiated from within the shock layer is not redistributed, as in the unrestricted flow, but rather is lost (absorbed) on the walls of the tube. The longitudinal limitation of the flow is also essential: the downstream flow is not really the unrestricted region of constant flow, where the radiation is at equilibrium with the matter, and the black-body radiation with temperature T_2 maintains dynamic equilibrium in the shock layer. Instead the shock region abuts on the region of radiative cooling, where the gas experiences cooling and contraction at nearly constant pressure. This region borders on the piston which drives the shock wave and is represented by the cool driver gas in the ordinary shock tube, the hot combustion products in the explosion shock tube, or the current sheet in the electromagnetic shock tube. The "piston" may act as a heat sink, as in the former case, or as a heat source, as in the latter two, but it never occurs in thermal equilibrium with the region of radiative cooling. Consequently, the structure of the shock-heated region in the experimental ionizing shock waves is, in fact, essentially nonequilibrium, and depends greatly on the type of experimental setup.

The existence of bulk energy losses implies that the law of conservation of energy flux, which serves as a basis for the Rankine-Hugoniot boundary conditions, in real experimental conditions is satisfied to a limited accuracy. Hence we can use the concept of shock transition from one steady state to another, and identify a region in a non-stationary flow with a shock front only up to this limited degree of accuracy. The small parameter, which justifies the use of the chosen approximation, is the ratio of the characteristic scale of the shock width $\Delta_{ion}(2)$ to the characteristic length of the region of radiative cooling Δ_{rc}. The value of Δ_{rc} can be estimated with the aid of the Kramers-Unsöld formula for the bulk energy loss in the continuous spectrum Q_c [2.48,57,59]:

$$Q_c = \frac{e^6 n^2 (\hbar\omega_c + T_e) Z_{eff}^2 \bar{G}}{(6\pi)^{3/2} \varepsilon_0^3 \hbar m_e^{3/2} c^3 T_e^{1/2}} , \tag{2.3.46}$$

where ω_c is the effective cutoff frequency [2.49] (commonly, $\hbar\omega_c \sim 2$ eV), Z_{eff} is the effective charge of the ion, $\bar{G}$ is the Gaunt factor averaged over the Maxwellian distribution of electrons. The energy loss by radiation in lines is of the same order of magnitude (see below). The characteristic time of irradiation of plasma thermal energy is $t_{rc} = E/2Q_c$, where

$$E = N\left(\frac{3}{2} T + \frac{3}{2} \alpha T_e + \alpha J\right) .$$

Eventually, the characteristic length of the region of radiative cooling of plasma with the density N_2, temperature T_2 and degree of ionization α_2 is given by

$$\Delta_{rc} = \frac{(6\pi)^{3/2} \hbar m_e^{3/2} \varepsilon_0^3 c^3 u_{x0}}{e^6 Z_{eff}^2 \bar{G}} \; \frac{T_2^{1/2}\left[(3/2)T_2(1 + \alpha_2) + \alpha_2 J\right]}{N_2 \alpha_2^2 (\hbar\omega_c + T_2)} . \tag{2.3.47}$$

Substituting in (2.3.47) the typical experimental values for argon [2.60] ($u_{x0} = 7.7 \cdot 10^5$ cm/s, $N_2 = 12 N_0 = 3.8 \cdot 10^{17}$ cm^{-3}, $\alpha_2 = 0.4$, $T_2 = 13{,}700$ K, $Z_{eff} = 1.5$ [2.49], $\bar{G} = 1$), we obtain $\Delta_{rc} = 60$ cm. As a matter of fact, the temperature and the degree of ionization in the region of radiative cooling substantially decline over a distance of about 20 cm; however, the characteristic scale of the shock layer Δ remains small in comparison with Δ_{rc}. Accordingly, with the accuracy of Δ/Δ_{rc} we may assume that the collisional processes of relaxation in the shock front eventually lead to the establishment of equilibrium, corresponding to the Rankine-Hugoniot conditions.

These results are illustrated in Fig.2.10. We see that the agreement between theory and experiment is not spoiled by neglecting radiative cooling, which allows us to calculate the structure of the collision-shaped relaxation region regardless of the effects of radiative cooling.

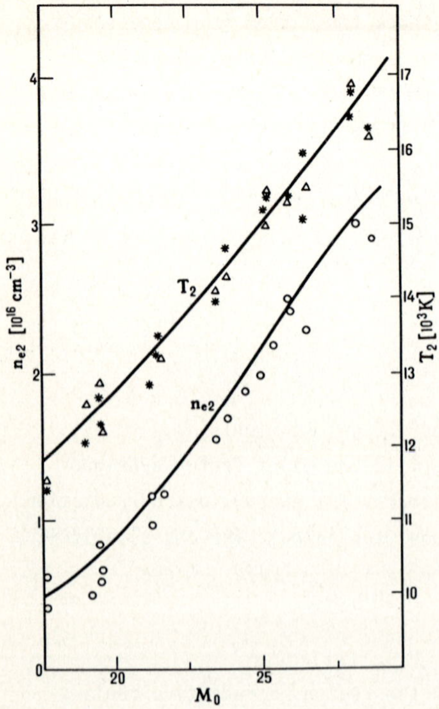

Fig.2.10. Shock adiabat for argon ($p_0 = 1$ Torr, $T_0 = 300$ K), ignoring radiative cooling: calculated curves and experimental points [2.48]. (○): maximal electron density behind the shock wave; (Δ and *): maximal temperature behind the shock wave measured by radiation in continuum (Δ) and in line $\lambda = 4158$ Å (*) (by courtesy of M. Pinêgre, P. Valentin)

Gasdynamic calculations of the structure of the relaxation region yield reliable results for ionizing shock waves of moderate intensity, when the precursor phenomena can be neglected, and therefore, the boundary condition supplementing the conditions (1.6.29-31) is $\alpha(0) = 0$. At low levels of ionization the electronic heat conduction on the scales of temperature and ionization relaxation is also negligible. Then the equations of continuity of fluxes of momentum and energy become simple algebraic relations, from which we get, in dimensionless variables,

$$(u_x - 1)(u_x - u_{GD}) = \frac{3I\alpha}{10M_0^2} , \tag{2.3.48}$$

$$T + \alpha T_e = \frac{5M_0^2}{3} u_x\left(1 - u_x + \frac{3}{5M_0^2}\right) . \tag{2.3.49}$$

Recall that $u_{GD} = (M_0^2 + 3)/4M_0^2$.

The equations (2.3.48,49) can be used to reduce the order of the set of shock-layer equations, expressing u_x and T in terms of α and T_e. The value of u_x must be taken as the lower root of the quadratic equation (2.3.48),

$u_x < (1 + u_{GD})/2$, since the flow in the relaxation region is subsonic, and $u_x = u_{GD}$ at $x = 0$, $\alpha = 0$. The velocity derivative, which enters the equation of electron heat conduction, can be expressed from (2.3.48):

$$\left(\frac{1 + u_{GD}}{2} - u_x\right)\frac{du_x}{dx} = -\frac{3I}{20M_0^2}\frac{d\alpha}{dx} \quad . \tag{2.3.50}$$

From (2.3.50) it is clear that compression of plasma exhibits a monotonic increase with the monotonic rise in the degree of ionization.

The structure of the relaxation region is described by the equations of ionization kinetics and electronic heat conduction, where the variables u_x and T are expressed by α and T_e. Assuming that the ionization takes place only via the excited level A^* with energy E, reactions (2.3.9,14,19,23), the first two being the limiting steps, we may write the kinetic equation of ionization in the form (2.2.6).

With due account taken of (2.3.50), the equation of electronic heat conduction assumes the form

$$\frac{3}{2}\frac{dT_e}{dx} = \frac{3IT_e}{20M_0^2u_x}\left(\frac{1 + u_{GD}}{2} - u_x\right)^{-1}\frac{d\alpha}{dx} - \left(\frac{3}{2}T_e + I\right)\frac{1}{\alpha}\left(\frac{d\alpha}{dx}\right)_e + \frac{T - T_e}{\Delta_r} \quad , \tag{2.3.51}$$

where Δ_r was defined elsewhere, see (2.2.17).

To avoid misunderstanding we emphasize that (2.2.6,3.51) are not exact. In (2.2.6) the ionization kinetics is actually determined by the excitation kinetics, which is not related to the kinetics of triple recombination with capture onto the ground level by the principle of detailed equilibrium. This equation gives a fair description of the ionization kinetics; however, in the region where recombination is important it must be revised. Strictly speaking, (2.2.6) should not merely be modified, but rather supplemented by the kinetic equation of excitation, which describes rapid ionization from the excited levels at the head of the relaxation region, and the approach to the equilibrium occupancy of excited levels towards the end of the relaxation region [2.38,40]. However, recombination gains importance only towards the very end of the region of ionization relaxation, near the state of equilibrium, and improved precision of the relevant kinetic coefficients adds little to the knowledge of the structure of the relaxation region. Accordingly, the expected results are hardly worth the efforts spent. Note also that (2.2.6) does not account for ionization by ion impact, since this contribution is small—at a low degree of ionization—compared with the contribution of the reaction (2.3.9). For $\alpha \gtrsim 10^{-3}$ the leading role is taken by reaction (2.3.19) (the electron avalanche).

The second term on the right-hand side of (2.3.51) accounts for the energy expenditure of the electron gas on impact ionization $-(3T_e/2+I)$ per each newly produced electron. This description accounts only for ionization from the excited levels and excitation by electron impact. Obviously, at the head of the relaxation region the excited atoms are ionized in collisions with other atoms. As a result, the gas of heavy particles imparts energy I to each electron; each of the newly produced electrons also acquires the energy of $3T_e/2$ from the electron gas, and a certain amount of energy of about T from the gas of heavy particles, not accounted for in (2.3.51). This contribution to the heating of electrons can be assessed only by straightforwardly taking into account the kinetics of excitation and deexcitation of atoms. However, this correction would be of minor importance: where there are few electrons, their exact temperature is of little relevance, and where there are many, ionization by atom impact is not essential.

Numerical integration of (2.3.49-51,2.6) poses no problems. Since the temperature relaxation comes to completion sooner than the ionization relaxation, $T_e = T$ near the singular point 2, corresponding to the terminal state. Consequently, the approach to this singular point is described by the single equation (2.2.6), the variables u_x, T and $T_e = T$ being expressed in terms of α with the aid of (2.3.49,50). In the vicinity of point 2 this equation can be written in the form

$$\frac{d\alpha}{dx} = -\left(\frac{\alpha_{eq}(T)}{\Delta_{ion}^{(e)}} + \frac{1}{\Delta_{ion}^{(a)}}\right)\frac{(1-\alpha_{eq}(T))}{\alpha_{eq}(T)}\left[\alpha - \alpha_{eq}(T)\right] , \tag{2.3.52}$$

i.e., the integral curve, as $x \to +\infty$, enters point 2. This enables one to start integrating (2.3.49-51,2.6) directly from $x = 0$, and the numerical solution of this set approximately describes the structure of the relaxation region.

The numerical solution is especially sensitive to the choice of the constant C_a^*, which characterizes the effective cross section of excitation by atomic collisions, see (2.7,10). The value of C_a^* determines the time of relatively slow progress of ionization to the onset of the electron avalanche, i.e., ultimately, the entire structure of the relaxation region. Independent figures for C_a^* are not available, and in practice one has to deal with a converse problem: find C_a^* from the measured profile of ionization in an ionizing shock wave. The reported values of C_a^* are highly divergent, from

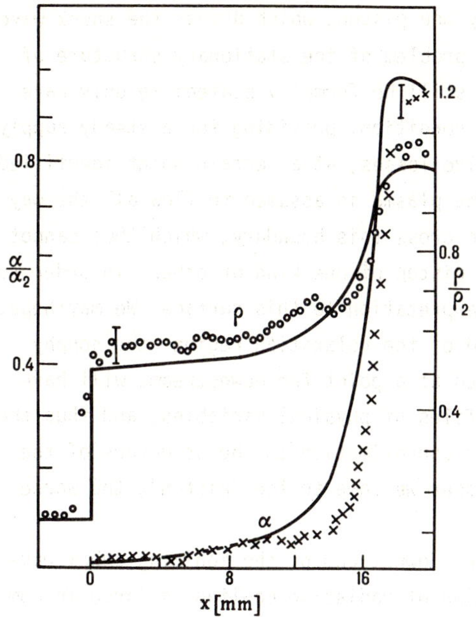

Fig.2.11. Calculated and experimental structure of the region of ionization relaxation for the ionizing shock wave in argon for $M_0 = 13.6$ [2.62] (by courtesy of I.I. Glass)

$1.2 \cdot 10^{-19} cm^2/eV$ in [2.26] to $2.5 \cdot 10^{-20} cm^2/eV$ in [2.61], a sixfold difference. *McLaren* and *Hobson* [2.61] ascribed this difference to the effects of the boundary layer, disregarded in [2.26] (see below). The latest results seem to confirm the value quoted in [2.62]: adjusting C_a^* so as to obtain the best possible fit between theoretical profiles of the relaxation region and experimental findings, *Glass* and *Lin* [2.62] found $C_a^* = 10^{-19} cm^2/eV$.

Theoretical curves and experimental points from [2.62] are reproduced in Fig.2.11. The calculated structure of the relaxation region for a shock wave in argon ($p_0 = 5.12$ Torr, $T_0 = 296.6$ K, $M_0 = 16.5$) with $C_a^* = 10^{-19} cm^2/eV$ fits experiment well. However, in another experiment on the same installation [2.63] with slightly different settings ($p_0 = 5.01$ Torr, $M_0 = 13.0$), the best-fit value of C_a^* was $2 \cdot 10^{-19} cm^2/eV$. The origin of this difference is not yet clear. Perhaps the cause really hides in the boundary effects on the walls of the shock tube, which cannot be included in the framework of the unidimensional gasdynamic model (Sect.2.4.1).

The structure of the region of radiative cooling can be described by stationary unidimensional gasdynamic equations only to a limited accuracy. Indeed, the existence of radiative bulk plasma energy losses points to the nonstationarity of the flow, since the volume of the radiating layer builds up with time, and thus the outflow of energy cannot be permanently compensated at the expense of work done by the piston, which drives the shock wave with a constant speed. However, the problem of the stationary structure of the region of radiative cooling can still be formally stated: we only have to establish an additional boundary condition, providing for a steady supply of energy to make up for the radiative losses, at a certain point identified with the end of the region. Since the plasma is assumed to flow all the way along the x axis, it must be free to cross this boundary, which thus cannot be identified with the surface of a piston of one kind or other, in order to give a straightforward physical interpretation to this surface. We may hope, however, that—at least near the end of the relaxation region—the nonphysical boundary condition, established at a point far downstream, will have reasonably little effect on the profiles of physical variables, and thus the stationary solution will give a fair approximation to the structure of the region of radiative cooling. The closer we come to the "piston", the worse the accuracy.

With the typical plasma parameters downstream of the ionizing shock wave, the characteristic scale of the region of radiative cooling is large in comparison with the scales of temperature and ionization relaxation Δ_r and Δ_{ion}. Consequently, the plasma in this region may be considered isothermal and being in the state of local ionization equilibrium:

$$T_e = T \quad , \tag{2.3.53}$$

$$\alpha = \alpha_{eq}(T) \quad , \tag{2.3.54}$$

where $\alpha_{eq}(T)$ is determined by the Saha formula (2.2.2).

The equations for the continuity of fluxes of mass and momentum retain their validity, thus allowing us to express the dimensionless variables T and $\alpha = \alpha_{eq}(T)$ via the plasma velocity u_x by virtue of the Saha formula and (2.3.49):

$$T = \frac{5M_0}{3} \frac{u_x(1 - u_x + 3/5M_0^2)}{1 + \alpha_{eq}(T, u_x)} \quad . \tag{2.3.55}$$

Expression (2.3.55) is a transcendental equation which connects T and u_x and can be solved numerically. It is easily proved that at $u_x < u_{GD}$

$$\frac{dt}{du_x} > 0 \quad , \qquad \frac{d\alpha_{eq}(T)}{dT} > 0 \quad . \tag{2.3.56}$$

The equation which describes the stationary structure of the region of radiative cooling in dimensional variables has the form

$$\frac{d}{dx}\left[\frac{1}{2}m_a N u_x^3 + \frac{5}{2} N u_x (T + \alpha T_e) + \alpha N J\right] = -(Q_c + Q_\ell) \quad ,$$

where Q_c accounts for the radiative losses due to emission in a continuous spectrum, see (2.3.46), and Q_ℓ accounts for the radiation in lines. Making use of equations for continuity of fluxes of mass and momentum, this equation can be written in dimensionless form:

$$\left[\left(\frac{1 + u_{GD}}{2} - u_x\right) + \frac{3I}{20M_0^2}\frac{d\alpha_{eq}(T)}{dT}\frac{dT}{du_x}\right]\frac{du_x}{dx} = -\frac{Q_c + Q_\ell}{4m_a\bar{N}_0\bar{u}_{x0}^3} \quad . \tag{2.3.57}$$

Since the flow in the region of radiative cooling is subsonic, the first term in brackets on the left-hand side of (2.3.57) is positive; by virtue of (2.3.56) the second term is also positive. Thus, the plasma in this region continues to be compressed. Its dimensional pressure

$$p = p_0\frac{5M_0^2}{3}\left(1 - u_x + \frac{3}{5M_0^2}\right) \tag{2.3.58}$$

remains virtually constant, since u_x is much less than unity directly behind the ionizing shock front (e.g., in argon $u_2 < 0.1$ at $p_0 = 1$ Torr, $M_0 > 18$). A further decrease in u_x causes only a slight rise in pressure. The temperature falls off somewhat slower than the compression builds up, since the equilibrium ionization $\alpha_{eq}(T)$ decreases with the cooling of the plasma, and on the right-hand side of (2.3.55) the term in brackets decreases together with the denominator. Especially sharp is the change in the degree of ionization of the plasma, the ionization being exponentially dependent on temperature. The density of electrons $n_e = \alpha/u_x$ also falls off very rapidly, since α decreases much more steeply than u_x.

The profiles of these variables were calculated in [2.60] for the region of radiative cooling in the wake of the shock in argon ($p_0 = 1$ Torr, $T_0 = 300$ K, $M_0 = 22.9$). The radiative losses in the continuous spectrum were estimated according to the Kramers-Unsöld formula (2.3.46) with $Z_{eff} = 1.5$, $\hbar\omega_c = 2.84$ eV, $\bar{G} = 1$. The radiative losses in lines were calculated under the assumption of Lorentzian line shapes.

Calculations [2.48] indicate that resonant transitions to the ground level do not contribute to the radiative cooling, since the radiation in these lines is trapped in the plasma owing to the high value of the absorption coefficient. The energy losses by radiation in lines are due mainly to transitions to the higher-lying levels 1s and 2p. With the typical parameters of the shock-heated argon plasma downstream ($N_2 = 2.8 \cdot 10^{17}\,cm^{-3}$, $\alpha_2 = 0.5$,

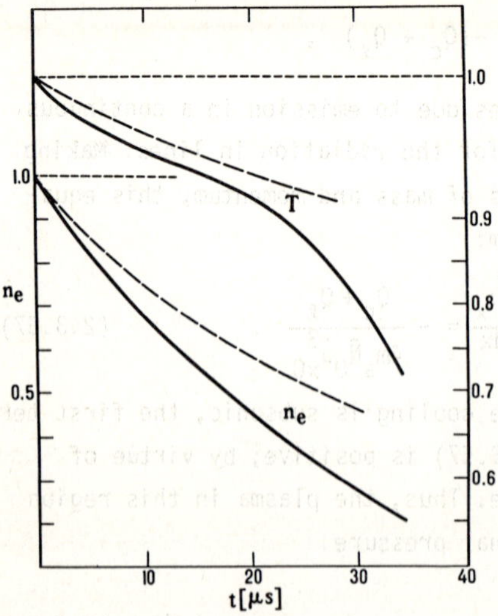

Fig.2.12. Calculated (----) and measured (——) profiles of temperature and electron density in the region of radiative cooling behind the shock wave in argon [2.60] (by courtesy of M. Pinègre, P. Valentin)

$T_2 = 13{,}700$ K), the energy loss in the continuous spectrum Q_c constitutes $1.52 \cdot 10^3$ W/cm^3, whereas the losses due to transitions to 1s, 2p and higher-lying levels constitute, respectively, $4.9 \cdot 10^2$, $1.7 \cdot 10^2$ and 5.1 W/cm^3 [2.48]. Comparing the calculated structure of the region of radiative cooling in argon ($p_0 = 1$ Torr, $T_0 = 300$ K, $M_0 = 19.9$) with the experimental results [2.60] (Fig.2.12) reveals good agreement near the shock, and increasing discrepancy with increasing distance from the shock.

With higher values of M_0 the agreement is worse. Apparently, this is due to the increasing importance of the effects associated with the restricted extent of the radiating volume of plasma and the nonunidimensionality of the flow, the increase in the optical thickness of the boundary layer on the walls of the shock tube, and the mixing of the shock-heated plasma with the cold driver gas.

2.4 Effects of Plasma Flow Nonunidimensionality in Ionizing Shock Waves

2.4.1 Effects of the Wall Boundary Layer in a Shock Tube on the Structure of the Relaxation Region

Deviations from idealized unidimensional flow patterns in experiments with shock tubes have long been observed and investigated [2.64-68]. The effect mainly consists of that the shock front velocity is lower and the velocity of the contact surface, separating the driver gas from the shock-compressed gas, is higher than their respective values predicted by the unidimensional theory (Sect.1.6.4). In the course of time, the former velocity decreases, and the latter increases, until (asymptotically) the shock front and the contact surface travel at the same velocity and at a certain (maximum) distance from one another ℓ_m. This is because the piston (the contact surface), which drives the shock wave, is not quite impenetrable: the flow of mass coming to the shock front escapes from the shock-compressed region to the wall boundary layer, and thus the mass, contained between the shock front and the contact surface, remains constant [2.64-66]. This effect is more pronounced, the greater the length-to-radius ratio of the shock tube L/R and the lower the initial gas pressure p_0.

The quantitative theory of this effect in the quasi-unidimensional approximation (which assumes all the variables to be functions of the coordinate x along the axis of the shock tube) was developed in [2.67,68]. According to this theory, the motion of the shock-compressed downstream gas $(x > 0)$ in the shock-fixed coordinate system can be described as a steady subsonic flow in a diffuser with variable cross section A(x):

$$\frac{A(x)}{A(0)} = \frac{1}{1-(x/\ell_m)^q} \equiv \frac{1}{f(x)} \quad , \tag{2.4.1}$$

where $q = 1/2$ or 4/5 depending on whether the wall boundary layer is laminar or turbulent, respectively. The expressions for ℓ_m are obtained by solving the boundary layer equations [2.67-69]. The condition of applicability of the quasi-unidimensional approximation is given by

$$\delta \ll R \ll \ell_m \quad , \tag{2.4.2}$$

δ being the thickness of the boundary layer at $x = \ell_m$. In the shock tube experiments it is generally satisfied.

By virtue of (2.4.1) the unidimensional continuity equation $\rho uA = \text{const}$ can be written in the form

$$\rho(x)u_x(x) = \rho_2 u_2 f(x) \quad , \tag{2.4.3}$$

or, in dimensionless variables N and u_x, normalized to the downstream values,

$$Nu_x = f(x) \quad . \tag{2.4.4}$$

At $x = \ell_m$ the velocity u_x tends to zero together with f(x): the mass flux across the contact surface vanishes. Profiles of variables, characterizing the flow of the shock-compressed gas between the shock front and the contact surface, are obtained from the adiabaticity integral. For an ideal gas with constant heat capacity we have

$$\frac{\gamma}{\gamma - 1}\frac{p}{\rho} + \frac{u_x^2}{2} = \text{const} \quad , \tag{2.4.5}$$

whence — in dimensionless variables — we readily obtain

$$N = \left(\frac{2 + (\gamma - 1)M_2^2}{2 + (\gamma - 1)M^2}\right)^{1/(\gamma-1)} \quad , \tag{2.4.6}$$

where the local Mach number M can be expressed with the aid of (2.4.4):

$$M = M_2 N^{-(\gamma+1)/2} f(x) \quad . \tag{2.4.7}$$

The sought-for profiles are found by solving (2.4.6,7) together [2.68]. The values of variables on the contact surface ($x = \ell_m$, $M = f(x) = 0$) are easily found from (2.4.6) and (1.6.30):

$$T_c = \frac{(\gamma + 1)^2 M_0^2}{2(2\gamma M_0^2 - \gamma + 1)} \quad ; \quad N_c = T_c^{1/(\gamma-1)} . \tag{2.4.8}$$

In particular, with the weak ($M_0 - 1 \ll 1$) and the strong ($M_0 \gg 1$) shock waves the boundary-layer effects result in the rise of the dimensionless downstream temperature from unity to the values

$$T_c(M_0 \to 1) = \frac{\gamma + 1}{2} + O(M_0 - 1) \quad ; \quad T_c(M_0 \to \infty) = \frac{(\gamma + 1)^2}{4\gamma} + O(M_0^{-2}) \quad . \tag{2.4.9}$$

With $\gamma = 5/3$ the temperature is raised by a factor of 4/3 for a weak shock, and by a factor of 16/15 for a strong shock. According to (2.4.8), this rise in temperature is accompanied by compression, its ratio being 1.54 for a weak shock and 1.10 for a strong shock. We see that the effect is more pronounced if M_0 is not too high. However, even with $M_0 \sim 10$ a slight rise in temperature may strongly affect the structure of the ionization-relaxation region of the ionizing shock wave, its width being an exponential function of temperature.

This approximation was used for calculating the structure of the ionization relaxation region in argon in [2.23]. The set of shock-layer equations accounts for the finite radius of the shock tube by using the continuity equation (2.4.4) in place of $Nu_x = 1$. Accordingly, wherever u_x^{-1} denotes dimensionless density, it is replaced by $f(x)u_x^{-1}$. The calculations were carried out under the assumption of a laminar boundary layer [$q = 1/2$ in (2.4.1)].

The results obtained in [2.23] under experimental conditions typical of moderate-intensity ionizing shock waves in ordinary shock tubes indicate that the compression and heating of the plasma, due to the existence of the boundary layer, can reduce the width of the relaxation region substantially. For instance, with $p_0 = 1$ Torr, $T_0 = 300$ K, $M_0 = 12$ the standard calculation yields a width of 50 cm, whereas the values obtained with due account of the effects of the boundary layer are 30, 20 and 10 cm for the radius of shock tubes of 5, 2.5, 1.25 cm, respectively. Simultaneously, the degrees of ionization and density upon reaching the point of ionization equilibrium exhibit an increase (by 5%-6% with $R = 5$ cm). The quantity $p_0\tau_{eq}$ characterizing the time of establishment of ionization equilibrium, which in purely unidimensional approximations displays a marginal dependence on p_0, turns out to be quite sensitive to the initial pressure with not too high values of R and M_0. The boundary-layer effects become insubstantial with $p_0 \gtrsim 1$ Torr, $R \gtrsim 5$ cm, $M_0 \gtrsim 15$. These conclusions are consistent with the results of numerous experiments [2.25,37,56,69-73].

It must be noted that if the value of the constant C_a^* of excitation by atomic collisions is itself estimated from the measured width of the relaxation region, then its value can be considerably exaggerated when the boundary-layer effects are neglected [2.61]. However, under the experimental conditions in [2.62] the required correction is insignificant and cannot be held responsible for the discrepancy (Sect.2.3.4) between the values of C_a^*, as estimated in [2.26,62] (and, incidentally, in [2.62] from the experiments with $M_0 = 13$ and $M_0 \geqslant 15.9$). This question is far from being entirely clear. Recent measurements of the transverse structure of the flow in ionizing shock waves [2.62] indicate that the quasi-unidimensional approximation is adequate for the shock wave in argon with $M_0 = 13.1$, $p_0 = 5.16$ Torr. The measured densities of electrons and heavy particles towards the end of the relaxation region are found to change drastically in a thin boundary layer (~ 2 mm) next to the walls of the shock tube; with increasing distance the densities cease being dependent on y. The experimental transverse profiles fit in well with the results of theoretical calculations. Hence it follows that with $M_0 \leqslant 13$ the downstream flow may be safely considered homogeneous and thus describable by the quasi-unidimensional gasdynamic equations, which for the large-diameter shock tubes reduce to the conventional unidimensional equations. However, at $M = 15.9$ ($p_0 = 5.1$ Torr) the transverse structure becomes essentially nonuniform beyond the limits of the boundary layer. *Glass* and *Lin* [2.62] attribute this to the manifestation of radiative cool-

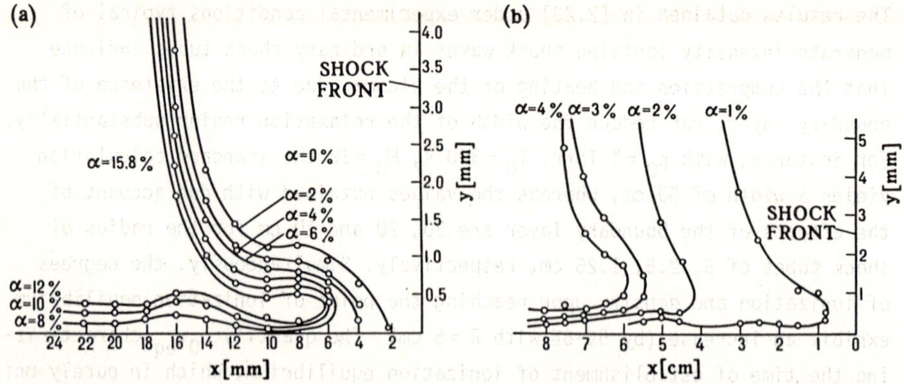

Fig.2.13a,b. Transverse structure of the region of ionization relaxation near the shock-tube wall for ionizing shocks in argon: (**a**) $M_0 = 16.5$, $p_0 = 5.12$ Torr, $T_0 = 296.6$ K; (**b**) $M_0 = 13.6$, $p_0 = 5.09$ Torr, $T_0 = 296.7$ K [2.62] (by courtesy of I.I. Glass)

ing. Whatever the real cause, under these conditions deviations from the theoretical results of [2.23] are quite possible, and perhaps these circumstances are responsible for the high variance of the values of C_a^*.

The experimental result, inexplicable within the framework of quasi-unidimensional approximation, is the displacement of the electron avalanche towards the gasdynamic shock front in the wall boundary layer [2.62] (Fig. 2.13). Note that this effect is not associated with the well-known bending of plane shock fronts in channels, observed in shock tubes with an initial pressure of about one atmosphere [2.74,75] and discussed in detail in [2.76]. Here the shock front remains planar, while nonuniformity is confined to the relaxation region. Specially designed experiments indicate that this displacement phenomenon can hardly be attributed to the interaction of the shock-heated gas with the material of the tube walls or with impurities. Coating the tube walls with a good insulator (cellulosic coating), with a metal with a different work function (tungsten), with water, whose vapor becomes the major impurity in argon, or with a solution of common salt, which contains a readily ionized component (Na), does not change the observed effect. This effect seems to be due to the yet unknown peculiarities of ionization relaxation in the wall boundary layer. No adequate theory has yet been evolved: even with a laminar boundary layer one has to deal with the problem of two-dimensional structure of the ionization-relaxation region; this gives rise to immense mathematical complications. Besides, the effect observed may be associated with the intrinsic instability of ionizing shock waves.

2.4.2 Instability of Ionizing Shock Waves

Instabilities which disturb the pattern of the steady, unidimensional plasma flow on ionizing shock waves, were manifest in many experiments [2.62,77,78]. The theory of these phenomena is not advanced, and does not definitely answer the question whether these effects are associated with the intrinsic instability (under certain conditions) of the plane ionzing shock wave front, or the downstream plasma flow is disturbed because of the violation of spatial uniformity on its boundaries in the wall boundary layer or on the contact surface. Besides, the uniformity of the flow is affected by the emission and absorption of radiation. Apparently, all these factors contribute to the instabilities observed.

Instability of the plane shock front, due to the peculiar form of the shock adiabat of the ionizing shock wave, may arise when one of the inequalities (1.7.27-29) is satisfied. Calculations indicate that under conditions typical for experiments with ionizing shock waves, the conditions of instability (1.7.27,28) are never satisfied, so the ionizing shock waves do not split into two-wave configurations, (Sect.1.7.4). Inequality (1.7.29) is usually not satisfied either, but its right- and left-hand sides occasionally are almost equal to each other. Under real flow conditions in the shock wave, confined by the boundary layer and the contact surface, the criterion (1.7.29) may perhaps be relaxed, the sign of strict inequality being replaced by $\gtrsim$. Then the splitting instability may be observed in shock waves in dissociative and ionizable gases with the shock velocities corresponding to bendings of shock adiabats associated with dissociation and ionization.

This phenomenon was investigated in [2.77], where CO_2 was used as the dissociative medium, and the ionizing shock waves were studied in argon. Nonuniformity of the flow, interpreted as a manifestation of instability, was monitored by the transverse configuration of shifts of interference bands. In a spatially uniform plasma the corresponding shift is proportional to the path length of the beam within the boundaries of the shock tube, i.e., to $(R^2 - r^2)^{1/2}$ for the beam which is offset by r from the axis of the shock tube. This elliptic dependence of the shift δf on r was actually observed with stable shock waves, whereas the dependence $\delta f(r)$ with the shock waves classified as instable was quite irregular. In some cases the flow pattern was intermediate between definitely stable and definitely unstable. Instability was actually detected near bends of shock adiabats (with the velocity of $5 \cdot 10^5$cm/s for CO_2 and $1.1 \cdot 10^6$cm/s for Ar). Although the rigid criterion for the splitting instability here apparently is not satisfied, the instability threshold is certainly approached. The smallness of the characteristic scales of shock fronts in comparison with the shock tube radius ($\Delta \lesssim 5$ mm in CO_2, $\Delta \lesssim 2$ mm in Ar; R = 2.5 cm) allows us to conclude that we are dealing exactly with the instability on a narrow plane shock, considered as a discontinuity in the idealized gasdynamic approximation. The spatial resolution of the diagnostic equipment [2.77] allowed positive discrimination between the above-mentioned instability and the manifestations of contact surface instability, especially connected with the relatively low velocities of shock waves [2.62]. The contact surface instability, not associated directly with the properties of an ionizing shock wave, also disturbs the uniformity of the downstream flow.

Instability of a different kind, due to the finite width of the relaxation region, was observed in [2.62,63] with the ionizing shock waves in argon. The experiments were performed with shock waves of moderate intensity, for which the instability criterion (1.7.29) is not satisfied even in its slackened form. Deviations from the uniform plasma flow were also detected by the irregularity of the interference pattern. Adding a small quantity of impurity to the argon (0.4% of hydrogen) markedly raises the ionization rate at the initial portion of the ionization-relaxation region, dominated by atomic collisions, and its width is reduced to one-third of the previous value. In this case an instability is not observed; its mechanism remains unclear. As demonstrated in [2.63], this instability (towards the end of the relaxation region, and further in the region of radiative cooling) is accompanied by a deviation from uniform flow, associated with emission and absorption of radiation. On the whole, the flow pattern becomes quite complicated; there are good reasons to suppose that instabilities give rise to turbulence.

Turbulence in ionizing shock waves in argon was detected in [2.78]. The experiment was carried out in an electromagnetic shock tube (R = 2.5 cm) at low pressures ($p_0 = 0.25$-0.025 Torr) in a wide range of Mach numbers ($M_0 = 16$-42). Diagnostic equipment included an array of electrostatic probes connected to the negative lead of the voltage source and serving as collectors of ionic current. The advantage of this technique is that it enables one to measure the local parameters of the plasma, while the interference techniques necessarily involve integration of the measured parameters over the volume of the ray tube. The probe signals are reproduced in Fig.2.14; they all clearly follow the same pattern. The peak, corresponding to the arrival of the gasdynamic shock, is followed by rapid irregular oscillations.

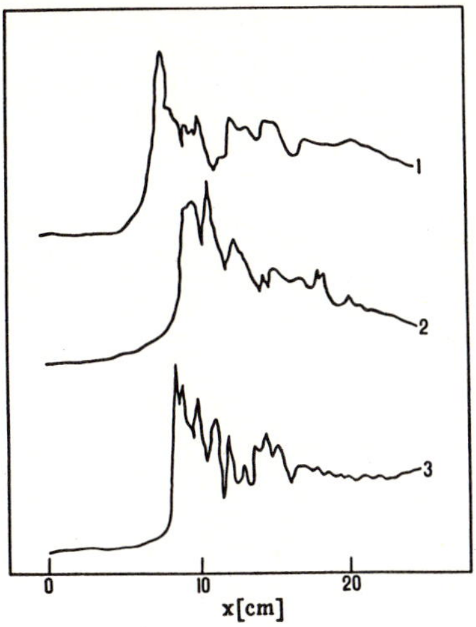

Fig.2.14. Signals of electrostatic probes in the plasma downstream of ionizing shock waves in argon: (1) $p_0 = 100$ mTorr, $M_0 = 16$; (2) $p_0 = 60$ mTorr, $M_0 = 26$; (3) $p_0 = 40$ mTorr, $M_0 = 36$; [2.78]
(by courtesy of J.A. Johnson III)

The region of oscillations is 12-15 cm long and is virtually independent of the Mach number. Fourier analysis of these signals allows the clear-cut principal mode and a number of lower harmonics to be singled out with good reproducibility. This pattern is consistent with the theoretical concept of oscillations near the turbulence threshold, and can be observed in a large number of weakly turbulent flows of different kinds [2.79]. The conclusion regarding the turbulization of the downstream flow in the ionizing shock wave in argon for $M_0 \geqslant 16$ is also consistent with the observations of the transverse structure of the relaxation region (Sect.2.4.1) [2.62]. The similarity of the results throughout the entire range of Mach numbers allows one to assume that the development of turbulence is not associated with the splitting instability, occurring on the bend of the shock adiabat. Apparently, the turbulence arises mainly due to the action of purely gasdynamic mechanisms in the relaxation region; for one, this is confirmed by the fact that the wavelength of the dominant mode is close to the radius of the shock tube.

We see that the actual structures of ionizing shock waves, which generally are well described within the framework of unidimensional or quasi-unidimensional approximations, upon close examination exhibit certain deviations from the predictions of the unidimensional treatment. Theoretical description of such structures requires the use of at least a two-dimensional gasdynamic approximation, accounting for the existence of fluctuations, and is further complicated by the necessity of detailed calculations of ionization kinetics in the turbulent wall boundary layer. Such a problem effectively defies solution; it cannot even be stated correctly, since no first-principle theoretical model is yet available capable of handling such flows, and the existing phenomenological theories are open to criticism [2.80,81].

Prior to constructing a consistent theory for weakly turbulent structures, one has to solve a large number of problems pertaining to fluctuations in gasdynamics and the theory of turbulence. It must be acknowledged, however, that the ionizing shock waves, along with the familiar aims of turbulence theory like the Couette flow or thermal convection, provide good grounds of investigation. The reproducibility of results obtained with shock tubes, the possibility to pinpoint the beginning of turbulence on the Mach numbers, e.g., between 13 and 16 under experimental conditions of [2.62], and, finally, the availability of a reasonable first approximation (unidimensional gasdynamic model), encourages hope for a major theoretical breakthrough in this direction in the not too distant future. In particular, the approach to the description of weak turbulence, recently proposed in [2.82] on the basis of the technique developed in [2.83] seems very promising. It has already been successfully employed to obtain interesting results regarding the turbulent flows in shock tubes [2.84].

3. Magnetohydrodynamic Shock Waves in Plasmas

The flow of plasma in shock waves in an external magnetic field is described by electromagnetic equations and equations of fluid mechanics. A most simple example of such a description is offered by one-fluid MHD equations (Sect. 3.1.1), whose application is restricted to large-scale and low-frequency flow patterns. The MHD equations describe the flow far upstream and downstream from the shock front, and therefore may be used to classify shock waves and discontinuities in plasma, to assure evolutionarity etc. The study of shock structure calls for a more detailed description, offered by the two-fluid transfer equations (Sect.3.1.2).

In Sect.3.2 we consider linear and nonlinear waves in a conducting fluid in a magnetic field. The dispersion equations for fast and slow magnetosonic and Alfvén waves are derived in Sect.3.2.1 in the context of the MHD approximation. Damping and dispersion of these waves due to physical processes, characterized by finite length scales, are analyzed in Sect.3.2.2 on the basis of a two-fluid model. In Sect.3.2.3 we describe MHD flows in the form of simple Riemann waves. The nonlinear distortion of the running-wave profiles, leading to the generation of shock waves, is shown to pertain to the fast and slow waves, but not to the Alfvén waves.

Discontinuities and shock waves in magnetohydrodynamics are examined in Sect.3.3, and classified in Sect.3.3.1. Similarly to gasdynamics (Sect. 1.6.1), all kinds of discontinuities except shock waves appear as stepwise profiles of simple finite-amplitude waves, without nonlinear steepening. The boundary conditions on the MHD shock fronts, evolving from the conservation laws, Maxwell equations and the requirement of increasing entropy across the shock, allow the existence of six different types of shock transitions. However, inspecting evolutionarity criteria for the MHD shock waves reveals that only two types of shocks are evolutionary and may be actually realized (Sect.3.3.3). These shock waves (fast and slow) may be viewed as the result of nonlinear superposition of the fast and slow finite-amplitude magnetosonic waves, respectively, consistent with the results of Sect.3.2.3. The fast and the slow shock-wave structures in the MHD approximation are illustrated in Sect.3.3.4 for the extreme cases of small and large magnetic Prandtl numbers (Pm); and analogy is drawn with the gasdynamic shock waves with small and large Pr. In Sect.3.3.5 we discuss the evolutionarity criterion for the singular (switch-on and switch-off) MHD shocks. Analyzing the evolutionarity requires including the so-called diffusion waves, intermediate between undamped and dissipative waves.

In Sects.3.4-6 we describe structures of shock waves in plasma for different orientations of the magnetic field with respect to the shock plane. Section 3.4 is devoted to the most important practical case of transverse shock waves, which classify as fast ones. In Sect.3.4.2 we analyze structures of transverse shock waves in magnetized plasma; here the action of transverse electron heat conduction is suppressed by the magnetic field ($Pr \gg 1$), and viscosity dominates among the dissipative mechanisms. The dispersive processes (inertia of electrons and/or charge separation), which give rise to oscillatory structures, are manifested in the moderate-intensity shock waves in cold plasma (Sect.3.4.4). In nonmagnetized plasma (small Pm; Sect.3.4.3) the main dissipative mechanism is represented by Joule heating, which, as indicated in Sect.3.3.4, is capable of shaping the structures of only moderately strong shock waves. The structures of strong shock waves contain isomagnetic subshocks, whose internal structure resembles that of the shocks in nonmagnetized plasma and may include electron-isothermal subshocks. Electron heat conduction and Joule dissipations dominate in partly magnetized plasma. The structures of strong shock waves contain viscous subshocks, which are at the same time isomagnetic and electron-isothermal. In Sect.3.4.4 we examine the role of longitudinal electric field of plasma polarization and its effect on the conduction anisotropy in transverse shock waves. In Sect.3.4.5 we discuss the experimental data on transverse MHD shocks in plasma.

In Sect.3.5 we consider the structures of the switch-on shock waves, which in nonmagnetized plasma (Sect.3.5.2) are shaped by Joule dissipations, while the structures of strong enough shocks contain isomagnetic subshocks. In magnetized plasma (Sect.3.5.3) an important role is played by the Hall currents, which are responsible for the oscillatory structure of moderately weak shock waves. The structures of strong shock waves are found to contain viscous subshocks, which are isomagnetic and isothermal in electron temperature.

The structures of the switch-off shock waves, dealt with in Sect.3.6, are described by the same equations as those of the switch-on shocks, since they have much in common. However, the respective boundary conditions exhibit essential differences (the switch-off shock waves are slow), which are responsible for certain peculiar features of the switch-off shock waves.

3.1 Basic Equations

3.1.1 Magnetohydrodynamic Equations

A flow of a conducting fluid in an external magnetic field is accompanied by interconnected electromagnetic and hydrodynamic phenomena. Accordingly, the flow is described by coupled equations of the electromagnetic field and the equations of motion of the fluid. A most simple and clear example of such a description is given by the magnetohydrodynamic (MHD) equations.

Magnetohydrodynamics deals with electrically neutral conducting fluid without internal degrees of freedom, whose state, as in conventional gasdynamics, is described by two thermodynamic parameters, e.g., ρ and S, and the mean velocity $\mathbf{u}$. The electromagnetic field in this approximation is wholly determined by the magnetic field $\mathbf{B}$. The magnetic field satisfies the equation

$$\operatorname{div} \mathbf{B} = 0 \quad , \tag{3.1.1}$$

and the equation of induction, derived from other Maxwell's equations by eliminating the electric field $\mathbf{E}$. To obtain the desired equation, we take the Maxwell equation

$$\operatorname{curl} \mathbf{B} = \mu_0\left(\varepsilon_0 \frac{\partial \mathbf{E}}{\partial t} + \mathbf{j}\right) \tag{3.1.2}$$

and disregard the displacement current (the first term in parentheses) as small compared with the conduction current $\mathbf{j}$. The electric field and current are coupled by the familiar Ohm's law

$$\mathbf{j} = \sigma \mathbf{E}^* \quad , \tag{3.1.3}$$

where $\mathbf{E}^*$ is the electric field in the particle-fixed coordinate system, σ is the conductivity of the fluid

$$\mathbf{E}^* = \mathbf{E} + \mathbf{u} \times \mathbf{B} \quad . \tag{3.1.4}$$

Into the Maxwell equation

$$\frac{\partial \mathbf{B}}{\partial t} = - \operatorname{curl} \mathbf{E} \tag{3.1.5}$$

we substitute the expression for **E** in terms of **j** and **B**, obtained under the stated assumptions from (3.1.2-4). Assuming also $\sigma(\mathbf{r},t) = \text{const}$, we come to the induction equation

$$\frac{\partial \mathbf{B}}{\partial t} = \operatorname{curl} \mathbf{u} \times \mathbf{B} + \nu_m \Delta \mathbf{B} \quad , \tag{3.1.6}$$

where

$$\nu_m = 1/\mu_0\sigma \tag{3.1.7}$$

is called the magnetic viscosity (by analogy with the kinematic viscosity) and has the dimension of the coefficient of diffusion. If the second term on the right-hand side of (3.1.6) can be neglected, this equation becomes the induction equation for an ideally conducting fluid

$$\frac{\partial \mathbf{B}}{\partial t} = \operatorname{curl} \mathbf{u} \times \mathbf{B} \quad , \tag{3.1.8}$$

corresponding to the limit

$$\sigma \to \infty \quad , \quad E^* \to 0 \tag{3.1.9}$$

with finite $j(\neq 0)$.

The MHD equations of motion are similar to the conventional gasdynamic equations but for the terms accounting for the magnetic-field effects [3.1]. The continuity equation remains the same:

$$\frac{\partial \rho}{\partial t} + \operatorname{div} \rho \mathbf{u} = 0 \quad . \tag{3.1.10}$$

In the momentum transfer equation the gasdynamic stress tensor (1.1.24) is supplemented by the Maxwellian magnetic stress tensor:

$$\Pi_{ik} \to \Pi_{ik} + (1/\mu_0)\left[(1/2)B^2\delta_{ik} - B_iB_k\right] \quad . \tag{3.1.11}$$

In the heat transfer equations the expressions for density and flux of the fluid's energy are supplemented, respectively, by density and flux of the electromagnetic field energy, expressed via **B** with the aid of (3.1.2-4):

$$\rho\varepsilon \to \rho\varepsilon + B^2/2\mu_0 \quad , \tag{3.1.12}$$

$$\mathbf{q} \to \mathbf{q} + (1/\mu_0)\mathbf{B} \times (\mathbf{u} \times \mathbf{B}) - (1/\mu_0^2\sigma)\mathbf{B} \times \operatorname{curl} \mathbf{B} \quad . \tag{3.1.13}$$

Let us convert the transfer equations into the more customary Navier-Stokes equations, assuming the coefficients of viscosity η, ζ and thermal conductivity κ to be constant:

$$\rho\left[\frac{\partial \mathbf{u}}{\partial t} + (\mathbf{u}\nabla)\mathbf{u}\right] = -\nabla p + \eta\Delta\mathbf{u} + (\eta/3 + \zeta)\ \mathrm{grad\ div}\ \mathbf{u}$$

$$- (1/\mu_0)\mathbf{B} \times \mathrm{curl}\ \mathbf{B} \quad , \tag{3.1.14}$$

$$\rho T\left[\frac{\partial S}{\partial t} + (\mathbf{u}\nabla)S\right] = \sigma_{ik}\frac{\partial u_i}{\partial x_k} + \kappa\Delta T + (1/\mu_0^2\sigma)(\mathrm{curl}\ \mathbf{B})^2 \quad . \tag{3.1.15}$$

The last terms on the right-hand sides of (3.1.14,15) account, respectively, for the density of electromagnetic forces and the entropy production at the expense of current dissipation (Joule heat).

Let us analyze the physical meaning of the two terms on the right-hand side of (3.1.6). In the ideally conducting medium the second term is absent. With due account of (3.1.1), (3.1.8) can be converted into

$$\left[\frac{\partial}{\partial t} + (\mathbf{u}\nabla)\right]\mathbf{B} = (\mathbf{B}\nabla)\mathbf{u} - \mathbf{B}\ \mathrm{div}\ \mathbf{u} \quad . \tag{3.1.16}$$

Expressing div $\mathbf{u}$ with the aid of (3.1.10)

$$\mathrm{div}\ \mathbf{u} = -\frac{1}{\rho}\frac{\partial\rho}{\partial t} - \frac{\mathbf{u}}{\rho}\nabla\rho \tag{3.1.17}$$

and substituting (3.1.17) into (3.1.16) we get

$$\left[\frac{\partial}{\partial t} + (\mathbf{u}\nabla)\right]\frac{\mathbf{B}}{\rho} \equiv \frac{d\mathbf{B}}{dt} = \left(\frac{\mathbf{B}}{\rho}\nabla\right)\mathbf{u} \quad . \tag{3.1.18}$$

Consider a small vector $\boldsymbol{\delta}\mathbf{L}$, drawn between two closely spaced fluid particles. If $\mathbf{u}$ is the velocity of one of these, then the velocity of the other is $\mathbf{u} + (\boldsymbol{\delta}\mathbf{L}\nabla)\mathbf{u}$, i.e., these particles will move during the time lapse dt by $\mathbf{u}\,dt$ and $\mathbf{u}\,dt + (\boldsymbol{\delta}\mathbf{L}\nabla)\mathbf{u}\,dt$, respectively, and the vector $\boldsymbol{\delta}\mathbf{L}$ will be changed by $(\boldsymbol{\delta}\mathbf{L}\nabla)\mathbf{u}\,dt$. Hence we get the equation in $\boldsymbol{\delta}\mathbf{L}$

$$\frac{d}{dt}\,\delta\mathbf{L} = (\boldsymbol{\delta}\mathbf{L}\nabla)\mathbf{u} \quad . \tag{3.1.19}$$

It follows that if the two particles were initially located on the same magnetic force line ($\boldsymbol{\delta}\mathbf{L} \parallel \mathbf{B}$), then, in accordance with (3.1.18,19), they will keep on this line. The same is easily demonstrated for particles located at a finite distance from one another. In this way the induction equation describes the flow in which the magnetic force lines are "glued" to the particles of the fluid, or frozen into the fluid.

At the other extreme, e.g., quiescent fluid, the right-hand side of (3.1.6) is dominated by the second term. Then (3.1.6) is reduced to the diffusion equation, in which the role of the diffusion coefficient is played by magnetic viscosity ν_m. This equation describes the penetration of magnetic field into the fluid due to the latter's finite conductivity.

The magnetic-field "freeze-in" criterion is given by the ratio of the two terms on the right-hand side of (3.1.6). If L is the characteristic length scale, then the magnetic-field freeze-in is characterized by a dimensionless magnetic Reynolds number Rm,

$$Rm = \frac{uL}{\nu_m} \quad . \tag{3.1.20}$$

For $Rm \gg 1$ the magnetic field is frozen into the fluid; for $Rm \ll 1$ it freely permeates the conducting fluid.

Now let us discuss the conditions of applicability of the MHD approximation for shock waves in a plasma with magnetic field. It is easily proved that with typical strengths of the magnetic field and the plasma parameters in collisional shock waves the quasistationarity condition of electromagnetic processes is satisfied and thus the displacement current may be neglected. Indeed, taking into account that according to (3.1.4,8,9) $E \sim uB$, $|\mathrm{curl}\,\mathbf{B}| \sim B/L$, $\partial E/\partial t \sim E/\tau$, τ being the characteristic time scale, from the inequality

$$|\mathrm{curl}\ \mathbf{B}| \gg \left|\varepsilon_0\mu_0 \frac{\partial \mathbf{E}}{\partial t}\right| \equiv \frac{1}{c^2}\left|\frac{\partial \mathbf{E}}{\partial t}\right| \tag{3.1.21}$$

we obtain

$$uL/c^2\tau \ll 1 \quad . \tag{3.1.22}$$

Since $L/\tau \sim u$, (3.1.22) is equivalent to the condition of applicability of the nonrelativistic theory ($u \ll c$), which is not violated even with the strongest laboratory shock waves. Condition (3.1.22) can also be expressed as a limitation imposed on the magnetic field, by estimating, with the aid of (3.1.14), the acceleration of plasma via the magnetic pressure gradient: $\rho u/\tau \sim B^2/\mu_0 L$; i.e., from (3.1.22) we get

$$B^2/\mu_0 \ll \rho c^2 \quad . \tag{3.1.23}$$

This inequality is satisfied with a good margin in the range $B \lesssim 10$ T, $n \gtrsim 10^{14}\,\mathrm{cm}^{-3}$, being consistent with the conditions of collisional shock waves in the laboratory.

Less realistic, as far as the plasma in a shock wave is concerned, is the magnetohydrodynamic one-fluid model (which describes the flow in terms of only ρ, $\mathbf{u}$, S or T) with the Ohm's law (3.1.3). As indicated in Sect.2.1, the two-fluid effects (separation of charges, unequal temperatures of electrons and ions, etc.) are clearly manifest even in moderately strong shock waves in the plasma without external fields. In an external magnetic field the components of moving plasma experience the action of oppositely directed electromagnetic forces; the transfer coefficients depend differently on the magnetic field (Sect.3.1.2), and thus the two-fluid effects are more pronounced.

It follows that studying the structure of shocks in a magnetized plasma within the framework of a Navier-Stokes approximation cannot be based on the MHD model; preference should be given to the two-fluid transfer equations. Notice, however, that two-fluid and nonequilibrium effects in shocks in a plasma are characterized by finite-length scales, such as the Debye radius, the length of temperature relaxation, etc. If the scale L we are concerned with is large compared to these, then using the one-fluid approximation is justified. Ohm's law (3.1.3) still holds in the limit $L\to\infty$, since then, according to (3.1.20), $Rm\to\infty$ and the plasma may be considered a perfectly conducting fluid. Thus, the MHD equations of ideally conducting dissipationless fluid are valid as long as we are dealing with the equilibrium states in the upstream and downstream infinities, or with long enough running waves considered in determining the evolutionarity condition, or with nonstationary plasma flows characterized by large-scale and low-frequency changes of all the variables, in which the shocks are treated as discontinuities.

3.1.2 Two-Fluid Transfer Equations for a Plasma

Now let us write the two-fluid transfer equations which will be used further for studying the shock structures in plasma with a magnetic field. These equations were derived and the transfer coefficients were calculated in [3.2].

The set of the Navier-Stokes transfer equations for a simple two-component plasma ($\gamma_e = \gamma_i = 5/3$; $Z = 1$) has the form

$$\frac{\partial n_e}{\partial t} + \mathrm{div}(n_e \mathbf{u}^e) = 0 \quad , \tag{3.1.24}$$

$$\frac{\partial n_i}{\partial t} + \mathrm{div}(n_i \mathbf{u}^i) = 0 \quad , \tag{3.1.25}$$

$$m_e n_e\left(\frac{\partial}{\partial t} + u^e_\beta \frac{\partial}{\partial x_\beta}\right)u^e_\alpha = -\frac{\partial p_e}{\partial x_\alpha} - \frac{\partial \Pi^e_{\alpha\beta}}{\partial x_\beta} - en_e(\mathbf{E} + \mathbf{u}^e \times \mathbf{B})_\alpha + \mathbf{R}_\alpha \quad , \tag{3.1.26}$$

$$m_i n_i\left(\frac{\partial}{\partial t} + u^i_\beta \frac{\partial}{\partial x_\beta}\right)u^i_\alpha = -\frac{\partial p_i}{\partial x_\alpha} - \frac{\partial \Pi^i_{\alpha\beta}}{\partial x_\beta} + en_i(\mathbf{E} + \mathbf{u}^i \times \mathbf{B})_\alpha - \mathbf{R}_\alpha \quad , \tag{3.1.27}$$

$$\frac{3}{2} n_e\left(\frac{\partial}{\partial t} + u^e_\beta \frac{\partial}{\partial x_\beta}\right)T_e + p_e \frac{\partial u^e_\beta}{\partial x_\beta} = - \frac{\partial q_{e\beta}}{\partial x_\beta} - \Pi^e_{\alpha\beta} \frac{\partial u^e_\alpha}{\partial x_\beta} + Q_e \quad , \tag{3.1.28}$$

$$\frac{3}{2} n_i\left(\frac{\partial}{\partial t} + u^i_\beta \frac{\partial}{\partial x_\beta}\right)T_i + p_i \frac{\partial u^i_\beta}{\partial x_\beta} = - \frac{\partial q_{i\beta}}{\partial x_\beta} - \Pi^i_{\alpha\beta} \frac{\partial u^i_\alpha}{\partial x_\beta} + Q_i \quad . \tag{3.1.29}$$

Here the subscripts e,i pertain to electrons and ions, respectively.

The force of friction (the transfer of momentum from ions to electrons by elastic collisions) is denoted by $\mathbf{R}$; $\Pi^s_{\alpha\beta}$, Q_s stand for the tensor of viscous tension and the heat resulting from the elastic collisions for the particles of kind s (s = e,i); the remaining designations are customary. The partial pressures for the electron and ion components are

$$p_e = n_e T_e \quad , \quad p_i = n_i T_i \quad . \tag{3.1.30}$$

The equations (3.1.24-30) are supplemented by Maxwell equations — the Poisson equation (2.1.7), (3.1.1,5), as well as (3.1.2) in the quasi-stationary approximation —

$$\mathrm{curl}\, \mathbf{B} = \mu_0 e(n_i \mathbf{u}_i - n_e \mathbf{u}_e) \quad . \tag{3.1.31}$$

The above-listed equations in the adopted Navier-Stokes approximation completely describe plasma dynamics in electromagnetic fields, whether created by the external sources or resulting from internal currents and charge separation in the plasma.

The presence of a magnetic field is responsible for the anisotropy of transfer phenomena in plasma. The between-collisions path of a particle is $\lambda = v_{Ts}\tau_s$ along the force lines, and $\min(\lambda, r_s)$ in the transverse sense, $r_s = v_{Ts}/\Omega_s$ being the cyclotron radius, $v_{Ts} = (8T_s/\pi m_s)^{\frac{1}{2}}$ the mean thermal velocity, and $\Omega_s = eB/m_s$ the cyclotron frequency. The degree of anisotropy of transfer phenomena is measured by the parameter

$$\lambda/r_s = \Omega_s \tau_s \quad . \tag{3.1.32}$$

For $\Omega_s\tau_s \ll 1$ the anisotropy is absent. On the other hand, for $\Omega_s\tau_s \gg 1$ the anisotropy is highly pronounced, and the characteristic lengths of transfer in the longitudinal and the transverse directions are quite different. Particles of kind s are called magnetized if $\Omega_s\tau_s \gg 1$, and nonmagnetized at the opposite extreme. In a simple plasma

$$\frac{\Omega_e \tau_e}{\Omega_i \tau_i} = \left(\frac{m_i}{2m_e}\right)^{\frac{1}{2}} \left(\frac{T_e}{T_i}\right)^{\frac{3}{2}} \quad , \tag{3.1.33}$$

so $\Omega_e\tau_e \gg \Omega_i\tau_i$ for $T_e \gtrsim T_i$. Accordingly, the transfer anisotropy may be featured by the plasma as a whole, i.e., both its components, $1 \ll \Omega_i\tau_i \ll \Omega_e\tau_e$, or by the electron component alone.

The force $\mathbf{R}$ is comprised of two components: the force of mutual friction $\mathbf{R}_u$, arising from the relative motion of electrons and ions with velocity $\mathbf{U} = \mathbf{u}^e - \mathbf{u}^i$, and the thermal force $\mathbf{R}_T$, associated with the electron temperature gradient

$$\mathbf{R} = \mathbf{R}_u + \mathbf{R}_T \quad . \tag{3.1.34}$$

For nonmagnetized electrons

$$\mathbf{R}_u = -0.51\, m_e n_e \mathbf{U}/\tau_e \quad ; \quad \mathbf{R}_T = -0.71\, n_e \nabla T_e \quad ; \tag{3.1.35}$$

and for magnetized electrons

$$\begin{aligned} \mathbf{R}_u &= -\, m_e n_e (0.51\, \mathbf{U}_\parallel + \mathbf{U}_\perp) \quad ; \\ \mathbf{R}_T &= -\, 0.71\, n_e \nabla_\parallel T_e + \frac{3}{2}\, (n_e/\Omega_e\tau_e) \nabla T_e \times \mathbf{b} \quad , \end{aligned} \tag{3.1.36}$$

where the subscripts $\parallel$ and $\perp$ mark the vector components parallel and normal to the magnetic force lines, and $\mathbf{b} = \mathbf{B}/B$.

In nonmagnetized plasma the viscosity tensor is expressed via the stress tensor

$$W_{\alpha\beta} = \frac{\partial u_\alpha}{\partial x_\beta} + \frac{\partial u_\beta}{\partial x_\alpha} - \frac{2}{3}\, \delta_{\alpha\beta} \frac{\partial u_\alpha}{\partial x_\alpha}$$

as

$$\Pi^s_{\alpha\beta} = -\eta^s_0 W^s_{\alpha\beta} \quad . \tag{3.1.37}$$

The viscosity coefficients for electrons and ions have the form

$$\eta^i_0 = 0.96\, n_i T_i \tau_i \quad , \quad \eta^e_0 = 0.73\, n_e T_e \tau_e \quad . \tag{3.1.38}$$

In magnetized plasma the oblique components of the viscosity tensor contain a small parameter $(\Omega_s\tau_s)^{-1}$ or $(\Omega_s\tau_s)^{-2}$); the transfer of momentum even along the magnetic field is reduced by a factor of $\Omega_s\tau_s$. However, should the plasma be compressed normal to the magnetic field, the transfer of momentum is about the same as in the absence of the field. The compression of the magnetic field raises the energy of the transverse motion of plasma particles, which then is evenly redistributed through collisions. Hence, the diagonal components of the stress tensor are of the same order as in the absence of magnetic field. For unidimensional motion

$$\Pi^s_{xx} = \Pi^s_{yy} = \Pi^s_{zz} = (\eta^s_0/3)\,\frac{\partial u^s_x}{\partial x} \quad . \tag{3.1.39}$$

The heat flux of electrons, like the force of friction, is composed of two terms, pertaining to the relative motion of plasma components and the electron temperature gradient:

$$\mathbf{q}_e = \mathbf{q}_{eu} + \mathbf{q}_{eT} \quad . \tag{3.1.40}$$

The heat flux of ions depends only on the ion temperature gradient. For non-magnetized plasma

$$\mathbf{q}_{eu} = 0.71\, n_e T_e \mathbf{U} \quad , \qquad \mathbf{q}_{eT} = -\kappa_{e\parallel} \nabla T_e \quad , \tag{3.1.41}$$

$$\mathbf{q}_i = -\kappa_{i\parallel} \nabla T_i \quad , \qquad \text{where} \tag{3.1.42}$$

$$\kappa_{e_\parallel} = 3.16\, n_e T_e \tau_e / m_e \quad , \qquad \kappa_{i_\parallel} = 3.9\, n_i T_i \tau_i / m_i \quad . \tag{3.1.43}$$

For magnetized plasma

$$\mathbf{q}_{eu} = 0.71 n_e T_e \mathbf{U} + \frac{3}{2}\,(n_e T_e / \Omega_e \tau_e)\mathbf{b} \times \mathbf{U} \quad , \tag{3.1.44}$$

$$\mathbf{q}_{eT} = -\,\kappa_{e_\parallel} \nabla_\parallel T_e - \kappa_{e_\perp} \nabla_\perp T_e - \frac{5}{2}\,(n_e T_e / eB)\mathbf{b} \times \nabla T_e \quad , \tag{3.1.45}$$

$$\mathbf{q}_i = -\,\kappa_{i_\parallel} \nabla_\parallel T_i - \kappa_{i_\perp} \nabla_\perp T_i + \frac{5}{2}\,(n_i T_i / eB)\mathbf{b} \times \nabla T_i \quad , \tag{3.1.46}$$

where

$$\kappa_{e_\perp} = 4.66 n_e T_e \tau_e / m_e (\Omega_e \tau_e)^2 \quad ,$$
$$\kappa_{i_\perp} = 2 n_i T_i \tau_i / m_i (\Omega_i \tau_i)^2 \quad . \tag{3.1.47}$$

The amount of heat acquired by ions in collisions with electrons is given by

$$Q_i = 3\,\frac{m_e}{m_i}\,\frac{n_e}{\tau_e}\,(T_e - T_i) \quad . \tag{3.1.48}$$

The heat acquired by electrons is

$$Q_e = -\,Q_i - \mathbf{RU} \quad . \tag{3.1.49}$$

To illustrate the use of the two-fluid transfer equations, let us employ them to derive Ohm's law. We neglect the inertia of electrons and the charge separation in the plasma, i.e., set the left-hand side of the equation of motion of electrons (3.1.26) equal to zero, and $n_e = n_i = n$. Then for $\Omega_e \tau_e \gg 1$,

(3.1.26) can be transformed into

$$\mathbf{j} = \sigma_{\parallel}\mathbf{E}'_{\parallel} + \frac{\sigma_{\perp}}{1 + (\Omega_e\zeta_e)^2}(\mathbf{E}'_{\perp} + \Omega_e\tau_e\mathbf{b}\times\mathbf{E}'_{\perp}) \quad , \qquad (3.1.50)$$

where

$$\mathbf{E}' = \mathbf{E} + \mathbf{u}^i\times\mathbf{B} + (\nabla p_e - \mathbf{R}_T)/en \quad , \qquad (3.1.51)$$

$$\sigma_{\perp} = e^2 n\tau_e/m_e \quad , \qquad \sigma_{\parallel} = 1.96\sigma_{\perp} \quad . \qquad (3.1.52)$$

In the limit $\Omega_e\tau_e \ll 1$, in (3.1.50) $\sigma_{\perp}$ must be replaced by $\sigma_{\parallel}$. As follows from (3.1.50-52), Ohm's law in its most simple form (3.1.3) is useful only for a plasma with nonmagnetized electrons, and then only for some definite directions of the current, which bring the last term on the right-hand side of (3.1.51) to zero.

3.2 Magnetohydrodynamic Waves

3.2.1 Linear MHD Waves

Consider the propagation of small perturbations in a quiescent homogeneous ionized gas in the constant magnetic field $\mathbf{B}_0$. As indicated in Sect.3.1.1, the study of sufficiently large-scale and low-frequency perturbations, i.e., $k\to 0$, $\omega\to 0$, may be based in the MHD equations of an ideal medium.

Like a common gas (Sect.1.4.1), a conducting fluid may display entropy waves, which carry perturbations of entropy S and density ρ at constant pressure p. Entropy waves do not propagate; they are carried along by the fluid. In the unidimensional case the entropy wave appears as an arbitrary stationary profile $\rho = \rho_0 + \rho'(x)$, $S = S_0 + S'(x)$, $|\rho'| \ll \rho_0$, at $p(x) = \text{const}$. With other kinds of linear waves in the adopted adiabatic approximation the perturbations of pressure p' and density ρ' are coupled by

$$p' = a^2\rho' \quad . \qquad (3.2.1)$$

Now we linearize the MHD equations, assuming the perturbations of velocity $\mathbf{u}$, density ρ' and magnetic field $\mathbf{B}'$ to be of the first order of smallness: $\rho = \rho_0 + \rho'$, $\mathbf{B} = \mathbf{B}_0 + \mathbf{B}'$. We shall consider the plane monochromatic waves, where all the perturbations are proportional to $\exp(-i\omega t + i\mathbf{kr})$. The coordinate system is chosen so that vector $\mathbf{k}$ is aligned with the x axis, and the unperturbed magnetic field lies in the (x,y) plane. Then from (3.1.1) we get $kB'_x = 0$, i.e., only the magnetic field components normal to the direction of propagation of the wave are perturbed. Introducing the phase velocity $\tilde{c} = \omega/k$, from the continuity equation (3.1.10) we find

$$\tilde{c}\rho' = \rho_0 u_x \quad , \tag{3.2.2}$$

from the induction equation (3.1.8)

$$\tilde{c}B'_y + B_{0x}u_y - B_{0y}u_x = 0 \quad , \tag{3.2.3}$$

$$\tilde{c}B'_z + B_{0x}u_z = 0 \quad , \tag{3.2.4}$$

and from the equation of motion (3.1.14)

$$\tilde{c}u_x - a^2\frac{\rho'}{\rho_0} - \frac{B_{0y}}{\mu_0\rho_0}B'_y = 0 \quad , \tag{3.2.5}$$

$$\tilde{c}u_y + \frac{B_{0x}}{\mu_0\rho_0}B'_y = 0 \quad , \tag{3.2.6}$$

$$\tilde{c}u_z + \frac{B_{0x}}{\mu_0\rho_0}B'_z = 0 \quad . \tag{3.2.7}$$

We see that the set (3.2.2-7) falls into two sets, one of which, (3.2.4,7), contains only the perturbations u_z and B'_z, while the other contains the perturbations u_x, u_y, ρ' and B'_y. It follows that these perturbations propagate independently. The condition of compatibility of (3.2.4,7) gives us

$$\tilde{c} = \pm c_a = \pm\frac{B_{0x}}{(\mu_0\rho_0)^{\frac{1}{2}}} \quad . \tag{3.2.8}$$

The waves carrying perturbations of the magnetic field normal to the direction of the constant magnetic field $\mathbf{B}_0$ are called Alfvén waves. Rewriting the dispersion equation (3.2.8) in the form

$$\omega = \pm\mathbf{k}\mathbf{B}_0/(\mu_0\rho_0)^{\frac{1}{2}} \quad , \tag{3.2.9}$$

we find that the group velocity of Alfvén waves

$$\mathbf{v}_g = \frac{\partial\omega}{\partial\mathbf{k}} = \pm\frac{\mathbf{B}_0}{(\mu_0\rho_0)^{\frac{1}{2}}} \tag{3.2.10}$$

does not depend on the wave vector and is always directed along the magnetic force lines. Alfvén waves may therefore be viewed as transverse vibrations of frozen-in magnetic force lines, considered as taut strings.

Now turn to the set of equations (3.2.2,3,5,6) for the perturbations u_x, u_y, ρ', B'_y. The condition of solvability of this set brings us to

$$(\tilde{c}^2 - a^2)(\tilde{c}^2 - c_a^2) - \tilde{c}^2 c_a'^2 = 0 \quad , \tag{3.2.11}$$

where $c_a' = B_{0y}/(\mu_0\rho_0)^{\frac{1}{2}}$. The larger and smaller (in terms of their absolute values) roots of (3.2.11) in $\tilde{c}^2$ give us, respectively, the squares of the velocities of the fast and slow magnetosonic waves c_f^2 and c_s^2. Obviously,

$$c_s \leqslant \min(a,c_a) \quad , \tag{3.2.12}$$

$$\min(a,c_a) \leqslant \max(a,c_a) \quad , \tag{3.2.13}$$

$$\max(a,c_a) \leqslant c_f \quad , \tag{3.2.14}$$

the equality signs in (3.2.12,14) corresponding to the propagation of waves along the magnetic field, when $B_{0y} = 0$, and in (3.2.13) to the case when, additionally, $a = c_a$, which at a given value of B_{0x} is realized only for one definite density $\rho_0 = \tilde{\rho}$. For $\rho_0 < \tilde{\rho}$ and $B_{0y} = 0$ $c_s = a$, $c_f = c_a$; the reverse holds for $\rho_0 > \tilde{\rho}$.

As follows from (3.2.11), the propagation of magnetosonic waves depends on both the elasticity of the medium and the elasticity of the magnetic field. Their relative importance depends on the quotient a^2/c_a^2, which (accurately up to a factor of the order of unity) coincides with the dimensionless parameter

$$\beta = \frac{2\mu_0 p_0}{B_0^2} \quad , \tag{3.2.15}$$

which expresses the ratio of thermal pressure to magnetic pressure.

Thermal pressure dominates for $\beta \gg 1$. In this case the fast magnetosonic wave becomes a common sonic wave ($c_f = a$), which, according to (3.2.5,6), carries the perturbations of the transverse velocity u_y and the magnetic field B_y' along with the density perturbations. The dispersion equation for the slow magnetosonic wave reduces to (3.2.8). In this extreme the Alfvén waves and the slow magnetosonic waves are similar to the gasdynamic vortex waves (Sect.1.4.1), which transfer the perturbations of the transverse velocity components. Owing to the existence of the initial magnetic field, these perturbations turn out to be associated with the magnetic-field perturbations, and their rate of propagation with respect to the medium is nonzero, if small compared to the speed of sound (as $\beta^{-\frac{1}{2}}$).

For $\beta \ll 1$ (cold plasma) the magnetic pressure dominates. The dispersion equation for fast magnetosonic waves yields

$$c_f = (c_a^2 + c_a'^2)^{\frac{1}{2}} \quad . \tag{3.2.16}$$

The group and phase velocities are equal, do not depend on k and are directed along the vector **k**, as in common sonic waves. The speed (3.2.16) can be viewed as the speed of sound $(\gamma p_0/\rho_0)^{1/2}$ in a medium dominated by the magnetic pressure $p_H = B^2/2\mu_0$, and the adiabatic exponent $\gamma = 2$, i.e., $p_H \propto \rho^2$, which is due to the freeze-in of magnetic field into the plasma, $B \propto \rho$, cf. (3.1.18). Such waves in a cold plasma are called magnetoacoustic waves. The dispersion equation for the slow magnetosonic waves in this limit has the form

$$c_s = a \cos\theta \quad , \tag{3.2.17}$$

where $\theta = \arctan(B_{0y}/B_{0x})$ is the angle between the original direction of magnetic force lines and the direction of propagation of the wave. By analogy with (3.2.8) the group velocity here is also $\mathbf{v}_g = a\mathbf{b}$, with $\mathbf{b} = \mathbf{B}_0/B_0$. In other words, the slow magnetosonic wave is a common sonic wave, propagated similarly to Alfvén's waves, only along the magnetic field. Such waves are called ion-acoustic waves.

Observe finally that the direction of propagation at right angles with the initial magnetic force lines ($\theta = \pi/2$) is singular for MHD waves. Here, as follows from (3.2.11),

$$c_f = (a^2 + c_a'^2)^{1/2} \quad , \tag{3.2.18}$$

whereas $c_a = 0$, $c_s = 0$. The nonpropagating Alfvén and the slow magnetosonic waves are then entirely similar to the common gasdynamic vortex waves. From (3.2.4,7) we find that two Alfvén waves, traveling in positive and negative directions, become two independent stationary profiles $u_z = u_z(x)$, $B_z' = B_z'(x)$. Accordingly, instead of two slow magnetosonic waves we come to deal with independent stationary profiles $u_y = u_y(x)$ and $\rho' = \rho'(x)$, $B_y' = B_y'(x)$ at $p(\rho) + B_y^2/8\pi = \mathrm{const}$.

3.2.2 Damping and Dispersion of Linear MHD Waves

The dispersion equations considered in Sect.3.2.1 represent the first terms of an expansion of $\omega(k)$ in small k. The account for effects described by finite spatial scales brings us to the expansion similar to (1.4.14):

$$\omega = \pm\tilde{c}k - i\mu k^2 \pm \gamma k^3 + \dots \quad . \tag{3.2.19}$$

Here $\tilde{c} > 0$ is one of the velocities c_a, c_f, c_s; $\mu > 0$ is the coefficient of damping; plus and minus signs relate to the direction of propagation of the wave. The last term in (3.2.19) accounts for the dependence of phase velocity on k (the effects of dispersion); γ is easily proved to be real-valued and may have either sign. As follows from (3.2.19), with small enough values of k the dispersion effects can be neglected as being small compared to the effects of dissipations. On the contrary, with finite values of k the dispersion phenomena may be considerable and even dominate over the dissipative effects due to both the running waves ($\omega \neq 0$) and the dissipative waves, which shape the shock structures ($\omega = 0$, Sect.1.7.2).

The explicit form of the expansion (3.2.19) can be derived in the same way as the first term of it in Sect.3.2.1, by linearizing either the dissipative MHD equations or the complete set of two-fluid transfer equations. The pertinent calculations can be found in [3.3]. If we confine ourselves to expanding (3.2.19) to second order in k, the damping coefficient μ can easily be shown to add up from the relevant coefficients, pertaining to all dissipative processes. In the MHD approximation for an Alfvén wave we get

$$\mu = \frac{1}{2}\left(\frac{1}{\mu_0\sigma} + \frac{\eta}{\rho_0}\right) \tag{3.2.20}$$

(since Alfvén waves do not carry over the perturbations of density and temperature, the heat conduction makes no contribution to damping). For the fast and slow magnetosonic waves

$$\mu_{f,s} = \frac{1}{2}\left\{(c_{f,s}^2 - c_a^2 \cos\theta)\left[\frac{\kappa c_s^2}{\rho_0}\left(\frac{1}{C_V} - \frac{1}{C_p}\right) + \frac{c_{f,s}^2}{\rho_0}\left(\frac{4}{3}\eta + \zeta\right)\right]\right.$$

$$\left. + (c_{f,s}^2 - c_s^2)\left(\frac{c_a^2}{\rho_0}\eta\cos\theta + \frac{c_{f,s}^2}{\mu_0\sigma}\right)\right\}(c_{f,s}^4 - c_a^2 c_s^2 \cos\theta)^{-1} \quad . \tag{3.2.21}$$

The contribution of Joule losses to the damping coefficient μ has the order of magnetic viscosity ν_m. The joint contribution of viscosity and heat conduction has the order of kinematic viscosity $\nu \cong v_{Ti}^2\tau_i$. Hence, we find that the relatively low damping of MHD waves due to viscosity and heat conduction is ensured by the inequality (1.2.16), which legitimizes the use of a "continuous" treatment (free path $v_{Ti}\tau_i$, the characteristic scale is the wavelength $L = 2\pi/k$). The damping due to Joule losses turns out to be small, as long as the magnetic field may be considered in a first approximation to be frozen-in, i.e.,

$$Rm \gg 1 \quad , \tag{3.2.22}$$

where in (3.1.20) $u = \tilde{c}$, $L = 2\pi/k$. Following the lines set down in Sect.1.5.2, we establish that the dissipative effects dominate over the nonlinear ones, and the linearization of MHD equations is justified if the dimensionless amplitude of the wave M_0, defined as the ratio of the amplitude of velocity oscillation in the wave to the velocity of the wave itself, satisfies — in addition to the inequality (1.5.21) — inequality

$$M_0 \ll 1/Rm \ll 1 \quad . \tag{3.2.23}$$

For assessing the characteristic scales of dispersion effects we can use the same procedure and find the third-order terms in k in (3.2.19). However, a simpler and more clear-cut approach consists in stating the exact form of the dispersion relation $\omega(k)$, neglecting all the dissipative terms in two-fluid transfer equations, then finding the limits of applicability of such an approximation. We shall not go into the details of wave dispersion in a plasma [3.4] and discuss only a few particular cases.

With transversely propagating magnetoacoustic waves ($\beta \ll 1$) the main physical cause of dispersion is the inertia of electrons. The dispersion equation has the form

$$\omega^2 = c_a'^2 k^2 \frac{\omega_{pe}^2}{\omega_{pe}^2 + c^2k^2} \quad , \tag{3.2.24}$$

where $\omega_{pe} = (e^2 n_0/\varepsilon_0 m_e)^{1/2}$ is the Langmuir frequency, and n_0 is the density of particles in the plasma. The characteristic scale of dispersion has the order of c/ω_{pe}. If this scale is large compared with the dissipative scales, then dispersion dominates over dissipation.

In a rarefied plasma in a strong magnetic field, for

$$\frac{B^2}{2\mu_0} \gg n_0 m_e c^2 \tag{3.2.25}$$

the mechanism of dispersion is grounded in the violation of plasma quasineutrality (separation of charges). Here the dispersion arises from electrostatic vibrations of ions with respect to electrons, which are glued to magnetic force lines (freeze-in condition). The dispersion equation has the form

$$\omega^2 = c_a'^2 k^2 \frac{\omega_{pi}^2}{\omega_{pi}^2 + c_a'^2 k^2} \quad , \tag{3.2.26}$$

where $\omega_{pi} = (e^2 n_0/\varepsilon_0 m_i)^{1/2}$. The characteristic scale of dispersion is c_a'/ω_{pi}. Providing that (3.2.25) is satisfied, this scale is superior to the scale c/ω_{pe}, and it is this dispersion mechanism that is dominant.

In an oblique wave ($\theta \neq \pi/2$) the dispersion is characterized by a much larger spatial scale. The physical mechanism of dispersion here is based on the Hall effect. For the near-transverse propagation ($|\theta - \pi/2| \ll 1$) the dispersion equation has the form

$$\omega^2 = c_a'^2 k^2 \left(1 + \frac{c_a'^2 k^2}{\Omega_i^2} \cos^2\theta\right) \quad . \tag{3.2.27}$$

The characteristic scale of dispersion is $c_a' \cos\theta/\Omega_i$, large compared to c/ω_{pe} as long as

$$|\theta - \pi/2| > (m_e/m_i)^{\frac{1}{2}} \quad . \tag{3.2.28}$$

3.2.3 Nonlinear Simple MHD Waves

Consider now the propagation of MHD perturbations of finite amplitude, having the form of simple Riemann waves. The simplest of these is the entropy wave; obviously, any profile

$$\rho = \rho(x - u_x t), \quad S = S(x - u_x t), \quad \mathbf{u} = \text{const}, \quad p = \text{const}, \quad \mathbf{B} = \text{const} \tag{3.2.29}$$

is an exact solution of the set of MHD equations of an ideal fluid.

For propagating simple waves all the variables do not depend on x and t separately, but rather on a certain combination of these $\eta = \eta(x,t)$. The curve $\eta(x,t) = \text{const}$ in the plane (x,t) corresponds to the constant phase of the wave (constant values of all variables). Evidently, the speed of propagation of a fixed phase point is

$$\frac{dx}{dt} = -\frac{\partial\eta}{\partial t} \Big/ \frac{\partial\eta}{\partial x} \quad . \tag{3.2.30}$$

Let us express in (3.1.8,10,14) the derivatives with respect to x and t in terms of derivatives with respect to η. For the rate of propagation of the wave with respect to the fluid we introduce the designation $\tilde{c} = dx/dt - u_x$. Observing that, according to (3.1.1) and in a finite-amplitude wave, $B_x = \text{const} = B_{0x}$, we come to the set of equations

$$\tilde{c}\,\frac{d\rho}{d\eta} = \rho\,\frac{du_x}{d\eta} \quad , \tag{3.2.31}$$

$$\tilde{c}\,\frac{dB_y}{d\eta} + B_{0x}\,\frac{du_y}{d\eta} - B_y\,\frac{du_x}{d\eta} = 0 \quad , \tag{3.2.32}$$

$$\tilde{c}\,\frac{dB_z}{d\eta} + B_{0x}\,\frac{du_z}{d\eta} - B_z\,\frac{du_x}{d\eta} = 0 \quad , \tag{3.2.33}$$

$$\tilde{c}\,\frac{du_x}{d\eta} - \frac{a^2}{\rho}\,\frac{d\rho}{d\eta} - \frac{1}{\mu_0\rho}\left(B_y\,\frac{dB_y}{d\eta} + B_z\,\frac{dB_z}{d\eta}\right) = 0 \quad , \tag{3.2.34}$$

$$\tilde{c}\,\frac{du_y}{d\eta} + \frac{B_{0x}}{\mu_0\rho}\,\frac{dB_y}{d\eta} = 0 \quad , \tag{3.2.35}$$

$$\tilde{c}\,\frac{du_z}{d\eta} + \frac{B_{0x}}{\mu_0\rho}\,\frac{dB_z}{d\eta} = 0 \quad . \tag{3.2.36}$$

The set of homogeneous linear equations, in terms of derivatives over η of the variables ρ, u_x, u_y, u_z, B_y, B_z, (3.2.31-36), in each point (x,t) coincides with the set of equations in small perturbations of the same variables against a uniform background (3.2.2-7) (a perfect fit can be achieved by rotating the y,z axes). For this reason the solvability conditions are the same: they require that $\tilde{c}$ be equal to one of the values $\pm c_a$, $\pm c_f$, $\pm c_s$. A natural way of separating the simple waves into the Alfvén and the fast and slow magnetosonic waves exists. Let us emphasize that presently the characteristic velocities c_a, c_f, c_s are not fixed but rather are functions of η.

Consider first the simple Alfvén waves. Setting $\tilde{c}^2 = c_a^2 = B_{0x}^2/\mu_0\rho$, from (3.2.32,33,35,36) we find $du_x/d\eta = 0$. Hence, making use of (3.2.31,34-36), for simple waves propagating in the positive direction we get

$$u_x = \text{const}, \quad \rho = \text{const}, \quad B_y^2 + B_z^2 = \text{const}, \quad \tilde{c} = c_a = \text{const},$$
$$u_y + \frac{B_y}{(\mu_0\rho_0)^{\frac{1}{2}}} = \text{const}, \quad u_z + \frac{B_z}{(\mu_0\rho_0)^{\frac{1}{2}}} = \text{const} \quad . \tag{3.2.37}$$

As follows from (3.2.37), a coordinate system exists such that

$$\mathbf{u} = -\frac{\mathbf{B}}{(\mu_0\rho_0)^{\frac{1}{2}}} \,. \tag{3.2.38}$$

In this coordinate system $\mathbf{u} \times \mathbf{B} = 0$, i.e., according to (3.1.8) **B** does not depend on t, and the simple Alfvén wave is stationary. Its structure is represented by arbitrary stationary profiles of transverse-velocity components and magnetic-field components, which satisfy (3.2.37,38). This implies that the transverse velocity $u_\perp = (u_y^2 + u_z^2)^{\frac{1}{2}}$ and magnetic field $B_\perp = (B_y^2 + B_z^2)^{\frac{1}{2}}$ rotate without changing their absolute values (which is why simple Alfvén waves are also called rotational waves). For the variable η it would be natural to choose the angle of rotation $\varphi = \arctan(B_z/B_y)$, which may be an arbitrary function of a new coordinate x'; $\varphi = f(x')$. Going back to the old coordinate x we obtain the general expression for the simple Alfvén wave profile:

$$\varphi = f[x - (u_x + c_a)t] \quad . \tag{3.2.39}$$

Observe that all the particles of the fluid travel at the same speed u_x, and therefore the profile does not suffer any deformations which might result in breakdown.

Now turn to the fast and slow magnetosonic simple waves. Note that at $\tilde{c}^2 \neq c_a^2$ a simple wave can always be considered plane-polarized, setting $u_z = 0$, $B_z = 0$. Then (3.2.33,36) become identities. If $\tilde{c}$ satisfies (3.2.11), then for the variable η one may choose the density ρ and rewrite the set (3.2.31,32, 34,35) in the form

$$\frac{dB_y^2}{d\rho} = 2\mu_0(\tilde{c}^2 - a^2) \quad , \tag{3.2.40}$$

$$\frac{du_x}{d\rho} = \frac{\tilde{c}}{\rho} \quad , \tag{3.2.41}$$

$$\frac{du_y}{d\rho} = -\frac{(\tilde{c}^2 - a^2)B_{x0}}{\rho\tilde{c}B_y} \quad . \tag{3.2.42}$$

Hence we realize the sense of variations in simple magnetosonic waves, traveling in the positive direction ($\tilde{c} > 0$): B_y^2 increases (decreases) with ρ in the fast (slow) waves; u_x increases as ρ increases; in the fast waves u_y and B_y are antibatic at $B_{x0} > 0$ and symbatic at $B_{x0} < 0$; the converse is true for the slow waves.

Assuming the heat capacity of the medium to be constant, the flow adiabaticity gives $p \propto \rho^\gamma$. Then the analysis of the set of equations (3.2.40-42) is facilitated by the dimensionless variables $\xi = a^2/c_a^2 \propto p$, $q = \tilde{c}^2/a^2$. From (3.2.40) we derive

$$\frac{d}{d\xi}\frac{B_y^2}{B_{x0}^2} = \frac{2}{\gamma}(q - 1) \quad . \tag{3.2.43}$$

At the same time, (3.2.11) is equivalent to

$$\frac{B_y^2}{B_{x0}^2} = \frac{q - 1}{q}(\xi q - 1) \quad . \tag{3.2.44}$$

Substituting (3.2.44) into (3.2.43), we obtain the basic equation describing the properties of the fast and slow simple waves [3.5-7]:

$$\frac{dq}{d\xi} = \left(\frac{2}{\gamma} - 1\right) \frac{(1 - q)q^2}{1 - \xi q^2} . \qquad (3.2.45)$$

This equation is easily integrable for certain values of γ, including $\gamma = 5/3$, 4/3. However, a qualitative notion about the behavior of its solutions is readily gained by examining the integral curves of (3.2.45) in the (ξ,q) plane, where the quadrant $\xi > 0$, $q > 0$ corresponds to the simple waves running in the positive direction. The phase diagram of this equation is reproduced in Fig.3.1; it shows no considerable change with $1 < \gamma < 2$.

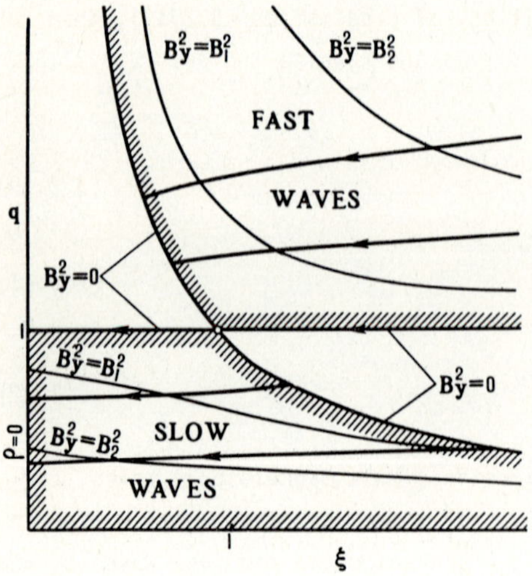

Fig.3.1. Integral curves of (3.2.45), representing, in the (ξ,q) plane, the change in plasma state in the simple MHD expansion waves (arrows indicate the direction of variation)

The fast magnetosonic waves in the plane (ξ,q) correspond to the region restricted from below by the curve $B_y^2 = 0$, which, according to (3.2.44), incorporates the ray $q = 1$, $\xi > 1$ and the hyperbolic portion $q = 1/\xi$, $\xi < 1$; the integral curves in this region correspond to $\tilde{c} = c_f$ in (3.2.40-42) and in the definition of q. The region of the slow magnetosonic waves is restricted from above by the same curve (straight-line segment $q = 1$, $0 \leqslant \xi < 1$ and hyperbolic portion $q = 1/\xi$, $\xi > 1$), by the segment $\xi = 0$, $0 < q \leqslant 1$ (which corresponds to $\rho = 0$) and by the ray $q = 0$, $\xi > 0$ (which corresponds to the limit $B_y^2/B_{x0}^2 \to \infty$) from below; here $\tilde{c} = c_s$. Curves $B_y^2 = \text{const}$ are plotted for two nonzero values B_1^2 and $B_2^2 > B_1^2$. Lines $\rho = \text{const}$ correspond to the verticals $\xi = \text{const}$.

Note that in both these regions the right-hand side of (3.2.45) is positive, i.e., $dq/d\xi > 0$. Hence with the aid of (3.2.41) we find at once that for the simple magnetosonic waves

$$\frac{d}{d\rho}(\tilde{c} + u_x) > 0 \quad . \tag{3.2.46}$$

This means that the compressed portions in such waves race ahead, which distorts the wave profile, and ultimately leads to a breakdown or a shock. As in conventional gasdynamics, discontinuities inevitably arise in simple compression waves, and thus centered simple compression waves do not exist. Further, we shall be concerned mainly with simple magnetosonic expansion waves, and the arrows on the integral curves in Fig.3.1 indicate the direction of variation of ξ and ρ for this type of wave.

From Fig.3.1 it is clear that in the slow expansion waves the density may drop to zero with simultaneous increase in B_y^2. A special case is represented by the slow expansion wave with $B_y = 0$, in which the density starts going down from values below $\tilde{\rho}$. From (3.2.44,45) we easily get

$$\frac{d}{d\xi}\left(\frac{B_y^2}{2\mu_0} + p\right) = \frac{B_{x0}^2}{\mu_0 \gamma} q \quad , \tag{3.2.47}$$

i.e., with small q (large B_y/B_{x0}) the sum of magnetic and thermal pressure in a slow expansion wave is approximately constant.

In fast expansion waves the magnetic field drops to zero, while the density decreases to nonzero values below $\tilde{\rho}$, i.e., the zero-density region cannot be formed behind a fast expansion wave. A special case is represented by the fast expansion wave with $B_y = 0$, in which the density decreases to $\rho = \tilde{\rho}$.

The singular point of (3.2.45) in the (ξ,q) plane $\xi = 1$, $q = 1$, i.e., $B_y = 0$, $\rho = \tilde{\rho}$, corresponds to the unique condition in a gas with a magnetic field, when neither the fast nor the slow expansion waves can exist.

Substitution of (3.2.44,45) into (3.2.41,42) yields

$$\frac{du_x}{d\xi} = \frac{a_0}{\gamma \xi_0} (\xi/\xi_0)^{-(\gamma+1)/\gamma} q^{\frac{1}{2}} \quad , \tag{3.2.48}$$

$$\frac{du_y}{d\xi} = -\frac{a_0}{\gamma \xi_0} (\xi/\xi_0)^{-(\gamma+1)/2\gamma} \left(\frac{q-1}{q\xi - 1}\right)^{\frac{1}{2}} \quad , \tag{3.2.49}$$

where the subscript 0 denotes the starting point in the plane (ξ,q). By integrating these equations, u_x and u_y can be expressed in terms of ξ, i.e., in terms of ρ.

The spatial profiles of flow variables are found from

$$x = [c(\rho) + u_x(\rho)]t + F(\rho) \quad , \tag{3.2.50}$$

where the function $F(\rho)$ depends on the initial and boundary conditions; (3.2.50) implicitly defines $\rho(x,t)$, which makes it possible to express all the other variables via x and t. If the initial and boundary conditions do not contain length-like or time-like parameters, the function $F(\rho)$ in (3.2.50) must be identical to zero. Then (3.2.50) describes a centered expansion wave, in which ρ (and hence all the other variables) depend only on the ratio x/t. The centered MHD expansion waves, like their gasdynamic counterparts, divide the regions of constant flow; weak discontinuities, which represent their boundaries, travel with respect to the gas at the speeds of dominant magnetosonic waves. Between these limits the flow variables are defined by the solution of (3.2.44,45,48-50) and keep constant along any line $x/t = \text{const}$.

For simple MHD waves, like for linear waves, the direction of propagation strictly perpendicular to the magnetic field is singular. With fast magnetosonic waves this direction has nothing special about it, but for the slow and Alfvén waves the speeds $\tilde{c}$ turn to zero. Setting in (3.2.31-36) $B_{x0} = 0$, $\tilde{c} = 0$, in lieu of the two Alfvén and two slow magnetosonic simple waves traveling in the positive and negative directions, we get arbitrary stationary profiles

$$\rho = \rho(\tilde{x}), \quad u_y = u_y(\tilde{x}), \quad u_z = u_z(\tilde{x}), \quad B_y = B_y(\tilde{x}), \quad B_z = B_z(\tilde{x}) \tag{3.2.51}$$

($\tilde{x} = x - u_x t$), governed by the condition

$$p + \frac{B_y^2 + B_z^2}{2\mu_0} = \text{const} \quad , \qquad u_x = \text{const} \quad . \tag{3.2.52}$$

These profiles are entirely similar to the nonpropagating waves, which in conventional gasdynamics correspond to finite-amplitude vortex perturbations.

3.3 Discontinuities and Shock Waves in Magnetohydrodynamics

3.3.1 Classification of Discontinuities

Let us find out the boundary conditions which ought to be satisfied on a MHD discontinuity surface [3.1,8,9].

We use the coordinate system fixed on an element of a (moving) discontinuity surface and let the x axis run normal to this element, in the direction of material flow across the discontinuity, only if the latter is not at rest with respect to the particles of the fluid. Obviously, the flux of mass across the discontinuity surface is continuous, i.e.,

$$[\rho u_x] = 0 \quad . \tag{3.3.1}$$

The continuity of density of momentum flux is a corollary of the law of conservation of momentum:

$$[\Pi_{\alpha x}] = 0 \quad , \quad \alpha = x, y, z \quad . \tag{3.3.2}$$

Making use of (3.1.11), we rewrite (3.3.2) in the form of conditions of continuity of the normal and tangential components of the momentum flux:

$$\left[p + \rho u_x^2 + \frac{1}{2\mu_0}(\mathbf{B}_\perp^2 - B_x^2)\right] = 0 \quad , \tag{3.3.3}$$

$$\left[\rho u_x \mathbf{u}_\perp - \frac{1}{\mu_0} B_x \mathbf{B}_\perp\right] = 0 \quad . \tag{3.3.4}$$

Subscript $\perp$ here marks the transverse, i.e., lying in the (y,z) plane, components of vectors. Finally, from the condition of continuity of energy flux we obtain with the aid of (3.1.13)

$$\left[\rho u_x(w + \mathbf{u}^2/2) + \frac{1}{\mu_0}\left(u_x B^2 - B_x(\mathbf{uB})\right)\right] = 0 \quad . \tag{3.3.5}$$

In addition, continuous on the discontinuity surface are the normal component of the magnetic field and the tangential component of the electric field:

$$[B_x] = 0 \quad , \tag{3.3.6}$$

$$[\mathbf{u}_\perp B_x - u_x \mathbf{B}_\perp] = 0 \quad . \tag{3.3.7}$$

First of all, let us consider those discontinuities in which the flux of matter across the discontinuity surface is zero, $\rho u_x = 0$. In such discontinuities transport of matter occurs only parallel to the discontinuity surface. If the magnetic force lines do not run parallel to the discontinuity surface ($B_x \neq 0$), then from (3.3.3-7) it follows that the magnetic field, velocity and pressure are continuous, while a jump is observed in the density, temperature and entropy, called a contact discontinuity. The discontinuity surface is just an interface between two stationary media, which display different densities at the same pressure. Evidently, the contact discontinuity represents a particular case of a simple entropy wave, when the profiles $\rho(x)$ and $S(x)$, appearing in (3.2.29), are step-shaped.

In a special case when the magnetic field lies in the plane of the discontinuity ($B_x = 0$), (3.3.4-7) are satisfied identically. In such a discontinuity the jump is displayed by the density and tangential components of the velocity and magnetic field; at the same time $[p + B^2/2\mu_0] = 0$. By analogy with gasdynamics this type of discontinuity is called tangential. A tangential discontinuity represents a particular case of simple Alfvén and slow magnetosonic waves, nonpropagating in the transverse direction when the profiles of ρ, u_y, u_z, B_y, B_z in (3.2.51) are step-shaped.

Another type of discontinuity is characterized by nonzero flux of matter across the discontinuity surface, although the density jump is zero: $[\rho] = 0$ and thus $[u_x]=0$. The velocity of propagation of such discontinuities can be found from the condition of existence of the nontrivial solution of (3.3.4,7), whence

$$u_x = \frac{B_x}{(\mu_0\rho)^{\frac{1}{2}}} \quad . \tag{3.3.8}$$

We see that the rate of propagation of discontinuity equals the velocity of the Alfvén wave; for this reason such discontinuities are called Alfvén (rotational) discontinuities. An Alfvén discontinuity represents a special case of a simple Alfvén wave (3.2.29), when the function has the form of a step. Hence the conclusion immediately follows about the continuous nature of entropy and pressure in Alfvén discontinuities.

Finally, the discontinuities in which the flux of mass across the surface and the density jump $[\rho]$ are simultaneously nonzero represent the shock waves. Note that MHD shock waves, similarly to the fast and slow magnetosonic simple waves, are always plane-polarized, i.e., the vector **B** and the normal to the shock front on either side of the front lie in the same plane, say, in the (x,y) plane. Indeed, for $B_x \neq 0$ from (3.3.4,7) we get

$$\left[\mathbf{B}\left(\frac{B_x^2}{\mu_0} - \rho u_x^2\right)\right] = 0 \quad , \tag{3.3.9}$$

whence it follows that the tangential component of the magnetic field in the shock wave does not change its direction. If, however, $B_x = 0$ (transverse shock wave), the same conclusion follows from (3.3.7): $[u_x \mathbf{B}_\perp] = 0$. From (3.3.4) we find that the velocity vectors ($\mathbf{u}$) on either side of the front lie in the same plane (x,y).

Let us emphasize an essential difference between MHD shocks and discontinuities of other kinds. It would be meaningless to investigate the structure of contact, tangential to Alfvén discontinuities: they are postulated discontinuous by choosing stepwise functions in the profiles (3.2.29,39, and 51), respectively; such discontinuities are stationary in the approximation of a dissipation-free medium. Accounting for dissipations results in diffusional smearing (broadening proportional to $t^{\frac{1}{2}}$) of these discontinuity profiles. On the contrary, the shock waves remain stationary even when the dissipative (and the dispersive) mechanisms are taken into account, which shape the finite-width structures of these shocks. Can this be taken as a strong indication that of all types of discontinuities only the shock waves have a direct physical meaning? This point is discussed below in Sect.3.3.5.

3.3.2 Boundary Conditions and the Shock Adiabat in Magnetohydrodynamics

Let us pay more attention to the boundary conditions on the discontinuity surface of MHD shock waves. Taking advantage of the fact that such shocks are plane-polarized, we choose a shock-fixed coordinate system in which the vectors of velocity and magnetic field lie in the plane (x,y); in accordance with (3.1.4,9) the electric-field vector is directed along the z axis. Marking the quantities pertaining to the uniform upstream and downstream flows by subscripts 0 and k, respectively, we obtain from (3.3.1-7) the following set of boundary conditions:

$$\rho_0 u_{x0} = \rho_k u_{xk} \quad , \tag{3.3.10}$$

$$p_0 + \rho_0 u_{x0}^2 + \frac{B_{y0}^2}{2\mu_0} = p_k + \rho_k u_{xk}^2 \quad \frac{B_{yk}^2}{2\mu_0} \quad , \tag{3.3.11}$$

$$\rho_0 u_{x0} u_{y0} - \frac{B_x B_{y0}}{\mu_0} = \rho_k u_{xk} u_{yk} - \frac{B_x B_{yk}}{\mu_0} \quad , \tag{3.3.12}$$

$$\begin{aligned} &\rho_0 u_{x0}\left(w_0 + \frac{1}{2} u_{x0}^2\right) + u_{x0} \frac{B_{y0}^2}{\mu_0} - \frac{B_x^2}{4\mu_0^2 \rho_0 u_{x0}} B_{y0}^2 \\ &= \rho \; u_{xk}\left(w_k + \frac{1}{2} u_{xk}^2\right) + u_{xk} \frac{B_{yk}^2}{\mu_0} - \frac{B_x^2}{4\mu_0^2 \rho_k u_{xk}} B_{yk}^2 \quad , \end{aligned} \tag{3.3.13}$$

$$B_x u_{y0} - B_{y0} u_{x0} = B_x u_{yk} - B_{yk} u_{xk} \quad . \tag{3.3.14}$$

In (3.3.13) the transverse components of velocity are expressed with the aid of (3.3.12). Note that B_x = const.

To obtain the equation for the shock adiabat, we express enthalpy in (3.3.13) in terms of internal energy and transform (3.3.13) with the aid of (3.3.10,14). After some straightforward calculation we get the equation for a MHD shock adiabat in the form similar to the gasdynamic equation (1.6.23):

$$\varepsilon_k(p_k,\rho_k) - \varepsilon_0(p_0,\rho_0) = \frac{1}{2}(p_0 + p_k)(\rho_k - \rho_0)/\rho_0\rho_k$$

$$+ \frac{1}{4\mu_0}(\rho_k - \rho_0)(B_{yk} - B_{y0})^2/\rho_0\rho_k \quad . \qquad (3.3.15)$$

For the ideal-gas plasma with the adiabatic exponent γ we obtain the equation for the shock adiabat with the aid of (1.1.21) in the form similar to (1.6.24):

$$\frac{p_k}{p_0} = \frac{(\gamma + 1)\rho_k - (\gamma - 1)\rho_0}{(\gamma + 1)\rho_0 - (\gamma - 1)\rho_k} + \frac{\rho_k - \rho_0}{(\gamma + 1)\rho_0 - (\gamma - 1)\rho_k} \times \frac{B_{yk}^2 - B_{y0}^2}{2\mu_0 p_0} \quad . \qquad (3.3.16)$$

Equation (3.3.16) differs from (1.6.24) by the last term, which represents the contribution of magnetic pressure. If magnetic pressure is small compared to thermal pressure ($1 \ll \beta$), then (3.3.16) is reduced to (1.6.24).

To investigate the form of MHD shock adiabats thoroughly, we go to dimensionless variables (as in Sect.1.6.1)

$$\rho = \frac{\bar{\rho}}{\bar{\rho}_0} \ , \quad u_x = \frac{\bar{u}_x}{\bar{u}_{x0}} \ , \quad u_y = \frac{\bar{u}_y}{\bar{u}_{x0}} \ , \quad B_y = \frac{\bar{B}_y}{\bar{B}_x} \ , \quad T = \frac{\bar{T}}{\bar{T}_0} \quad . \qquad (3.3.17)$$

Here, as before, we retain the same letters for the respective dimensionless variables, marking the dimensional variables with bars. We also introduce the sonic Mach number M and the Alfvén Mach number M_a, defined as

$$M = \bar{u}_x/a \quad , \qquad M_a = \bar{u}_x/c_a \quad . \qquad (3.3.18)$$

We consider ρ, u_x, ..., M and M_a as independent variables. Their values in the unperturbed flow are known: $\rho = u_x = \ldots = T = 1$; $M = M_0$, $M_a = M_{a0}$. Our task consists in finding the downstream values of these variables: $\rho(k) = \bar{\rho}(k)/\bar{\rho}(0)$, etc.

The condition of continuity of mass flux (3.3.10) for dimensionless variables can be written in the form

$$\rho u_x = 1 \quad . \qquad (3.3.19)$$

With the aid of (3.3.19) we easily get

$$M = M_0 u_x T^{-\frac{1}{2}} \quad , \qquad (3.3.20)$$

$$M_a = M_{a0} u_x^{\frac{1}{2}} \quad . \qquad (3.3.21)$$

Eliminating ρ, u_y, T by virtue of (3.3.11,12,20) we obtain a relation defining a curve in the plane (u_x, B_y), at each point of which the flow variables are linked to the initial upstream state by the conservation laws, i.e., the flow is characterized by the same values of fluxes of mass, momentum and energy:

$$F(u_x,B_y) \equiv \frac{(\gamma+1)}{(\gamma-1)}(u_x-1)(u_x-u_{GD}) + \frac{\gamma}{(\gamma-1)} u_x \frac{B_y^2 - B_{y0}^2}{M_{a0}^2}$$

$$- \frac{(B_y - B_{y0})^2}{M_{a0}^4} - \frac{2B_{y0}}{M_{a0}^2}(B_y - B_{y0}) = 0 \quad , \qquad (3.3.22)$$

where u_{GD} is defined by (1.8.8); $B_{y0} = \bar{B}_{y0}/\bar{B}_x = \tan\theta$ is the slant with respect to the upstream shock plane.

The downstream state is defined by the intercept of curve (3.3.22) with the hyperbola (3.3.14) whose equation, in dimensionless variables, is

$$B_y\left(u_x - \frac{1}{M_{a0}^2}\right) = B_{y0}\left(1 - \frac{1}{M_{a0}^2}\right) \quad . \qquad (3.3.23)$$

With the aid of (3.3.12) it is easy to prove that (3.3.23) expresses the condition of zero electric field in the fluid-fixed coordinate system ($\mathbf{u} \times \mathbf{B} = 0$), which has to be satisfied in the equilibrium states on either side of the shock ("zero field hyperbola"). As follows from (3.3.21), one of the branches of this hyperbola corresponds to a super-Alfvén flow ($M_a > 1$), while the other pertains to a sub-Alfvén flow ($M_a < 1$). Expressing B_y via u_x with the aid of (3.3.23) and substituting it into (3.3.22), we obtain a fourth-order equation in u_x, one of the roots being $u_x = 1$ and the remaining roots describing the possible shock transitions from the given initial state. Simple algebraic transformations demonstrate that two roots of this equation correspond to the intercepts of curve (3.3.22) with the super-Alfvén branch of (3.3.23), and the other two roots correspond to intercepts with the sub-Alfvén branch. Obviously, these pairs of roots may be nonexistent in the range of real values, or they may merge with one another in degenerate cases. But each pair vanishes or merges independently on its side of the straight line $u_x = 1/M_{a0}^2$, which separates the super-Alfvén region and the sub-Alfvén region in the plane (u_x,B_y). In particular, since one real-valued root $u_x = 1$ obviously exists, another root in the same region of the plane (u_x,B_y) must also exist.

Figure 3.2 shows the curve (3.3.22) and hyperbola (3.3.23) when all four roots exist and have different values. Let us enumerate them in order of decreasing velocity: points 1 and 2 are super-Alfvén, and points 3 and 4 are sub-Alfvén. To establish the relationship of velocities in these points with the velocities of magnetosonic waves, we divide (3.2.11) by $\bar{u}_x^{-2}$, denote the

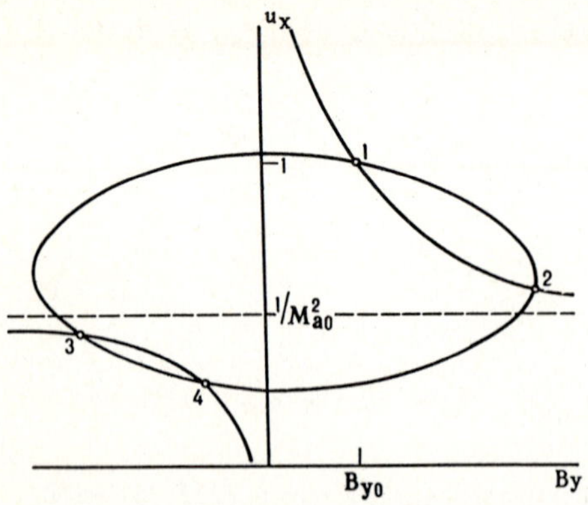

Fig.3.2. Intersections of closed curve (3.3.22) with the two branches of zero-field hyperbola (3.3.23) define the singular points 1-4, representing the states upstream and downstream of MHD shock waves

magnetosonic Mach number by $\tilde{M} = \bar{u}_x/\tilde{c}$, $\tilde{c} = c_f$ or c_s, and set $\tilde{M} = 1$, thus defining a curve in the plane (M^2, M_a^2), which corresponds to the equality of magnetosonic velocity and flow velocity

$$\left(1 - \frac{1}{M^2}\right)\left(1 - \frac{1}{M_a^2}\right) = B_y^2 \quad . \tag{3.3.24}$$

Hence the inequalities $\bar{u}_x > c_f$ and $\bar{u}_x > c_s$ are readily demonstrated to be equivalent to

$$M_a^2 > 1 + \frac{B_y^2 M^2}{M^2 - 1} \quad , \tag{3.3.25}$$

the conditions $\bar{u}_x > c_f$ and $\bar{u}_x > c_s$ being satisfied, respectively, for the values of M_a^2 lying in the plane (M^2, M_a^2) above the upper or the lower branch of the hyperbola, appearing on the right-hand side of (3.3.25).

Denote by u_F' and u_B' the derivatives of u_x with respect to B_y along the curves (3.3.22,23), respectively. In the point $u_x = 1$

$$u_F' = -\frac{B_{y0} M_0^2}{M_{a0}^2 (M_0^2 - 1)} \quad , \quad u_B' = -\frac{M_{a0}^2 - 1}{B_{y0} M_{a0}^2} \quad . \tag{3.3.26}$$

Choose for the starting point 0 the point 1 (Fig.3.2). As follows from the diagram, $0 > u_F' > u_B'$, whence with the aid of (3.3.25) we immediately find

$$c_f(1) \leqslant \bar{u}_x(1) \quad . \tag{3.3.27}$$

Bringing point 0 into coincidence with the points 2,3,4 in succession, we find that

$$c_a(2) \leqslant \bar{u}_x(2) \leqslant c_f(2) \quad , \tag{3.3.28}$$

$$c_s(3) \leq \bar{u}_x(3) \leq c_a(3) \quad , \tag{3.3.29}$$

$$\bar{u}_x(4) \leq c_s(4) \quad . \tag{3.3.30}$$

On the basis of (3.2.12,14) we conclude that point 1 is always supersonic [$M(1) > 1$], and point 4 subsonic [$M(4) < 1$]; points 2 and 3 may be either super- or subsonic.

Examining the change in the entropy along the curve (3.3.22), e.g., by comparing u'_F with the derivative of u_x over B_y along the isentrope $S(u_x, B_y) = \text{const}$, we can demonstrate similarly that

$$S(1) \leq S(2) \leq S(3) \leq S(4) \quad , \tag{3.3.31}$$

whence

$$T(1) \leq T(2) \leq T(3) \leq T(4) \tag{3.3.32}$$

(equality signs in (3.3.27-32) correspond to conceivable degenerate cases and bear no relation to Fig.3.2). A rigorous derivation of (3.3.27-32) can be found in [3.10-12].

Point k, which describes the state downstream, is one of the points 1-4; the number of point k must be higher than the number of point 0. The condition of increasing entropy is satisfied by six types of shock transitions: $1\to2$, $1\to3$, $1\to4$, $2\to3$, $2\to4$, $3\to4$. Below in Sect.3.3.3, we shall demonstrate that only two of these ($1\to2$ and $3\to4$) are evolutionary and physically realizable. In magnetohydrodynamics the distortion of profiles, resulting in discontinuities, is exhibited only by the fast and slow magnetosonic simple compression waves. In the resulting discontinuities the flow must go from the supersonic to the subsonic regime with respect to one of the characteristic velocities c_f, c_s. This condition is satisfied with the shock transitions $1\to2$ and $3\to4$. The corresponding MHD shock waves are appropriately called fast and slow.

Figure 3.3 depicts the regions in the plane (M^2, M_a^2) corresponding to the fast and slow MHD shock waves, respectively, for a fixed angle between the initial direction of magnetic field and the shock plane. Insofar as $M_a^2/M^2 = \gamma p_0 \mu_0 / B_x^2 = \text{const}$, a fixed initial upstream state of the medium corresponds to a definite straight ray, emerging from the origin of coordinates in the plane (M^2, M_a^2). The increase in the shock velocity corresponds to the motion along this ray in the sense of increasing Mach numbers. The fast and slow shock-wave regions are restricted by the inequalities (3.3.27-30), see also (3.3.25).

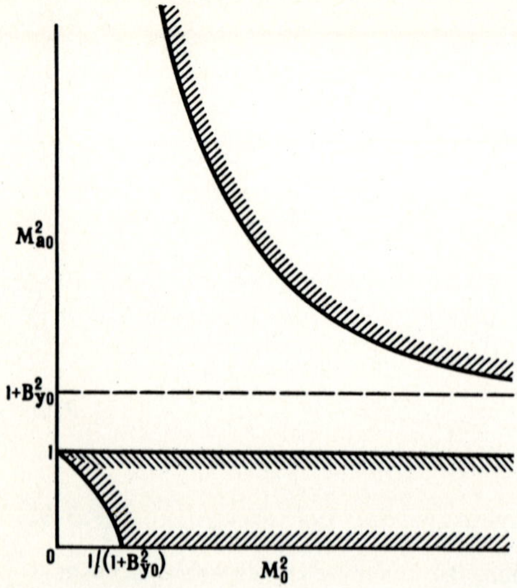

Fig.3.3. The region of existence of fast MHD shocks in the (M_0^2, M_{a0}^2) plane with fixed B_{y0} is restricted, from below, by the upper branch of hyperbola (3.3.25). The region of existence of slow MHD shocks is restricted, from below, by a portion of the lower branch of the same hyperbola and the straight line $M_{a0} = 0$, and from above by the straight line $M_{a0} = 1$

From Fig.3.3 it is clear that the strength of the slow shock waves is restricted, since their velocity obviously cannot rise above the Alfvēn velocity; on the contrary, the speed of a fast shock wave can be arbitrarily high. On the basis of (3.3.22,23) it is easy to show that with infinitely growing intensity the compression asymptotically approaches the limit of compression for the gasdynamic shock wave $(\gamma+1)/(\gamma-1)$. The relative magnetic-field variations in fast MHD shock waves always have the order of unity. The curves $u_x(k)$, $B_y(k)$, $T(k)$ downstream of the shock in the fluid with $\gamma = 5/3$ are plotted (as functions of M_3^2) in Fig.3.4 for the slow shock waves (here $k=4$) and as functions of M_1^2 in Fig.3.5 for the fast shock waves (here $k=2$) for two values of the angle of inclination of magnetic field: $\theta = 10^\circ$ (solid lines) and $\theta = 45^\circ$ (dashed lines). From Figs.3.4,5 it is clear that in both types of shocks compression and heating of the plasma occurs. The variations of magnetic field in the fast and slow shock waves (as in the respective simple waves) have opposing directions: in the fast waves B_y increases, while in the slow waves it decreases. The jumps of all variables across the shock can easily be expressed in terms of $u_x(k)$.

Consider, for instance, a shock wave traveling along the magnetic field (normal shock), $B_{y0} = 0$. In this case the hyperbola (3.3.23) degenerates into a pair of straight lines, $B_y = 0$ and $u_x = 1/M_{a0}^2$. If we assume $u_x(1) = 1$, then $u_x(2) = u_x(3) = 1/M_{a1}^2$; $u_x(4) = u_{GD}$. In this degenerate case the points 2 and 3 merge. The extreme case of the fast shock wave in the range $1 < M_{a1}^2 < u_{GD}^{-1}$ is

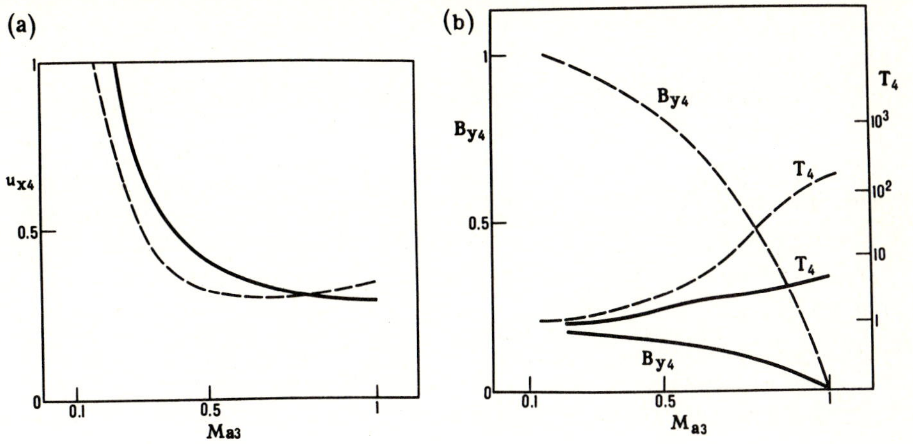

Fig.3.4a,b. Shock adiabats of slow MHD shock waves (*see text*)

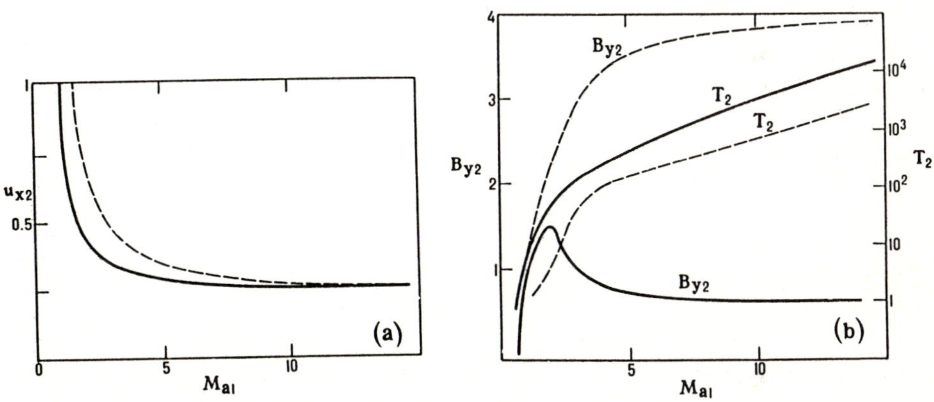

Fig.3.5a,b. Shock adiabats of fast MHD shock waves (*see text*)

the shock transition $1 \to 2 \equiv 3$, where the compression equals

$$u_x^{-1}(2 \equiv 3) = M_{a1}^2 \quad . \tag{3.3.33}$$

The transverse component of the magnetic field downstream becomes nonzero,

$$B_y^2(2 \equiv 3) = (\gamma + 1)(M_{a1}^2 - 1)(1 - M_{a1}^2 u_{GD}) \quad ; \tag{3.3.34}$$

accordingly, such shocks are called switch-on.

The same degenerate case is encountered when considering the strongest possible slow shock wave: the switch-off shock, for which $M_{a3} = 1$. Normalization to the downstream state 4, i.e., this time $u_x(4) = 1$, where the transverse magnetic field is switched off [$B_y(4) = 0$], enables one to employ the same expressions (3.3.33,34) with the transcription $1 \to 4$.

An important special case is represented by the transverse shocks with $B_x = 0$. Here the normalization procedure (3.3.17,18) cannot be applied directly; however, all the necessary expressions can be obtained through the transition to the limit ($B_{y0} \to \infty$, $M_{a0} \to \infty$), retaining the designations B_y and M_{a0} for the quantities B_y/B_{y0} and M_{a0}/B_{y0}, which in the transition to the limit remain finite [actually, now we have $B_y = \bar{B}_y/\bar{B}_{y0}$, $M_{a0} = \bar{u}_{x0}/c_a'$; see (3.2.11)]. From the discussion in Sect.3.2.3 it follows that the transverse MHD shock waves can only be fast, so the subscript 0 we replace by the subscript 1. The region of existence of these waves in the plane (M_1^2, M_{a1}^2) is found from (3.3.25)

$$M_{a1}^2 > \frac{M^2}{M^2 - 1} \quad , \tag{3.3.35}$$

which in the accepted notation is equivalent to $u_x(1) > c_f$. The equation in $u_x(2)$ is reduced to the quadratic equation

$$u_x^2 - u_x\left(u_{GD} + \frac{\gamma}{M_{a1}^2(\gamma + 1)}\right) - \frac{2 - \gamma}{M_{a1}^2(\gamma + 1)} = 0 \quad , \tag{3.3.36}$$

whose positive root is exactly the sought-for value of $u_x = u_x(2)$.

3.3.3 Evolutionarity Conditions for MHD Shock Waves

We shall now formulate the evolutionarity conditions for MHD shock waves on the basis of the general theory (Sect.1.7). It should be noted that since in the unidimensional case $B_x = \text{const}$ due to $\mathrm{div}\,\mathbf{B} = 0$, the flow is described by $n = 7$ variables: density ρ, entropy S, the three components of velocity $\mathbf{u}$, and the two transverse components of the magnetic field. For these variables there are $r = 7$ basic relations (3.3.1-7) on the discontinuity surface, which state the continuity of fluxes of mass of the three components of momentum and energy, and of the two tangential components of the electric field. In accordance with the results obtained in Sect.1.7.1, the equality $r = n$ assumes that for MHD shock waves the downstream state of the fluid is uniquely determined by the shock speed, which may take on any value within a certain range. The number of waves diverging from the shock for an evolutionary MHD shock wave ought to be equal to $s_1 + s_2 = r - 1 = 6$, according to (1.7.8). Insofar as for a uniform flow, corresponding to a MHD singularity of type k, an obvious relationship, see (3.3.27-30), exists

$$k = 1 + \begin{pmatrix}\text{the number of waves} \\ \text{whose velocities are} \\ \text{superior to velocity} \\ \text{of the flow}\end{pmatrix} \quad . \tag{3.3.37}$$

For the shock transition $k_1 \to k_2$ the number of waves diverging from the shock is

$$s_1 + s_2 = k_1 - 1 + 8 - k_2 = 7 + k_1 - k_2 \quad . \tag{3.3.38}$$

We see that the evolutionary shock transitions can be only the transitions $1 \to 2$, $2 \to 3$, $3 \to 4$. In fact, however, as we have already mentioned, it is only the transitions $1 \to 2$ (fast MHD shock wave) and $3 \to 4$ (slow MHD shock wave) that are realized in the evolutionary shocks, whereas the transitions $2 \to 3$ (called the trans-Alfvén MHD shock) are not evolutionary. The need for a more detailed consideration here arises from the fact that the matrix A of coefficients for the unknown amplitudes of outgoing waves in this case falls apart into separate blocks, and the fulfillment of the necessary condition of evolutionarity (1.7.8) must be checked individually for each group of variables pertaining to the same block, see remark following (1.7.8).

Indeed, in the coordinate system defined in Sect.3.3.2 we are dealing with the following situation. The entropy wave transmits perturbations of ρ, S; the fast and the slow magnetosonic waves transmit perturbations of ρ, u_x, u_y, B_y; the Alfvén wave transmits perturbations of u_z, B_z. Let us denote the entropy perturbations and those of the magnetic field in the fast and the slow magnetosonic waves and in Alfvén wave in the homogeneous flows upstream and downstream by S_ℓ^0, $F_\ell^\pm$, $S_\ell^\pm$, $A_\ell^\pm$, respectively. Here the values of the subscript $\ell = 0$ and $\ell = k$ correspond to the waves traveling ahead of and behind the shock, respectively; the superscripts "+" and "-" pertain to the propagation of the waves downstream and upstream, respectively. The perturbation of every variable on either side of the shock can be presented in the form of a linear combination of wave amplitudes:

$$\delta S_\ell = S_\ell^0 \quad , \tag{3.3.39}$$

$$\delta\rho_\ell = L_p(S_\ell^0, F_\ell^+, F_\ell^-, S_\ell^+, S_\ell^-) \quad , \tag{3.3.40}$$

$$\delta u_{x\ell} = L_x(F_\ell^+, F_\ell^-, S_\ell^+, S_\ell^-) \quad , \tag{3.3.41}$$

$$\delta u_{y\ell} = L_y(F_\ell^+, F_\ell^-, S_\ell^+, S_\ell^-) \quad , \tag{3.3.42}$$

$$\delta B_{y\ell} = F_\ell^+ - F_\ell^- + S_\ell^+ - S_\ell^- \quad , \tag{3.3.43}$$

$$\delta u_{z\ell} = L_z(A_\ell^+, A_\ell^-) \quad , \tag{3.3.44}$$

$$\delta B_{z\ell} = A_\ell^+ - A_\ell^- \quad . \tag{3.3.45}$$

Here L_p, L_x, L_y, L_z are linear quantics whose coefficients can easily be worked out with the aid of (3.2.2-7). However, we do not need them since it is sufficient to know that they are nonzero.

From the continuity equation we obtain (1.7.6), which couples the perturbation of the shock-front speed with the perturbation amplitudes of ρ and u_x on either side of the shock. Expressing the perturbation amplitudes in terms of wave amplitudes, we obtain

$$\delta D = L_D(S_0^0, S_k^0, F_0^+, F_0^-, F_k^+, F_k^-, S_0^+, S_0^-, S_k^+, S_k^-) \quad . \tag{3.3.46}$$

In the same fashion, making use of (3.3.46), we transform the relations between the perturbation amplitudes of variables to either side of the shock, which follow, like (1.7.6), from the boundary conditions, and the relations between amplitudes of propagating waves. It ought to be noted that it is only the relations between ρ, u_x, u_y, B_y, S that result from the continuity conditions of fluxes of the x and y components of momentum and energy and the z component of the electric field. The variables u_z and B_z enter the boundary conditions only quadratically, in the terms proportional to u_z^2 and B_z^2. Since in the adopted coordinate system $u_z = 0$, $B_z = 0$ both upstream and downstream, we may write

$$\delta u_z^2 = 2u_z \delta u_z = 0 \quad , \quad \delta B_z^2 = 0 \quad .$$

For this reason the said boundary conditions do not impose restrictions on the perturbed amplitudes of u_z and B_z. On the contrary, the continuity conditions of the z component of the momentum flux and the y component of electric field connect only the perturbed amplitudes of u_z and B_z.

Let us write the resulting relations for the amplitudes of different kinds of waves in shorthand. As far as we are concerned with the matrix of coefficients for the amplitudes of the outgoing waves, we assume the amplitudes of the incident waves to be zero, which in (3.3.39-45) enables the terms proportional to S_0^0, F_0^+, S_0^+, A_0^+, S_0^-, F_k^- to be dropped. All the remaining amplitudes in the shock transitions $1 \to 2$, $2 \to 4$, $3 \to 4$, may pertain to the outgoing waves (certainly, not all at once).

From the continuity conditions of the flux of the x components of momentum and energy we obtain the two relations

$$L_1(S_k^0, F_0^-, F_k^+, S_k^+, S_k^-) = 0 \quad , \tag{3.3.47}$$

$$L_2(S_k^0, F_0^-, F_k^+, S_k^+, S_k^-) = 0 \quad . \tag{3.3.48}$$

The continuity conditions of the y component of the flux of momentum and the z component of electric field give

$$L_3(F_0^-, F_k^+, S_k^+, S_k^-) = 0 \quad , \tag{3.3.49}$$

$$L_4(F_0^-, F_k^+, S_k^+, S_k^-) = 0 \quad . \tag{3.3.50}$$

Finally, from the conditions of continuity of the z component of the flux of momentum and y component of the electric field we get

$$L_5(A_0^-, A_k^+, A_k^-) = 0 \quad , \tag{3.3.51}$$

$$L_6(A_0^-, A_k^+, A_k^-) = 0 \quad . \tag{3.3.52}$$

Here, as before, $L_1, \ldots, L_6$ are independent linear quantities, whose exact form is irrelevant. Equation (3.3.47-52) illustrate the block structure of a matrix of coefficients for the unknown amplitudes of outgoing waves: (3.3.47-50) describe a 4×4 block, and (3.3.51,52) describe a 2×2 block. This implies that out of the seven amplitudes S_k^0, F_0^-, S_0^-, F_k^+, F_k^-, S_k^+, S_k^- only four correspond to the outgoing waves, while out of the three amplitudes A_0^-, A_k^+, A_k^- only two pertain to the outgoing waves. The waves with amplitudes S_k^0, F_k^+, S_k^+, A_k^+ run downstream and thus are always outgoing. Accordingly, the outgoing waves must include one of the waves F_0^-, S_k^- and one of the waves A_0^-, A_k^-, i.e., exactly one magnetosonic and one Alfvén wave out of those traveling upstream. We see that on both sides of the evolutionary MHD shock three outgoing magnetosonic waves and two outgoing Alfvén waves must exist. Including the outgoing entropy wave, which always exists downstream, there must be six outgoing waves, just as required by the evolutionarity condition obtained earlier.

The requirement that one and only one of the upstream Alfvén waves be outgoing implies the nonevolutionarity of the trans-Alfvén shock transitions $2 \to 3$, in which both such waves are incident on the shock. It follows that (3.3.51,52) provides two relations for the amplitude of a single outgoing wave A_k^+. Overdetermination of the problem of small perturbation of the trans-Alfvén shock wave indicates that the response to a small perturbation is not small, i.e., that such a shock is unstable and subject to spontaneous splitting into several finite-amplitude discontinuities, Sect.1.7.1.

For the fast MHD shock (transition $1 \to 2$), as easily proved with (3.3.37), the outgoing waves are the slow magnetosonic wave with the amplitude S_k^- and Alfvén's wave with the amplitude A_k^- downstream. For the slow MHD shock (transition $3 \to 4$) the outgoing waves are the fast magnetosonic wave with the amplitude F_0^- and Alfvén wave with the amplitude A_0^- upstream. These shock waves are thus evolutionary. We see that out of twelve MHD shock transitions allowed by the conservation laws, six comply with the condition of increasing entropy at the shock, but the evolutionarity condition leaves only two types of actually existing MHD shock waves, the fast and slow waves [3.13-16].

3.3.4 Shock Structures in the MHD Approximation

According to the general results obtained in Sect.1.7.2, only the fast and slow MHD shock waves are physically realizable; they must always exhibit a uniquely determined structure. Let us illustrate this point by a simple example, corresponding to interesting extreme cases.

Let us write the shock-layer equation in the one-fluid MHD approximation in dimensionless variables:

$$u_x - 1 + \frac{1}{\gamma M_0^2}\left(\frac{T}{u_x} - 1\right) - \Delta_v \frac{du_x}{dx} + \frac{B_y^2 + B_z^2 - B_{y0}^2}{2M_{a0}^2} = 0 \quad , \tag{3.3.53}$$

$$u_y - \Delta_v' \frac{du_y}{dx} - \frac{B_y - B_{y0}}{M_{a0}^2} = 0 \quad , \tag{3.3.54}$$

$$u_z - \Delta_v' \frac{du_z}{dx} - \frac{B_z}{M_{a0}^2} = 0 \quad , \tag{3.3.55}$$

$$u_x^2 + u_y^2 - 1 + \frac{2}{M_0^2(\gamma - 1)}(T - 1) + \frac{2B_{y0}(B_y - B_{y0})}{M_{a0}^2}$$
$$- 2\Delta_v u_x \frac{du_x}{dx} - 2\Delta_v' u_y \frac{du_y}{dx} - 2\Delta_T \frac{dT}{dx} = 0 \quad , \tag{3.3.56}$$

$$\frac{dB_y}{dx} = \frac{1}{\Delta_j}(u_x B_y - B_{y0} - u_y) \quad , \tag{3.3.57}$$

$$\frac{dB_z}{dx} = \frac{1}{\Delta_j}(u_x B_z - u_z) \quad . \tag{3.3.58}$$

Here the difference in the characteristic scales of viscosity Δ_v and Δ_v' depends on the relative values of the first and second viscosity coefficients η and ζ; with $\zeta = 0$ we have $\Delta_v' = \Delta_v/3$. The scale Δ_T is defined as in Sect.1.8, and

$$\Delta_j = \frac{1}{\mu_0 \sigma u_{x0}}$$

is the characteristic scale of diffusion of the magnetic field, σ being the conductivity of the gas.

Since MHD shock waves are plane-polarized, in the adopted coordinate system $u_z = B_z = 0$ both upstream and downstream of the shock. We shall try a solution which is plane-polarized itself, i.e., set $u_z \equiv B_z \equiv 0$ across the entire shock. Then (3.3.55,58) are satisfied identically.

The general structure of the MHD shock described by (3.3.53-58) depends, in the first place, on the relation of characteristic scales of gaskinetic transfer processes Δ_v and Δ_T to the scale of diffusion of the magnetic field Δ_j. For $Pr \sim 1$ (the temperature conductivity χ is of the order of the kinematic viscosity ν)

$$\frac{\Delta_v}{\Delta_j} \simeq \frac{\Delta_T}{\Delta_j} \simeq \frac{\nu}{\nu_m} = Pm \quad . \tag{3.3.59}$$

Here ν_m is the magnetic viscosity, see (3.1.7), and Pm is the magnetic Prandtl number. We see that the general appearance of the shock structure depends on the magnitude of the dimensionless parameter Pm. In accordance with (3.1.38, 39,52), in a simple plasma (no matter whatever its magnetization) the kinematic viscosity is of the order of $T\tau_i/m_i$, the magnetic viscosity is of the order of $m_e/\mu_0 e^2 n\tau_e$, and the magnetic Prandtl number can be defined as the ratio of these two quantities:

$$Pm = \frac{\mu_0 e^2 nT\tau_e\tau_i}{m_e m_i} \simeq \frac{2.6 \cdot 10^{11}}{(\lambda/10)^2} \cdot \frac{T^4[eV]}{n[cm^{-3}]} \quad . \tag{3.3.60}$$

From (3.3.60) it is clear that Pm depends mainly on the plasma temperature. With plasma densities typical of the experiments with collision-dominated shock waves, Pm is small if $T < 10$ eV and large if $T > 10$ eV.

Let us consider the cases when Pm is either small or large.

If $Pm \ll 1$, then the Joule mechanism of energy dissipation dominates, and the medium [to an accuracy of O(Pm)] may be considered an ideal gas without viscosity and heat conduction, having a finite conductivity. Disregarding, on the scale Δ_j, the effects of viscosity and heat conduction, we may reduce the set (3.3.53-58) to just two equations in u_x and B_y: (3.3.22) and

$$\Delta_j \frac{dB_y}{dx} = \left[B_y\left(u_x - \frac{1}{M_{a0}^2}\right) - B_{y0}\left(1 - \frac{1}{M_{a0}^2}\right)\right] \quad . \tag{3.3.61}$$

The point representing the state of flowing plasma in the (u_x, B_y) plane moves along the curve (3.3.22) in accordance with (3.3.61). Curve (3.3.22) is quadratic in u_x, i.e., for every admissible value of $B_y = B$ there exist, generally speaking, two values of u_x for which $F(u_x, B) = 0$: $u_{x+}(B)$ and $u_{x-}(B)$, $u_{x-} < u_{x+}$. It is clearly evident that the parameters of gas flow corresponding to $B_y = B$, $u_x = u_{x+}$ and $B_y = B$, $u_x = u_{x-}$ are linked together by relationships which coincide exactly with the boundary conditions for the gasdynamic shock wave. Indeed, by virtue of $\mathbf{B}_\perp = \text{const}$ and $\mathbf{u}_\perp = \text{const}$ these conditions are reduced to continuity conditions of the fluxes of mass, momentum and energy. It can be demonstrated that the upper branch of curve (3.3.22) corresponds to the supersonic gas flow, while the lower branch pertains to the subsonic flow (Chap.4). For this reason, as long as $Pm \ll 1$, the shock structure described by (3.3.22,61) is likely to display internal discontinuities, gasdynamic jumps, in which for fixed $B_y = B$ the speed u_x switches from u_{x+} to u_{x-}. Inasmuch as $B_y = \text{const}$ in such jumps [to an accuracy of $O(Pm)$], they are called isomagnetic. Within the same limits of accuracy the structure of an isomagnetic jump coincides with the structure of a gasdynamic shock.

Figure 3.6 illustrates typical MHD shock structures described by (3.3.22, 61) in the (u_x, B_y) phase plane. Bold lines pertain to the curve (3.3.22), the arrows indicate the directions of motion along this curve in accordance with (3.3.61) (motion in the same direction increases the entropy of the gas

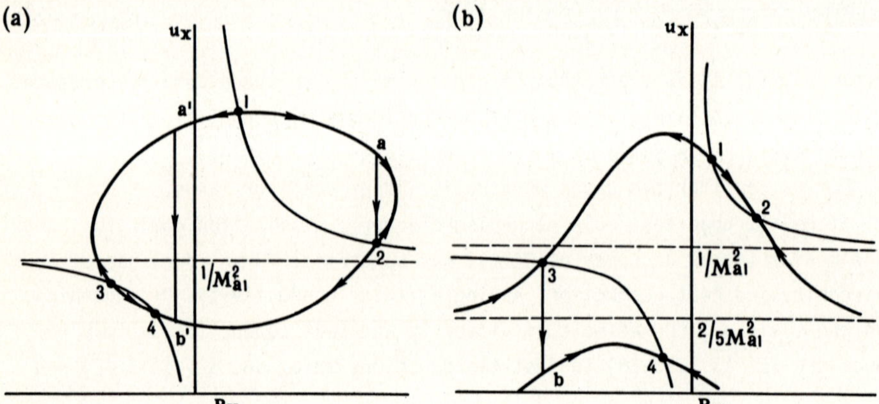

Fig.3.6a,b. Phase curves representing the structures of MHD shock waves in the (u_x, B_y) plane. For $Pm \ll 1$ they consist of resistive regions, corresponding to the movement along curve (3.3.22) as indicated by arrows, and isomagnetic subshocks (downward transitions along vertical arrows). (**a**): closed curve (3.3.22); (**b**): curve (3.3.22) consists of two branches

due to Joule dissipations). Hyperbolas with the asymptotes $B_y = 0$ and $u_x = 1/M_{a0}^2$ represent the zero-field hyperbola (3.2.23).

From Fig.3.6 it is clear that the structures of the fast and slow MHD shocks exist and are unique. The fast shock wave in Fig.3.6b has a purely resistive structure, shaped entirely by Joule dissipations. In Fig.3.6a the fast-shock wave structure is represented by the resistive region of the magnetic-field compression $1 \to a$, followed by the isomagnetic jump $a \to 2$. The structure of a slow MHD shock in Fig.3.6b starts with the isomagnetic jump $3 \to b$, followed by the resistive region of compression of magnetic field $b \to 4$. In Fig.3.6a the structure of the slow shock is purely resistive.

These curves illustrate the general appearance of the fast and slow MHD shock structures for $Pm \ll 1$. In particular, the isomagnetic jumps, if any, always appear in the head of a slow shock and in the tail of a fast shock. In fact, isomagnetic jumps arise if and only if they are evolutionary, which requires that the flow be supersonic ahead of the jump and subsonic behind the jump (Sect.1.7). As follows from Fig.3.6 (and can be proved rigorously), we cannot go over from the upper branch to the lower one by moving along the curve (3.3.22), as indicated by arrows (from supersonic to subsonic regimes): such a transition is possible only by means of an isomagnetic jump. By virtue of (3.3.27-30) point 1 is always supersonic, and point 4 subsonic. Therefore, the isomagnetic jump may occur in a fast shock wave when point 2 is subsonic (Fig.3.6a), and in a slow shock wave when point 3 is supersonic (Fig.3.6b). It can be demonstrated that the integral curves of (3.3.22,61) always come out of point 1 and enter point 4; they enter points 2 and 3 when these points are supersonic (Fig.3.6b) and come out of points 2 and 3 when they are subsonic (Fig.3.6a). Therefore the only possibility of entering the subsonic point 2, terminating the fast shock structure (Fig.3.6a), or coming out of the supersonic point 3, opening the slow shock structure (Fig.3.6b), consists in the isomagnetic jump from the upper (supersonic) to the lower (subsonic) branch of curve (3.3.22).

As follows from Fig.3.6, trans-Alfvén shock waves manifest no structure whatsoever.

For the shock transitions $1 \to 4$, $1 \to 3$, $2 \to 4$ the number of outgoing waves, according to (3.3.38), is less than required by the evolutionarity condition. The number of boundary conditions being fixed, this implies (Sect.1.7.2) that the corresponding structures, if any, are ambiguous. For instance, they are represented by the two-parameter set of structures of type $1 \to 4$ and one-parameter sets of structures of types $1 \to 3$ and $2 \to 4$. This circumstance is not reflected in Fig.3.6, which may create an illusion that the shock transition $1 \to 4$ corresponds to a one-parameter family of structures [$1 \to a' \to b' \to 4$, where point a' is any point in Fig.3.6a such that $B_y(4) < B_y(a') < B_y(1)$], whereas the shock transitions $1 \to 3$ and $2 \to 4$ seemingly display a uniquely defined structure, either purely resistive for $1 \to 3$ (Fig.3.6b) or for $2 \to 4$ (Fig.3.6a), starting with an isomagnetic jump ($2 \to 4$ in Fig.3.6b) or ending with an isomagnetic jump ($1 \to 3$ in Fig.3.6a). Actually, this false impression, which has found its way into some earlier works, is rooted in the simplifying assumption that $B_z = u_z \equiv 0$ throughout the entire structure. To obtain a correct picture we must admit that both transverse components of magnetic field are subject to variation within the limits of the shock. Then we get the proper families of structures, e.g., the transition $1 \to 3$ and $2 \to 4$ each correspond to a one-parameter set containing only one structure with $u_z = B_z \equiv 0$. These integral curves are portrayed in Chap.4.

In the other extreme ($Pm \gg 1$), the scales of viscosity and heat conduction are the largest, and the shock structure is shaped by these mechanisms. Accordingly, by virtue of (3.3.53-58), on the scale Δ_v to an accuracy of $O(Pm^{-1})$ the plasma is a perfect conductor. This means that in the (u_x, B_y) phase plane the point depicting the state of the gas moves along the zero-field hyperbola

$$B_y = B_{y0} \frac{1 - 1/M_{a0}^2}{u_x - 1/M_{a0}^2} \ , \qquad B_z = 0 \ . \tag{3.3.62}$$

The portions of hyperbola connect only two pairs of singular points: 1 with 2 and 3 with 4 (Fig.3.6). Consequently, only the fast and slow MHD shocks display uniquely determined structures. Trans-Alfvên shocks $2 \to 3$ and nonevolutionary shock transitions $1 \to 3$ and $2 \to 4$ in a high-conductivity gas ($Pm \gg 1$) do not show any structures.

Let us draw attention to the fact that the shocks with high (low) Pm are similar in nature to the shocks with high (low) Pr, (Sects.1.8.2,3). This similarity derives from the fact that the smallness of Pr or Pm indicates a relatively low viscosity; then the characteristic scales of heat conduction and Joule's heating, respectively, are the greatest. The Joule mechanism, however, is incapable of directly dissipating the momentum of the incident flow, and therefore cannot shape the strong-shock structures. Indeed, the density of the electromagnetic energy dissipated is of the order of $(B_k - B_0)^2/2\mu_0$, and, as indicated in Sect.3.3.2, the compression of magnetic field is of the order of unity, whereas the kinetic energy of the incident flow can be as high as desired. This is why the shock structures display discontinuities, or, to be more precise, narrow (with a width to the same order as Pr or Pm) isothermal or isomagnetic jumps, whose structures are shaped mainly by viscosity. In the other extreme (high Pr or Pm), the viscosity dominates throughout the entire shock, and there is no need to introduce internal discontinuities.

3.3.5 Evolutionarity of Singular MHD Shock Waves

Our discussion of the evolutionarity of MHD shocks omits the two extreme cases: the switch-on and the switch-off shock waves. Switch-on shock waves are realized by $B_{y1}=0$ in the range of Mach numbers $1 \leqslant M_{a1} \leqslant 2$ ($\gamma = 5/3$); $M_{a2}=1$ downstream (Sects.3.3.2,5.1). Switch-off shock waves can arise with any value of B_{y1}; in this case $M_{a1}=1$, Sects.3.3.2,6.1. In the classification of Sect.1.7 these shock waves are singular, inasmuch as the speed of the upstream flow in the switch-off shock wave and that of the downstream flow of the switch-on shock wave coincide with the local Alfvén speed (one of the characteristic velocities). The theory evolved in Sect.1.7 does not include the singular shock waves.

Obviously, the switch-on shock wave may be considered an extreme case of an evolutionary fast MHD shock for $1 \leqslant M_{a1} \leqslant 2$ and $B_{y1} \to 0$. The switch-off shock wave is an extreme case of a strong slow MHD shock for $M_{a1} \to 1$ and $B_{y1} = \mathrm{const}$. These circumstances impart physical meaning to the structures of the switch-on and switch-off waves, considered in Sects.3.5.6, i.e., evolutionary shock waves exist as close to these as desired. At the same time their evolutionarity cannot be checked by going to the limit, since this would imply an unavoidable violation of condition (1.7.3), which ensures the validity of the entire approach, see (1.7.4).

In view of this the early conclusions about nonevolutionarity of the switch-on and switch-off shock waves [3.12] are not substantiated.

The question of evolutionarity of singular MHD shocks had remained unsettled since the late 1950s, till the publication of [3.17] threw light on the subject. It was emphasized that inasmuch as in the singular shock waves the phase velocity of one of the waves with respect to the shock becomes zero, the dispersion equation in its simplest form

$$\omega = (\tilde{c} + u_x)k \tag{3.3.63}$$

is not valid. As a matter of fact, (3.3.63) represents the first term of the expansion of $\omega(k)$, see (3.2.19),

$$\omega = (\tilde{c} + u_x)k - \frac{1}{2}\, iDk^2 + \ldots \;, \qquad D > 0 \;, \tag{3.3.64}$$

which meets the needs as long as the inequalities (1.7.3) are satisfied. If $\tilde{c} = -u_x$, then the second inequality in (1.7.3) is incompatible with (3.3.63). The second term is to be considered, provided (1.7.3) is satisfied. In place of a single undamped wave with the characteristic speed $\tilde{c}$ we obtain two diffusive waves with the dispersion law

$$k_\pm = \pm\sqrt{\frac{\omega}{2D}}\,\frac{1 - i}{2} \;, \tag{3.3.65}$$

which is similar to (1.7.22). That the right-hand sides of (3.3.65) and (1.7.22) are complex conjugates is explained by the opposing directions of the flow in the two cases. Since $\mathrm{Im}\{k_+\} < 0$, $\mathrm{Im}\{k_-\} > 0$, one of these waves travels in the positive, and the other in the negative direction of the x axis. One of the two dissipative waves is "outgoing" with respect to the shock. This circumstance can be quite reasonably interpreted as diffusive smearing of the perturbation profile, which propagates at the phase velocity c with respect to the fluid, i.e., it is at rest with respect to the shock. The smearing, obviously enough, goes either way, both towards and away from the shock. It can be viewed as the propagation of waves, providing that the second inequality in (1.7.3) is satisfied, which requires that $\omega \ll D/\Delta^2$. If the shock front is shaped by the same dissipative mechanism responsible for damping the waves of the given type [the second term in (3.3.64)], and has a finite width $\Delta \sim D/\tilde{c}$, see (1.7.17), then this condition may be rewritten as

$$\omega \ll \tilde{c}^2/D \;. \tag{3.3.66}$$

Taking the diffusive waves into consideration, we may formulate the following rule: in a singular shock wave the undamped wave, whose phase velocity coincides with the characteristic velocity, corresponds to one outgoing diffusion wave.

Let us now apply this rule to singular MHD shock waves. Viscosity and finite conductivity taken into account, we come to the dispersion equation (3.3.65) with $2D = \nu + \nu_m$. For either type of singular shocks (switch-on and switch-off) we have, by virtue of the above-developed arguments, two outgoing Alfvén waves, one of which is the diffusive one (ahead of the front of the switch-off shock and behind the front of the switch-on shock, respectively). The total number of outgoing Alfvén waves thus satisfies the evolutionarity condition. However, having written explicitly the equations which connect amplitudes of the incident and the outgoing waves, we find that such a set of equations for the switch-on shock is degenerate: the equations are incompatible if the amplitude of the incident wave is nonzero, and have nontrivial solutions in the absence of incident waves. This implies that the switch-on shock wave is nonevolutionary: it is subject to splitting into a number of discontinuities and to spontaneously emitting Alfvén waves. The appropriate set of equations for the switch-off shock wave is nondegenerate, which implies evolutionarity of the shock [3.17,18].

The nature of splitting of a singular shock wave, nonevolutionary as far as the diffusion wave is concerned, has not been studied in detail. In this respect it would be natural to draw an analogy with the behavior of the contact, tangential, weak and rotational discontinuities, which (i) classify as singular, because their velocity with respect to the gas coincides with one of the characteristic velocities (the velocity of the entropy or Alfvén waves), and (ii) are thus nonevolutionary, as long as entropy waves or Alfvén waves are taken into account. In a one-dimensional problem such discontinuities display the diffusive kind of smearing, proportional to $\sqrt{t}$. Apparently, the decay of nonevolutionary singular shocks follows the same pattern: by gradually smearing out rather than by giving rise to other waves or discontinuities. Incidentally, such instability does not outlaw these types of shocks and discontinuities. For example, in the selfsimilar problem the distances between waves and discontinuities of various types grow in proportion with t, and so the relative width of regions occupied by the spreading discontinuities decreases as $t^{-\frac{1}{2}}$, i.e., in spite of the smearing, these discontinuities may still be considered as narrow. The pertinence of this argument to the on-switching shock waves can be rigorously proved only by a careful numerical computation, which will depend critically on the proper account for dissipations. The first computation of this kind, aimed at elucidating stability of the switch-on shock waves [3.19], indicates that the switch-on shock wave is stable when dissipations are small (in [3.19] the effect of dissipation is accounted for by a low numerical viscosity).
This result supplies an indirect proof of the assumption that the decay of the switch-on shock wave proceeds in a diffusive way due to dissipation.

3.4 Structures of Transverse Shocks

3.4.1 Boundary Conditions and the Shock Adiabat

The boundary conditions for the set of two-fluid transfer equations are most conveniently expressed in dimensionless variables. These variables here are defined as in Sect.2.1.1; definitions of the dimensionless magnetic field B_y and the Alfvén Mach number M_a for a transverse shock wave can be found in Sect.3.3.3. As demonstrated in Sect.3.3.3, the transverse shock wave realizes the fast-shock transition $1 \to 2$; accordingly, in this section the subscripts 1 and 2 pertain to the downstream and upstream states of the plasma, respectively. We confine ourselves to considering a simple plasma ($Z = 1$, $\gamma_i = \gamma_e = 5/3$). We choose a reference frame in which the transverse velocity of the plasma is zero both upstream and downstream. This option is guaranteed by virtue of (3.3.10,12) at $B_x = 0$.

Then for the upstream state ($x = -\infty$) we may write

$$u_x^i = u_x^e = n_i = n_e = T_e = T_i = B_y = 1 \quad ; \qquad E_x = \phi = 0 \quad ; \tag{3.4.1}$$

for the downstream state ($x = +\infty$), according to (3.3.36),

$$u_x^i = u_x^e = \frac{1}{8}\left\{1 + \frac{3}{M_1^2} + \frac{5}{2M_{a1}^2} + \left[\left(1 + \frac{3}{M_1^2} + \frac{5}{2M_{a1}^2}\right)^2 + \frac{8}{M_{a1}^2}\right]^{\frac{1}{2}}\right\} \equiv u_2 \quad . \tag{3.4.2}$$

Then from (3.3.10-14) we get

$$n_e = n_i = 1/u_2 \quad , \tag{3.4.3}$$

$$B_y = 1/u_2 \quad , \tag{3.4.4}$$

$$T_i = T_e = 1 + \frac{M_1^2}{3}(1 - u_2)\left(u_2 + 1 - \frac{3}{M_{a1}^2 u_2}\right) \equiv T_2 \quad . \tag{3.4.5}$$

From (2.1.7) follows that $E_x = 0$ downstream; the magnitude of the potential jump ϕ across the shock, as for a shock in nonmagnetized plasma (Sect.2.1.4), is not governed by the Rankine-Hugoniot conditions, but rather depends on the actual shock structure.

The equation of the shock adiabat in terms of pressure and relative volume has a simple form and is easily derived from (3.4.2-5):

$$\frac{\bar{p}_2}{\bar{p}_1} = \frac{1}{4u_2 - 1}\left(4 - u_2 + \frac{(1 - u_2)^3}{\beta_1 u_2^2}\right) \quad . \tag{3.4.6}$$

where $\beta_1 = 2\mu_0 \bar{p}_1 / \bar{B}_{y1}^2$.

The range of parameters M_1, M_{a1}, corresponding to the transverse shock waves, is defined by the inequality (3.3.35). From (3.3.35 and 4.2-5) we find

$$\frac{1}{4} < u_2 < 1 \quad , \quad 1 < T_2 \quad , \quad T_2/u_2 \quad , \quad T_2^{3/2}u_2 \quad , \quad T_2^2 u_2 < \infty \quad . \qquad (3.4.7)$$

The inequalities (3.4.7) indicate that in the transverse shock waves increases in the density of plasma (no more than fourfold), temperature, pressure, mean-free path length and time (indefinitely) are observed. In the limit of strong transverse shock waves ($M_1 \to \infty$, $M_{a1} \to \infty$) the compression of plasma, like that in gasdynamic shock waves, together with the compression of magnetic field, tends to 4.

As in Sect.2.1, here it may be convenient to normalize the variables to state 2 downstream. After the transcription $1 \leftrightarrow 2$ (3.4.1) and (3.4.2-5) will describe the downstream and upstream states, respectively. From the evolutionarity condition (3.3.28) [which is reduced to the inequality $\bar{u}_x(2) < c_f$, since $c_a = 0$ in the transverse direction], by analogy with (3.3.35) we find that the Mach numbers M_2 and M_{a2} downstream satisfy the inequality

$$M_{a2}^2 < \frac{M_2^2}{M_2^2 - 1} \quad . \qquad (3.4.8)$$

However, it is not the entire portion of the positive quadrant of the plane (M_2^2, M_{a2}^2) satisfying (3.4.8) that describes the possible states of plasma downstream of transverse shocks, just as not any value of $0 < M < 1$ may be assumed by the Mach number downstream of the gasdynamic shock (Sect.1.6.3). The region of interest, as illustrated in Fig.3.7, is restricted from below by a curve corresponding to the limit of strong shock waves ($M_1 \to \infty$, $T_2 \to \infty$ with a fixed finite-valued u_2). Parametric equation for this curve can easily

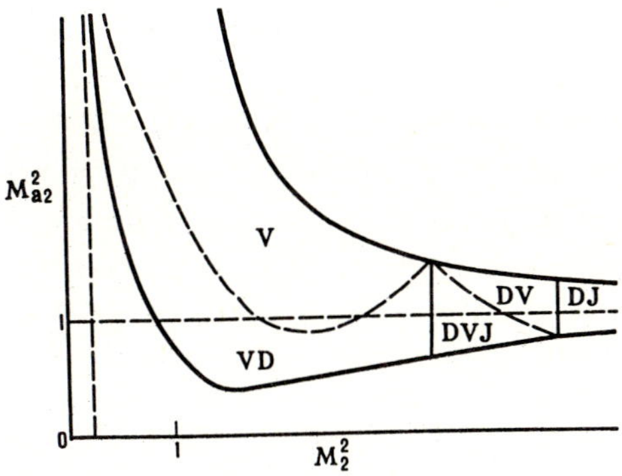

Fig.3.7. The region of existence of transverse shock waves in the (M_2^2, M_{a2}^2) plane (see text)

be deduced from (3.3.20,21 and 4.5) ($\frac{1}{4} < u_2 < 1$):

$$M_2^2 = \frac{3u_2^2(5u_2 + 1)}{5(1 - u_2)^3} , \qquad M_{a2}^2 = \frac{u_2^2(5u_2 + 1)}{2(4u_2 - 1)} . \tag{3.4.9}$$

As follows from (3.4.9), $1/5 < M_2^2 < \infty$, $2/5 < M_{a2}^2 < \infty$. In the limit of low magnetic pressure, i.e., when $M_{a2} \to \infty$, $\beta_2 = 6M_{a2}^2/5M_2^2 \to \infty$, this region becomes a finite interval $1/5 < M_2^2 < 1$, pertaining to shocks in gas or plasma without magnetic field. In the opposite extreme case of a cold plasma ($M_2 \to \infty$, $\beta_2 \to 0$) the interval of allowed values of M_{a2} is reduced to a point $M_{a2} = 1$. Then, even though the temperature jump across the shock may be large, the compression of plasma is about unity, since plasma is compressed together with magnetic field, and the thermal pressure is much smaller than the magnetic pressure.

3.4.2 Structure of Transverse Shock Waves in Magnetized Plasmas

The structures of transverse shock waves have been investigated by many researchers [3.20,21]. The earlier works [3.10,22-33] used more or less simplified idealized equations (MHD model and the like), which accounted for only some of the processes which shape the shock structure. The problem for a weak transverse shock wave with due account taken of all collisional mechanisms on the basis of 13-moment equations was studied in [3.34-36]. The structure of transverse shocks of arbitrary intensity was investigated with the aid of Navier-Stokes equations in [3.37-41]; here we shall chiefly base our discussion on these results. Let us start with the case of a magnetized plasma ($1 \ll \Omega_i\tau_i \ll \Omega_e\tau_e$).

We write the shock-layer equations directly in dimensionless variables, normalized to the equilibrium state of the plasma downstream (subscript 1). As in Sect.2.1.1, we employ $\varepsilon = (m_e/m_i)^{\frac{1}{2}}$, $\ell_1 = a_1\tau_{i1}$. We also introduce the designation

$$\delta_1 = 1/\Omega_{i1}\tau_{i1} , \tag{3.4.10}$$

where $\delta_1 \ll 1$ is a small parameter. Transverse components of electric field are assumed constant within the shock front, and in the chosen coordinate system $\bar{E}_y = 0$, $\bar{E}_z = -\bar{u}_{x1}\bar{B}_{y1}$. In the Poisson equation, as well as in other equations, the dimensionless concentrations of ions (n_i) and electrons (n_e) are expressed in terms of the dimensionless velocities u_x^i, u_x^e according to the continuity equation (2.1.14):

$$\frac{dE_x}{dx} = \frac{1}{r_{D1}}\left(\frac{1}{u_x^i} - \frac{1}{u_x^e}\right) . \quad (3.4.11)$$

The Maxwell equation (3.1.31) for B_y has the form

$$\frac{dB_y}{dx} = \frac{M_{a1}^2}{M_1\delta_1\ell_1}\left(\frac{u_z^i}{u_x^i} - \frac{u_z^e}{u_x^e}\right) . \quad (3.4.12)$$

In the two-fluid transfer equations we employ the expressions for the kinetic coefficients of magnetized plasma (Sect.3.1.2). If we disregard the skew components of the tensor of viscosity, then in the adopted coordinate system the solution for the shock structure can be sought in a planar form, assuming that $u_y^i = u_y^e = B_z = 0$ throughout the shock. Then the equations of motion of electrons and ions in the y coordinate and the z component of Maxwell equation (3.1.31) are satisfied identically. The shock structure is described — together with (3.4.11,12) — by equations of conservation of x and z components of momentum and energy of the plasma

$$u_x^i - 1 + \varepsilon^2(u_x^e - u_x^i) + \frac{3}{10M_1^2}\left(\frac{T_i}{u_x^i} + \frac{T_e}{u_x^e} - 2\right) + \frac{B_y^2 - 1}{2M_{a1}^2}$$
$$- \frac{3E_x^2}{20M_{a1}^2} - \Delta_{vi}\frac{du_x^i}{dx} - \Delta_{ve}\frac{du_x^e}{dx} = 0 , \quad (3.4.13)$$

$$u_z^i + \varepsilon^2(u_z^e - u_z^i) + \frac{r_{D1}}{M_1\delta_1\ell_1}E_x = 0 , \quad (3.4.14)$$

$$\frac{1}{2}\left[u_x^{i^2} + u_z^{i^2} + \varepsilon^2(u_x^{e^2} + u_z^{e^2} - u_x^{i^2} - u_z^{i^2})\right] + \frac{3}{4M_1^2}(T_e + T_i - 2)$$
$$+ \frac{B_y - 1}{M_{a1}^2} - \Delta_{vi}u_x^i\frac{du_x^i}{dx} - \Delta_{ve}u_x^e\frac{du_x^e}{dx} - \Delta_{Ti}\frac{dT_i}{dx} - \Delta_{Te}\frac{dT_e}{dx}$$
$$- \frac{9\sqrt{2}\varepsilon\delta_1(u_z^i - u_z^e)}{20M_1^2u_x^iu_x^eT_e^{3/2}} = 0 , \quad (3.4.15)$$

and equations of motion and heat balance of the electron component:

$$\varepsilon^2\frac{du_x^e}{dx} + \frac{3}{10M_1^2}\frac{d}{dx}\frac{T_e}{u_x^e} - \frac{u_z^eB_y}{M_1\delta_1\ell_1u_x^e} + \frac{3E_x}{10M_1^2r_{D1}u_x^e}$$

$$
-\frac{\varepsilon\sqrt{2}(u_x^i - u_x^e)}{M_1\ell_1 u_x^i u_x^e T_e^{3/2}} - \frac{d}{dx}\Delta_{ve}\frac{du_x^e}{dx} = 0 \quad , \tag{3.4.16}
$$

$$
\varepsilon^2\frac{du_z^e}{dx} + \frac{1}{M_1\delta_1\ell_1}\left(B_y - \frac{1}{u_x^e}\right) - \frac{\varepsilon\sqrt{2}(u_z^i - u_z^e)}{M_1\ell_1 u_x^i u_x^e T_e^{3/2}}
$$

$$
-\frac{9\sqrt{2}\varepsilon\delta_1}{20M_1^2 u_x^i u_x^e B_y T_e^{3/2}}\frac{dT_e}{dx} = 0 \quad , \tag{3.4.17}
$$

$$
\frac{3}{2}\frac{dT_e}{dx} + \frac{T_e}{u_x^e}\frac{du_x^e}{dx} - \frac{d}{dx}\left(\frac{10M_1^2}{3}\Delta_{Te}\frac{dT_e}{dx} + \frac{3\sqrt{2}\varepsilon\delta_1(u_z^i - u_z^e)}{2u_x^i u_x^e T_e^{3/2}}\right)
$$

$$
-\frac{10\sqrt{2}M_1\varepsilon}{3\ell_1 u_x^i u_x^e B_y T_e^{3/2}}\left[(u_z^i - u_z^e)^2 + (u_x^i - u_x^e)^2\right] + \frac{T_e - T_i}{\Delta_r}
$$

$$
+\frac{3\sqrt{2}\varepsilon\delta_1(u_z^e - u_z^i)}{2u_x^i u_x^e B_y T_e^{3/2}}\frac{dT_e}{dx} = 0 \quad . \tag{3.4.18}
$$

Here

$$
\Delta_{vi} = \Delta_{vi}^{(1)}T_i^{(5/2)} \quad , \quad \Delta_{vi}^{(1)} = 0.096\frac{\ell_1}{M_1}
$$

$$
\Delta_{ve} = \Delta_{ve}^{(1)}\frac{u_x^i}{u_x^e}T_e^{5/2} \quad , \quad \Delta_{ve}^{(1)} = 0.052\frac{\varepsilon\ell_1}{M_1} \quad , \tag{3.4.19}
$$

$$
\Delta_{Ti} = \Delta_{Ti}^{(1)}u_x^{i\,-2}B_y^{-2}T_i^{-\frac{1}{2}} \quad , \quad \Delta_{Ti}^{(1)} = 0.18\frac{\delta_1^2\ell_1}{M_1^3} \quad ,
$$

$$
\Delta_{Te} = \Delta_{Te}^{(1)}u_x^i u_x^{e\,-3}B_y^{-2}T_e^{-\frac{1}{2}} \quad , \quad \Delta_{Te}^{(1)} = 0.59\frac{\varepsilon\delta_1^2\ell_1}{M_1^3} \quad ,
$$

and Δ_r is defined as in (2.1.20) (transcription $0 \to 1$).

As in Sect.2.1, we consider the plasma to be quasi-neutral in the first approximation, assuming the Debye radius r_D to be the smallest among the pertinent characteristic scales, cf. estimation (2.1.26), and finding the profiles of electric field E_x and potential ϕ in the next approximation in the same way as in Sect.2.1.4. (The effects of charge separation with reference to transverse shock waves are discussed specifically in Sect.3.4.3.) In the adopted approximation $n_i = n_e$, i.e., in accordance with (2.1.14),

$$u_x^i = u_x^e \equiv u_x \quad . \tag{3.4.20}$$

With the aid of (3.4.20) we can estimate important characteristic scales contained in (3.4.11-18) besides those defined in (3.4.19). According to (3.4.14), u_z^i is small compared to u_z^e. Estimating u_z^e with the aid of (3.4.12) we find that the first term in the equation of motion of electrons (3.4.17) becomes significant (of the same order as the second term, which expresses the electric field in the frame of rest of electron component) on the scale

$$\Delta_d = \Delta_d^{(1)} u_x^{e^{1/2}} \quad , \quad \Delta_d^{(1)} = \frac{\varepsilon \delta_1 M_1 \ell_1}{M_{a1}} = \frac{c}{\omega_{pe}(1)} \quad , \tag{3.4.21}$$

equal to the collisionless skin depth, the characteristic length of dispersion of magnetoacoustic waves in a cold plasma, whose density corresponds to state 1, see (3.2.24). Similarly, using (3.4.12), we ascertain that the third term in (3.4.17), which accounts for intercomponential friction in plasma, has the same order of magnitude as the second term on the scale

$$\Delta_j = \Delta_j^{(1)} T_e^{-3/2} \quad , \quad \Delta_j^{(1)} = \frac{\sqrt{2}\varepsilon \delta_1^2 M_1 \ell_1}{M_{a1}^2} = \frac{\nu_m}{\bar{u}_{x1}} \quad , \tag{3.4.22}$$

which coincides with the diffusion length of the magnetic field, ν_m being the magnetic viscosity, see (3.1.7).

We shall also employ shock-layer equations, normalized to the downstream state. This requires a transcription $1 \to 2$ to be done in (3.4.10-22). According to (3.4.7), $\delta_2 < \delta_1 \ll 1$, i.e., δ_2 also is a small parameter.

To calculate the shock structure we employ the asymptotic multiscale technique, described in Sect.1.8.3. Some remarks must be made concerning the use of this technique in our present case. Two options for normalizing the variables exist: they may be related either to the equilibrium downstream values, or to the equilibrium upstream values. As we have seen, velocity, density and magnetic field in the shock wave vary within restricted limits, and their dimensionless values have the order of unity. However, the temperature jump may be arbitrarily high. For instance, we may take the scale characterizing isotropic viscosity in the form $\Delta_{vi}(1) \sim \ell_1/M_1$, or $\Delta_{vi}(2) \sim \ell_2/M_2$; then $\Delta_v(1)/\Delta_v(2) \sim T_2^{-5/2}$. Since the boundary conditions permit that $1 < T_2 < \infty$, the scale $\Delta_v(1)$ may be arbitrarily small compared with $\Delta_v(2)$. It follows that for a given physical process to dominate throughout the shock layer, its characteristic scale must be the largest of all throughout the entire shock. Otherwise, the competitive processes should be taken into account at the appropriate portions of the shock layer.

As follows from (3.4.19), the heat-conduction scales of ions and electrons are small compared to the respective viscosity scales ($1 \ll Pr$), since the heat conduction in the transverse directions is suppressed by the magnetic field. The scale of electron viscosity is small (in the parameter ε) compared to the ion viscosity. We see that the scales pertinent to the problem at hand are Δ_{vi}, Δ_r, Δ_α and Δ_j.

The largest of these four scales is always Δ_r. It is clear, however, that energy exchange between electrons and ions by itself does not produce the shock; it is responsible only for the region of temperature relaxation behind the shock front. Assuming in (3.4.12-18) the derivatives with respect to x to be of the order of $1/\Delta_r$, we obtain (with an accuracy down to terms that are small with respect to the parameters ε and δ):

$$u_z^i = u_z^e \quad , \tag{3.4.23}$$

$$u_x B_y = 1 \quad , \tag{3.4.24}$$

$$u_x = \text{const}, \quad B_y = \text{const}, \quad T_e + T_i = \text{const} \quad . \tag{3.4.25}$$

Taking into account that in point 2 (with appropriate normalization)

$$u_x = 1, \quad B_y = 1, \quad T_e + T_i = 2 \quad ,$$

we integrate (3.4.18) and find

$$\frac{x - x_0}{\Delta_r^{(2)}} = \frac{3}{2}\left(\frac{1}{2} \ln \frac{1 + T_e^{\frac{1}{2}}}{1 - T_e^{\frac{1}{2}}} - \frac{T_e^{3/2}}{3} - T_e^{\frac{1}{2}}\right) \quad . \tag{3.4.26}$$

This profile is similar to (2.1.40), although this time it describes the temperature relaxation behind the shock, not the nonequilibrium heating ahead of the shock. The integration constant is found from conditions on the interface between the shock layer and the relaxation region. At present we consider it infinitesimally thin, so far as Δ/Δ_r is small. Within the limits of this layer the relaxation effects are insignificant, and heating of electrons and ions occurs independently.

Among the effects which shape the shock structure in magnetized plasma the dominating ones are those of ion viscosity and inertia of electrons. For $M_2^2 \ll (\varepsilon\delta)^{-1}$ the ion viscosity dominates (region V in Fig.3.7), while for $M_2^2 \gg (\varepsilon\delta)^{-1}$ the main process is represented by dispersion due to inertia of electrons (region D in Fig.3.7). For $(\varepsilon\delta)^{-1} T_1^{5/2} \ll M_2^2 \ll (\varepsilon\delta)^{-1}$ dispersion dominates over viscosity at the head portion of the shock (region VD in Fig.3.7). In the dispersion region viscous dissipations dominate for $M_2^2 \ll (\varepsilon\delta^2)^{-1}$, while Joule dissipations prevail for $M_2^2 \ll (\varepsilon\delta^2)^{-1}$ (regions DV and DJ in Fig.3.7).

Consider the range of not too small values of β_2, where $M_2^2 \ll (\varepsilon\delta)^{-1}$. This region, as we know, is dominated by ion viscosity. Setting $\Delta = \Delta_{vi}$, we get the following equations, which describe the shock structure in the region indicated:

$$u_x - 1 + \frac{3}{10M_2^2}\left(\frac{T_e + T_i}{u_x} - 2\right) + \frac{B_y^2 - 1}{2M_{a2}^2} - \Delta_{vi}^{(2)} T_i^{5/2} \frac{du_x}{dx} = 0 \quad , \tag{3.4.27}$$

$$\frac{u_x^2 - 1}{2} + \frac{3}{4M_2^2}(T_e + T_i - 2) + \frac{B_y - 1}{M_{a2}^2} - \Delta_{vi}^{(2)} u_x T_i^{5/2} \frac{du_x}{dx} = 0 \quad , \tag{3.4.28}$$

$$B_y u_x = 1 \quad , \tag{3.4.29}$$

$$\frac{dT_e}{dx} + \frac{2}{3}\frac{T_e}{u_x}\frac{du_x}{dx} = 0 \quad . \tag{3.4.30}$$

In the entire region of interest the magnetic field (with an accuracy to small corrections of the order of ε and δ_2) is frozen, the heating of ions is due to ion viscosity, and the heating of electrons is the result of adiabatic compression. From (3.4.27-30) with boundary conditions (3.4.2-5) (with transcription $2 \to 1$) we get

$$T_e = T_1 (u_1/u_x)^{2/3} \quad , \tag{3.4.31}$$

$$T_i = 2 + \frac{4}{3}(1 - u_x) + \frac{10M_2^2}{9}(1 - u_x)^2 \left(1 - \frac{1}{u_x M_{a2}^2}\right) - T_e \quad , \tag{3.4.32}$$

$$\frac{x}{\Delta_{vi}^{(2)}} = \frac{3}{4} \int_{(1+u_1)/2}^{u_x} \frac{T_i^{5/2}(u) u^2 du}{(u - 1)(u - u_1)(u - u_{1-})} \quad , \tag{3.4.33}$$

where u_{1-} differs from (3.4.2) by the sign in front of the square bracket and transcription $1 \to 2$. The origin of coordinates is chosen so that $u_x(0) = (u_1 + 1)/2$. The transverse components of velocities of electrons and ions are small quantities of the order of δ_2. They are easily found from (3.4.12,14) if due account is taken in (3.4.14) of skew components, small in δ_2, of the viscosity tensor [3.39]:

$$u_z^i = \frac{3\delta_2 \ell_2 T_i}{20M_2}\frac{du_x}{dx} \quad , \qquad u_z^e = \delta_2 \ell_2 \left(\frac{3T_i}{20M_2} + \frac{M_2}{M_{a2}^2 u_x}\right)\frac{du_x}{dx} \quad . \tag{3.4.34}$$

In the next (first) approximation in r_D/Δ the magnitude of the electric field can be found from (3.4.16), and the deviation from quasineutrality from (3.4.11):

$$E_x = \frac{5r_{D2}}{3u_x}\left(T_e + \frac{3T_i}{10} + \frac{2M_2^2}{u_x M_{a2}^2}\right)\frac{du_x}{dx} , \tag{3.4.35}$$

$$n_i - n_e = r_{D2}\frac{dE_x}{dx} . \tag{3.4.36}$$

Equation (3.4.35) can be integrated with the aid of (3.4.31,32) to find the electric-potential profile and its jump across the shock in dimensional variables:

$$\begin{aligned} e(\varphi_2 - \varphi_1) = \frac{5}{9}\bar{T}_2 M_2^2 \Bigg[& \frac{u_1^2 - 1}{2} + \left(2 + \frac{6}{5M_2^2} + \frac{1}{M_{a2}^2}\right)(u_1 - 1) \\ & + \left(1 + \frac{2}{M_{a2}^2} + \frac{3}{M_2^2}\right)\ln u_1 + \frac{5}{M_{a2}^2}(1 - u_1^{-2}) \\ & + \frac{63T_1 u_1^{2/3}}{20M_2^2}\left(1 - u_1^{-2/3}\right)\Bigg] . \end{aligned} \tag{3.4.37}$$

As pointed out in Sect.2.1.4, the jump of potential is not predetermined by the boundary conditions, but depends on the structure of the shock. Consequently, measurements of the jump may supply valuable information about the structure of shocks in plasma.

Note that the absence of heat-conductive heating upstream in transverse shocks in magnetized plasma creates conditions for the appearance of peculiar structures when the unperturbed plasma flow is nonisothermal for $\bar{T}_{e1} > \bar{T}_{i1}$ (obviously, for $\bar{T}_{e1} < \bar{T}_{i1}$ the structure is similar to that described above). If $\bar{T}_{e1} \gg \bar{T}_{i1}$, or, more precisely,

$$\frac{\bar{T}_{e1}}{\bar{T}_{i1}} \gtrsim \left(\frac{\Delta_{vi}^{(1)}}{\Delta_{ve}^{(1)}}\right)^{2/3} \approx \left(\frac{m_i}{m_e}\right)^{1/3} , \tag{3.4.38}$$

then the viscous heating of the electron component may gain importance. Retaining only the terms which describe viscous and adiabatic heating in (3.4.18) and in the corresponding equation of heat balance of ions, we can integrate these equations, obtaining, in variables normalized to state 1,

$$\frac{T_e}{T_{e1}} = \left\{ \frac{\Delta_{ve}^{(1)}}{\Delta_{vi}^{(1)}} \left(\frac{T_{e1}}{T_{i1}}\right)^{3/2} + u_x\left[1 - \frac{\Delta_{ve}^{(1)}}{\Delta_{vi}^{(1)}} \left(\frac{T_{e1}}{T_{i1}}\right)^{3/2}\right]\right\}^{-2/3} \quad . \tag{3.4.39}$$

For $\bar{T}_{e1}/\bar{T}_{i1} \sim 1$ we return to (3.4.31). If, however, this ratio is large compared to the right-hand side of (3.4.38), then the Navier-Stokes equations, in accordance with (3.4.39), predict that it is the electrons that experience viscous heating, while the ions are heated adiabatically.

Consider now the range of small values of β_2, such that $M_2^2 \gg (\varepsilon\delta_2)^{-1}$. In this region the shock structure is shaped primarily by the inertia of electrons, and so the scale Δ_d is the largest. From (3.4.13-15,17) with an accuracy down to small quantities of the order of ε, δ_2^2, $1/M_2$ we obtain the set of equations

$$u_x - 1 + \frac{B_y^2 - 1}{2M_{a2}^2} = 0 \quad , \tag{3.4.40}$$

$$\frac{u_x^2 - 1}{2} + \frac{B_y - 1}{M_{a2}^2} + \frac{\Delta_d^{(2)^2}}{2M_{a2}^2}\left(u_x \frac{dB_y}{dx}\right)^2 = 0 \quad , \tag{3.4.41}$$

$$B_y u_x - 1 = \Delta_d^{(2)^2} u_x \frac{d}{dx}\left(u_x \frac{dB_y}{dx}\right) \quad . \tag{3.4.42}$$

Eliminating the dimensionless velocity u_x from these equations, rewriting them in dimensionless variables normalized to state 1 and switching to Lagrangian coordinates $u_x\Delta_d^{(2)} d/dx = d/d\xi$, we obtain the following equations for the dimensionless quantity B_y:

$$\frac{1}{2}\left(\frac{dB_y}{d\xi}\right)^2 + \frac{(B_y - 1)^2}{2}\left(\frac{(B_y + 1)^2}{4M_{a1}^2} - 1\right) = 0 \quad , \tag{3.4.43}$$

$$\frac{d^2B_y}{d\xi^2} + (B_y - 1)\left(\frac{B_y(B_y + 1)}{2M_{a1}^2} - 1\right) = 0 \quad . \tag{3.4.44}$$

These equations have a well-known mechanical analogy: the energy integral and the equation of motion of unit-mass point in a potential field

$$U(B_y) = \frac{(B_y - 1)^2}{2}\left(\frac{(B_y + 1)^2}{4M_{a1}^2} - 1\right) \quad .$$

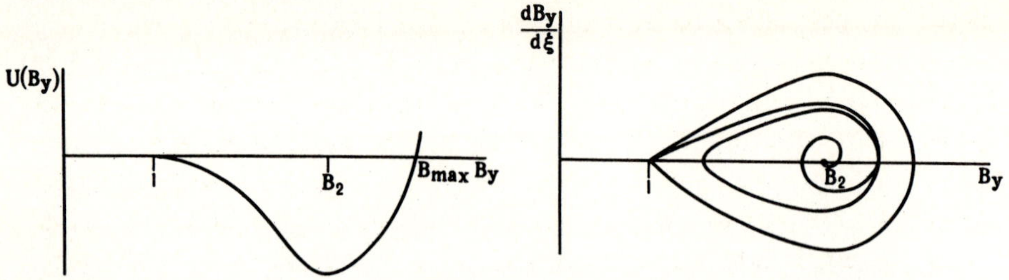

Fig.3.8. The potential $U(B_y)$ and phase curves of (3.4.44), describing soliton and periodic waves. Damping of the soliton by dissipations (asymptotic approach to $B_y = B_2$ as $\xi \to +\infty$) is described by the phase curve of (3.4.48)

The motion in the field of such a configuration has two points of equilibrium $B_1 = 1$ and $B_2 = [-1 + (1 + 8M_{a1}^2)^{\frac{1}{2}}]/2$, one of which corresponds to the equilibrium state upstream ($B = B_1$), and the other is close to the equilibrium state downstream $B_y = 1/u_2$, although it does not exactly coincide with the latter.

The phase diagram of (3.4.44) is illustrated in Fig.3.8. It is clear that in the absence of dissipations there is no way of passing from $B_y = B_1$ to $B_y = B_2$ and hence there is no solution representing a shock wave. This is why we have used normalization to state 1 in (3.4.43,44). The solution of (3.4.43,44) with the boundary condition $B_y(-\infty) = B_1 = 1$ is a soliton, namely

$$B_y(\xi) = 1 + \frac{2(M_{a1}^2 - 1)}{1 + M_{a1} \cosh(2\xi\sqrt{1 - M_{a1}^{-2}})} \quad . \tag{3.4.45}$$

The maximum value $B_{y\,max} = 2M_{a1} - 1$ corresponds to the maximum density $n_{max} = M_{a1}/(2 - M_{a1})$. The natural condition $n_{max} < \infty$ results in inequalities, familiar from the theory of collisionless shocks and expressing the criterion of laminar flow in the shock [3.42,43]:

$$M_{a1} < 2 \quad , \qquad B_{y\,max} < 3 \quad . \tag{3.4.46}$$

Let us analyze the meaning of inequality (3.4.46). For this purpose we use (3.4.16) to find the potential profile in the shock wave. With due account of the boundary conditions $\varphi(-\infty) = 0$ and $B_1 = B_y(-\infty) = 1$, we get in dimensional variables

$$e\varphi = \bar{T}_1 \frac{10M_1^2}{3M_{a1}^2} (B_y - 1) \quad .$$

By virtue of this the inequality (3.4.46) assumes the form

$$e\varphi_{max} < \frac{1}{2} m_i \bar{u}_{x1}^2 \quad . \tag{3.4.47}$$

Unless (3.4.47) is satisfied, the kinetic energy of ions in the incident plasma flow falls below the potential barrier $e\varphi_{max}$. This would create a counterflow of ions, reflected from the potential barrier, and the creation of multimode flow might result in the buildup of bunch instabilities and turbulization of the flow. The structures of collisionless shocks with due account of these essentially kinetic phenomena were considered in [3.44,45].

If, by way of a better approximation, we consider the terms accounting for viscous and Joule dissipations, small with respect to $1/\varepsilon\delta_2 M_2^2$ and δ_2, respectively, we obtain a damped soliton solution. If dissipations are small enough, the variables converge on their respective asymptotic downstream values in an oscillatory manner. Let us prove this point by linearizing the following equation [which represents a generalization of (3.4.44) with low-level dissipations taken into account]:

$$\frac{d^2B_y}{d\xi^2} + \frac{1}{u_x}\left(\frac{\Delta_{vi}^{(2)}T_i^{5/2}B_y}{\Delta_d^{(2)}M_{a2}^2} + \frac{\Delta_j^{(2)}T_e^{-3/2}}{\Delta_d^{(2)}}\right)\frac{dB_y}{d\xi}$$

$$+ (B_y - 1)\left(\frac{B_y(B_y + 1)}{2M_{a1}^2} - 1\right) = 0 \quad , \qquad (3.4.48)$$

in the vicinity of the singular point 2, with appropriate normalization of variables. As a result, we find the asymptotic profile of B_y as $x \to +\infty$:

$$B_y - 1 \propto \exp\left\{- [\gamma + i(\Omega^2 - \gamma^2)^{\frac{1}{2}}]\xi\right\} \quad ,$$

where

$$\Omega^2 = \left[3(1 + 8M_{a2}^2)^{\frac{1}{2}} - 8M_{a2}^2 - 1\right]/4M_{a2}^2 \approx 1 - M_{a2}^2$$

$$\gamma = \gamma_v + \gamma_j = \frac{\Delta_{vi}^{(2)}}{\Delta_d^{(2)}M_{a2}^2} + \frac{\Delta_j^{(2)}}{\Delta_d^{(2)}} \quad .$$

We see that for $\Omega > \gamma$ the solution describes oscillations which fade out downstream of the shock. The relative magnitude of γ_v and γ_j determines the role of ion viscosity and Joule dissipations in oscillation damping. For $(\varepsilon\delta^2)^{-1} \gg M_2^2 \gg (\varepsilon\delta)^{-1}$ the damping is due mainly to ion viscosity, and for $M_2^2 \gg (\varepsilon\delta^2)^{-1}$ the damping of oscillations is due primarily to Joule dissipations.

Note that weak shocks obviously may have only monotonic structure. Indeed, as already mentioned in Sect.3.2.2, on sufficiently large scales dissipations prevail over dispersion; as the strength of the shock decreases, the shock width, due to dissipations, tends to infinity, see, e.g., (1.8.12), whereas the characteristic dispersion scales remain unchanged. The setup of oscillations requires that $\Omega > \gamma_v$ or $\Omega > \gamma_j$. Since for $M_2 \gg 1$ also $M_{a2} \simeq 1$, we have $\Omega^2 = 1 - M_{a2}^2 \ll 1$. For a weak shock wave $T_i \simeq T_e \simeq 1$, and thus the condition $\Omega > \gamma$ bounds the intensity of the shock from below.

As indicated above, the coefficient of viscosity exhibits strong dependence on temperature, and $\Delta_{vi}(1)/\Delta_{vi}(2) \propto T^{5/2}$. Accordingly, if the temperature jump across the shock is large, the characteristic scale of ionic viscosity at the head of the shock layer may be much smaller than it its tail. On the contrary, the dispersion, due to inertia of electrons, is characterized by the scale $\Delta_d = c/\omega_{pe}$, which remains virtually constant throughout the shock. It follows that in a strong shock wave (when T_2 is large) the dispersion effects may dominate in the head of the shock layer (region VD in Fig.3.7). Then the head portion of the shock will display oscillations of velocity and magnetic field, whereas later all the variables will tend monotonically to their equilibrium downstream values. It can be demonstrated [3.40] that a necessary condition for the onset of oscillations is given by the inequality $M_2 > 1$.

3.4.3 Structures of Transverse Shock Waves in Nonmagnetized and Partly Magnetized Plasmas

We begin with an extreme case where the magnetic field is not too strong, and the plasma is not magnetized. In other words, we consider the range of variation of parameters where $\Omega_i\tau_i \ll \Omega_e\tau_e \ll 1$. Comparing the scales Δ_d and Δ_j, which characterize the inertia of electrons and the Joule dissipations, respectively, we conclude that the effects of current inertia are not significant for nonmagnetized plasma. The scales of electron heat conduction and ion viscosity are defined by (2.1.20); the scales of electron heat conduction and of electron-ion temperature relaxation are commensurable and differ from the scale of viscosity by a factor of ε^{-1}. According to (3.1.52), the scale of Joule dissipations differs from (3.4.22) only by a numerical coefficient, i.e.,

$$\Delta_j = \Delta_j^{(1)} T_e^{-3/2} \quad , \qquad \Delta_j^{(1)} = \frac{1.96\sqrt{2}\,\delta_1^2 M_1 \ell_1}{M_1^2} \quad . \tag{3.4.49}$$

Comparing (3.4.49) and (2.1.20) we find that the ratio of the scale of electron heat conduction to the scale of Joule losses is of the order of $\beta^2(\Omega_e\tau_e)^2$. We see that in nonmagnetized plasma with not too large values of $\beta < \Omega_e\tau_e$ the leading dissipative process is represented by Joule dissipations, and it is this scale that determines the width of the shock. For reference, let us specify the relevant restrictions for the numerical parameters of a hydrogen plasma:

$$1.6 \cdot 10^{-16} n T^{3/2} \gtrsim B \gtrsim 3 \cdot 10^{-3} T^{5/2} \quad , \qquad (3.4.50)$$

where T is expressed in electronvolts, B in teslas, and n in cm^{-3}. With the magnetic fields typical of experiments with collision-dominated shock waves, the inequality (3.4.50) is satisfied at relatively low plasma temperatures, as long as $Pm \ll 1$.

The shock width being large compared with Δ_r, the temperatures of electrons and ions are equal on the scale Δ_j:

$$T_i = T_e \equiv T \quad . \qquad (3.4.51)$$

The shock structure on the scales of Joule's dissipation is described by the equations of conservation of longitudinal momentum and energy of plasma, and the equation of z motion of electrons. Let us write these equations in dimensionless variables, normalized to state 2, disregarding, as in Sect. 3.4.2, the effects of charge separation and the action of all dissipative and dispersive mechanisms with the exception of the Joule mechanism:

$$u_x - 1 + \frac{3}{5M_2^2}\left(\frac{T}{u_x} - 1\right) + \frac{B_y^2 - 1}{2M_{a2}^2} = 0 \quad , \qquad (3.4.52)$$

$$\frac{u_x^2 - 1}{2} + \frac{3}{2M_2^2}(T - 1) + \frac{B_y - 1}{M_{a2}^2} = 0 \quad , \qquad (3.4.53)$$

$$\Delta_j^{(2)} T^{-3/2} \frac{dB_y}{dx} = u_x B_y - 1 \quad . \qquad (3.4.54)$$

From (3.4.52,53) we find B_y and T as functions of the dimensionless velocity u_x:

$$B_y = \frac{1}{u_x} + \frac{3}{5u_x}\left\{\left[1 + \frac{40M_{a2}^2}{9}(u_x - 1)(u_1 - u_x)(u_x - u_{1-})\right]^{\frac{1}{2}} - 1\right\} \quad , \qquad (3.4.55)$$

$$\begin{aligned} T = 1 &+ \frac{M_2^2}{3u_x}(1 - u_x)\left(u_x^2 + u_x - \frac{2}{M_{a2}^2}\right) \\ &- \frac{2M_2^2}{5u_x M_{a2}^2}\left\{\left[1 + \frac{40M_{a2}^2}{9}(u_x - 1)(u_1 - u_x)(u_x - u_{1-})\right]^{\frac{1}{2}} - 1\right\} \quad , \end{aligned} \qquad (3.4.56)$$

where u_{1-} is defined as in (3.4.33).

At the stationary point 2 downstream, the boundary conditions for dimensionless velocity, magnetic field strength and temperature $[B_y(2) = u_x(2) = T(2) = 1]$ are satisfied automatically. With the aid of (3.4.55,56) Eq.(3.4.54) is directly integrable, yielding the following expression for the structure of the shock layer:

$$\frac{x}{\Delta_j^{(2)}} = \frac{5}{3} \int_{(1+u_1)/2}^{u_x} \frac{dB_y(u)}{du} T^{-3/2}(u) \left\{\left[1 + \frac{40M_{a2}^2}{9} \right.\right.$$
$$\left.\left. \times (u - 1)(u_1 - u)(u - u_{1-})\right]^{\frac{1}{2}} - 1\right\}^{-1} du \quad . \tag{3.4.57}$$

In the next (first) approximation with respect to r_D/Δ_j, from the equation of longitudinal motion of electrons we find the electric field

$$E_x = -r_{D2}\left(1.71 \frac{dT}{du_x} - \frac{T}{u_x} + \frac{10M_2^2}{3M_{a2}^2} u_x B_y \frac{dB_y}{du_x}\right) \frac{du_x}{dx} \quad . \tag{3.4.58}$$

Integrating (3.4.58) we get the potential jump across the shock:

$$e(\varphi_2 - \varphi_1) = \bar{T}_2\left\{1.71(1 - T_1) + \frac{5M_2^2}{3M_{a2}^2}\left(1 - \frac{1}{u_1}\right)\right.$$
$$\left. + \int_1^{u_1} \left[\frac{T(u)}{u} + \frac{5M_2^2}{3M_{a2}^2} B_y^2(u)\right] du\right\} \quad . \tag{3.4.59}$$

As an example, the profiles of velocity, magnetic field, temperature and potential in a transverse shock in nonmagnetized plasma at $M_1 = 3$, $M_{a1} = 2$ are plotted in Fig.3.9.

Now let us return to (3.4.52-54). Rewrite (3.4.54) in the form

$$u_x B_y(u_x) - 1 = \Delta_j^{(2)} T^{-3/2} \frac{dB_y}{du_x} \frac{du_x}{dx} \quad . \tag{3.4.60}$$

From (3.4.55) it follows that $B_y \geqslant 1/u_x$, i.e., the left-hand side of (3.4.60) is nonnegative, and thus the sign of du_x/dx is predetermined by the sign of the derivative dB_y/du_x. Making use of (3.4.52,53), we obtain the expressions for dB_y/du_x and d^2B_y/du_x^2:

$$\frac{dB_y}{du_x} \frac{(2 - 5B_y u_x)}{M_{a2}^2} = 8u_x - 5 - \frac{3}{M_2^2} + \frac{5(B_y^2 - 1)}{2M_{a2}^2} \quad , \tag{3.4.61}$$

$$\frac{d^2B_y}{du_x^2} \frac{(2 - 5B_y u_x)}{M_{a2}^2} = 8 + \frac{5}{M_{a2}^2} \frac{dB_y^2}{du_x} + \frac{5u_x}{M_{a2}^2} \left(\frac{dB_y}{du_x}\right)^2 \quad . \tag{3.4.62}$$

◄ Fig.3.9

Fig.3.10 ▼

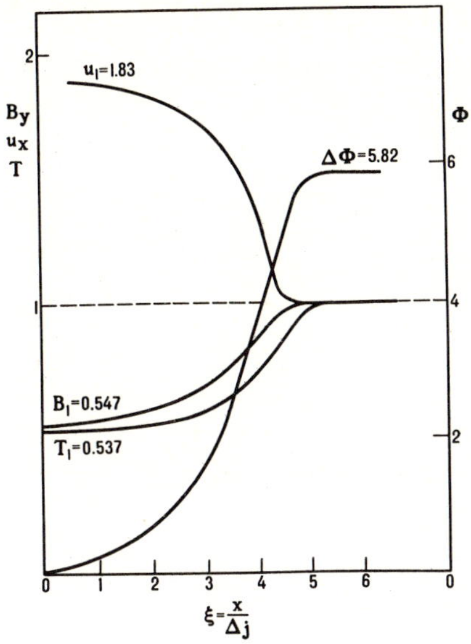

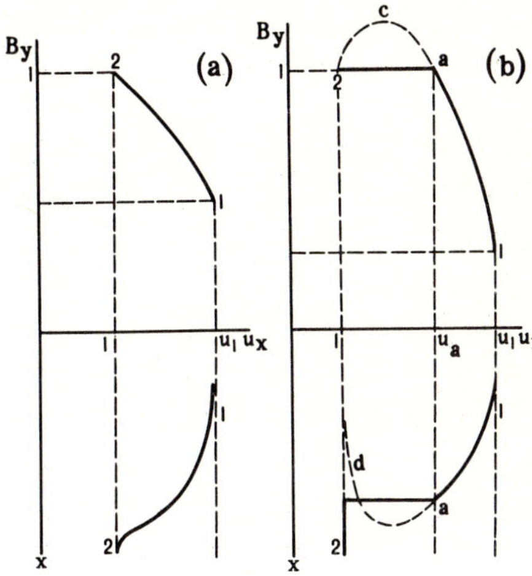

Fig.3.9. Structure of resistive shock wave in nonmagnetized plasma. The scale Δ_j pertains to the equilibrium downstream state

Fig.3.10a,b. Phase curves representing structures of transverse shock waves in nonmagnetized plasma: (**a**) purely resistive structure; (**b**) the structure containing isomagnetic subshock a →2

Rewriting (3.4.61) in dimensionless variables, normalized to the stationary points 1 and 2, respectively, and taking advantage of the fact that $B_y = u_x = 1$ at these points, we get the following expression for dB_y/du_x at points 1 and 2:

$$\left(\frac{dB_y}{du_x}\right)_{1,2} = -M^2_{a1,2}\left(1 - \frac{1}{M^2_{1,2}}\right) \quad . \tag{3.4.63}$$

Inasmuch as $M_1 > 1$, at point 1 $dB_y/du_x < 0$ always. On the contrary, at point 2 the inequality $dB_y/du_x < 0$ is satisfied only when $M_2 > 1$. Since at points 1 and 2 $dB_y/du_x < 0$ as long as $M_2 > 1$, and from (3.4.61,62) it follows that the conditions $dB_y/du_x = 0$ and d^2B_y/du_x^2 cannot be simultaneously satisfied, we see that the inequality $M_2 > 1$ is the necessary and sufficient condition for the fulfilment of inequality $dB_y/du_x < 0$ for $1 < u_x < u_1$. Then $B_y(u_x)$ is a monotonic function of u_x for all $1 < u_x < u_1$, whereas $u_x(x)$ is a monotonic function of x, as shown in Fig.3.10a. On the other hand, with $M_2 < 1$ the curve $B_y(u_x)$ exhibits a portion where $dB_y/du_x > 0$ (Fig.3.10b). It should be noted that, as we move along the curve $B_y(u_x)$ from the initial point 1 to the point 2, the entropy goes up on the leg $1 \to c$ and down on $c \to 2$. From Fig.3.10b it is clear that the corresponding solution $u(x)$ is physically meaningless, and necessitates introducing an internal discontinuity in the shock structure, as indicated in Sect.1.8, similar to an isothermal discontinuity in a gasdynamic shock wave at $Pr \ll 1$. At present, as follows from Fig.3.10b, the only natural way of introducing the required discontinuity is the solution $1 \to a \to 2$, which complies with the condition of increasing entropy.

We see that the internal discontinuity inevitably arises in the Joule shock wave for $M_2 < 1$, its structure and width being determined by dissipative processes, standing next in the hierarchy of scales [3.22]. Since the largest of these scales is still small compared to Δ_j [or, which is the same, $Pm \ll 1$; see (3.4.50)], the magnetic field within the discontinuity is constant [in accordance with (3.4.54)] with an accuracy to Δ_{Te}/Δ_j and equals its value at point 2. Such an internal discontinuity is called the isomagnetic jump.

Isomagnetic jumps were already discussed as structural components of MHD shocks at $Pm \ll 1$ (Sect.3.3.5). At that point we also emphasized the necessity of an internal discontinuity: the Joule mechanism is capable of dissipating a limited amount of incoming energy, whereas the energy carried by the incident flow may be as high as desired. Therefore, when the strength of the shock exceeds a certain threshold (the speed of the shock front is above a certain critical value for a given β_1), the Joule mechanism becomes inadequate, and one has to take into account other dissipative processes. The critical Mach number is determined by the intersection of curve $M_2^2 = 1$, see (3.3.20, 4.2-5), with the straight line $\beta_1 = \text{const}$. When the Mach number of a shock wave falls into the region between the hyperbola $M_1^{-2} + M_{a1}^{-2} = 1$ and the line $M_{a1}^B(\beta_1)$, defined by equation $M_2 = 1$, the shock structure is shaped solely by Joule dissipations. The structure of shocks with $M_{a1} > M_{a1}^B$ displays internal isomagnetic discontinuity. The limit of the Alfvén Mach number as $\beta_1 \to 0$ is $M_{a1}^B\ (\beta \to 0) = 2.76\ldots$.

Consider the structure of the isomagnetic jump for $M_{a1} > M_{a1}^B$. Inasmuch as the magnetic field within the internal discontinuity is constant, the discontinuity is of purely gasdynamic nature, i.e., the variations of all variables in the isomagnetic jump depend only on the magnitude of the sonic Mach number $M_2 < 1$. It follows that the isomagnetic-jump structure is similar to that of the shock in nonmagnetized plasma, considered in Sect.2.1.

Velocity and temperature of the plasma flowing into the isomagnetic jump (point a in Fig.3.10b) can easily be found from the equations of conservation of momentum and energy. From (3.452,53) at $B_y = 1$ (the magnetic field strength in the isomagnetic jump equals its final downstream value at point 2) we find

$$u_x(a) = \frac{M_2^2 + 3}{4M_2^2} , \quad T(a) = \frac{(M_2^2 + 3)(5M_2^2 - 1)}{16M_2^2} . \tag{3.4.64}$$

The local Mach number at point a can easily be found from (3.4.64):

$$M^2(a) = \frac{M_2^2 + 3}{5M_2^2 - 1} . \tag{3.4.65}$$

We see that $M(a) > 1$ for $M_2 < 1$, i.e., the isomagnetic discontinuity actually is a shock in a plasma without external magnetic field with the Mach number M(a) for the incoming flow. Taking advantage of the results obtained in Sects. 2.1.2,3, we find that the structure of the internal isomagnetic discontinuity for $1 > M_2^2 > 4/5$ is shaped by the processes of electron heat conduction and electron-ion heat exchange, whereas for $M_2^2 < 4/5$ we have to introduce yet another internal discontinuity, isothermal with respect to electronic temperature. The latter is indicated in Fig.3.11 by the line M_{a1}^T, described by $M_2^2 = 4/5$; it is easily proved that $M_{a1}^T(\beta_1 \to 0) = 3.02$ ... . Consequently, in general, the shock structure in nonmagnetized plasma may be viewed as a sequence of nested discontinuities, i.e., a viscous jump within the electron-isothermal jump at $M_{a1} > M_{a1}^T$, the electron-isothermal jump within the Joule shock for $M_{a1}^T > M_{a1} > M_{a1}^B$. The longitudinal electric field E_x and the potential jump in the head (Joule) portion of the shock containing an isomagnetic jump are given by (3.4.58,59), the integration in (3.4.59) being carried out to point a rather than to point 2. The electric-field profile and the potential

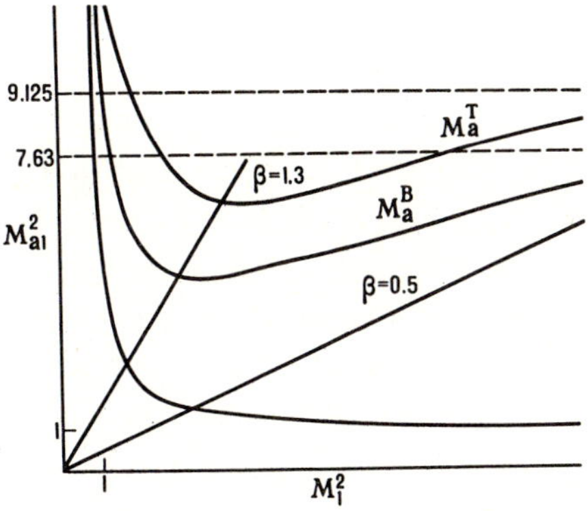

Fig.3.11. Regions in the (M_1^2, M_{a1}^2) plane pertaining to different types of structures of transverse shock waves in nonmagnetized plasma. Rays emanating from the origin of coordinates correspond to fixed initial states of a plasma with indicated values of β

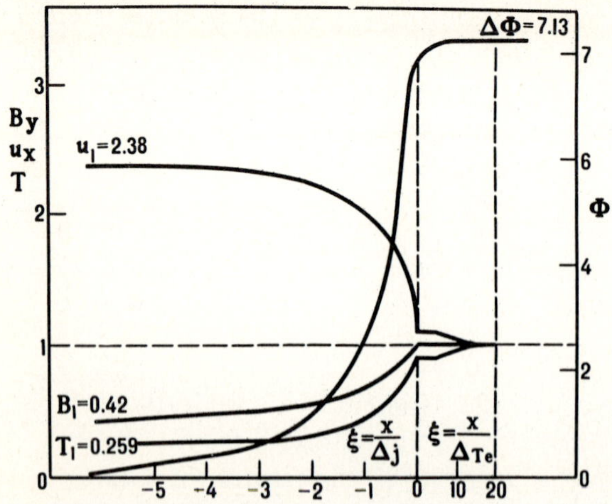

Fig.3.12. The structure of transverse shock wave with a weak isomagnetic subshock. The scales Δ_j, Δ_{Te} pertain to the equilibrium downstream state

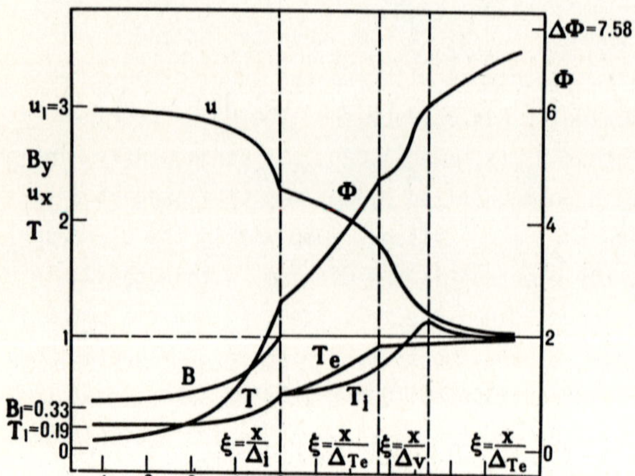

Fig.3.13. Structure of a transverse shock wave with a strong isomagnetic subshock. The scales Δ_j, Δ_{Te}, Δ_v pertain to the equilibrium downstream state

jump across the isomagnetic discontinuity are given by the formulas in Sect.2.1.4.

Figure 3.12 [3.40] illustrates the shock structure with a weak isomagnetic jump ($M^B_{a1} < M_{a1} < M^T_{a1}$) for $M_1^2 = 20$, $M_{a1}^2 = 7$. The structure with a strong isomagnetic jump is shown in Fig.3.13; here $M_1^2 = 17$, $M_{a1}^2 = 25$. Note that with increasing intensity of the shock wave the jumps of density and temperature tend to concentrate in the internal ion isomagnetic discontinuity (viscous discontinuity). If we also recall that with increasing temperature the plasma becomes magnetized no matter what the initial magnitude of the magnetic field, we may deduce that the structures of strong shock waves differ from those discussed in Sect.3.4.2 only in that they exhibit a small Joule portion in the head of the shock [3.40].

Finally, let us consider the case of partly magnetized plasma, when $\Omega_i\tau_i \ll 1 \ll \Omega_e\tau_e$. The shock structures here exhibit certain peculiarities providing the following condition is fulfilled:

$$\varepsilon^{-\frac{1}{2}} \ll \delta_1, \delta_2 \ll \varepsilon^{-1} \quad . \tag{3.4.66}$$

Then, the scales of electronic heat conduction and Joule dissipations are of the same order, and the scale of ion viscosity is comparatively small. As demonstrated in [3.40], the structures of relatively weak shocks are shaped by Joule dissipations and electronic heat conduction, whereas the heating of ions occurs adiabatically. In Sect.2.1 we indicated that Joule dissipations and electronic heat conduction, taken separately, are incapable of shaping the structures of arbitrarily strong shocks. The same is true for the two mechanisms acting together. The structures of strong shocks in partially magnetized plasma contain internal discontinuities, which are at once isomagnetic and electron-isothermal. In this case the necessary condition of internal discontinuity is given by the inequality $M_2^2 < M_{2\,cr}^2$, $M_{2\,cr}^2$ being a solution of the transcendental equation

$$M_2^2 = \frac{3}{5}\,(1 + u_1^{-2/3}T_1^{-1}) \quad . \tag{3.4.67}$$

In the limit $\beta_1 \to 0$ the critical value of the Alfvén Mach number is $M_{a1} \to 2\sqrt{3} = 3.46$.

3.4.4 Plasma Polarization in Transverse Shock Waves

So far, by way of first approximation, we have assumed the plasma to be quasi-neutral, ignoring (in view of the smallness of the ratio r_D/Δ, Δ being the characteristic scale of the problem) the contribution of the energy of the longitudinal electric field to the overall plasma energy and the effects of charge separation. Naturally, this does not imply that we neglect the effect of plasma polarization itself. On the contrary, the very concept of quasineutrality implies that the longitudinal electric field has a very considerable effect on the shock structure. To elucidate the role of a longitudinal polarization field, let us use the generalized Ohm's law (3.1.50). Expression (3.1.50) indicates that in a magnetized plasma the conduction perpendicular to the direction of the field is suppressed by magnetic field, as reflected by the small coefficient $[1+(\Omega_e\tau_e)^2]^{-1} \sim (\Omega_e\tau_e)^{-2}$. Physically speaking, this occurrence is due to the fact that in the absence of magnetic field the electron, acted upon by the static electric field E, assumes a longitudinal drift velocity $v_D^{(0)} = eE\tau_e/m_e$. In the crossed fields **E, B** this drift velocity, the collisions ignored, is normal to both fields and equals $v_D = E/B = v_D^{(0)}/\Omega_e\tau_e$. As follows from (3.1.50), accounting for collisions gives rise to the drift-velocity component $v_D^{(0)}/(\Omega_e\tau_e)^2$ in the direction of the effective field $\mathbf{E}_\perp'$.

Hence, the conclusion seems to follow that the characteristic length of conduction of a strongly magnetized plasma in the transverse direction is also multiplied by $(\Omega_e\tau_e)^2$. However, as demonstrated in Sect.3.4.2, in our present case of transverse magnetic field this conclusion is wrong: in fact, in a strongly magnetized plasma, where the transfer coefficients are highly anisotropic, the scale of Joule dissipations is of the same order as in the absence of the field, and is much smaller than the viscosity scale. As a result, the magnetic field in a viscous jump is frozen in. Should the Joule scale be multiplied by a large factor of $(\Omega_e\tau_e)^2$, the structure of a transverse shock in a magnetized plasma would have the form described in Sect. 3.4.3, which contradicts experimental findings (Sect.3.4.5).

The reason why the anisotropy of conduction is disguised becomes clear if we substitute the condition $j_x = 0$ into (3.1.50) (satisfied since the shock is stationary and unidimensional, Sect.2.1)

$$E_x' = -\Omega_e\tau_e E_y' \quad . \tag{3.4.68}$$

Substituting (3.4.68) into (3.1.50) we get

$$j_z = \sigma_\perp E_z' \quad . \tag{3.4.69}$$

Recall that $\sigma_\perp$ differs little from the conductivity of plasma in the absence of magnetic field, $\sigma_\parallel$, see (3.1.50). Equation (3.4.69) indicates that the magnetic field has little effect on the conductivity of plasma just because of the longitudinal Hall field E_x', which is due to plasma polarization and is normal to both magnetic field B_y and transverse current j_z. Although the separation of charges may be minor, the field E_x' itself is not small: in accordance with (3.4.68), the effective longitudinal field in magnetized plasma is strong in comparison with the transverse field.

From arguments developed above it follows that the unidimensionality condition of the flow in magnetized plasma is much more restrictive than in gasdynamics. Here the unidimensionality is not a local characteristic of the flow, pertaining to the vicinity of some point of the shock, but rather it characterizes the flow in general, owing to the presence of long-range electromagnetic fields. If this condition is violated, the longitudinal current may be nonzero, just as observed experimentally [3.46]. At $j_x \neq 0$ condition (3.4.68) also becomes invalid, and the longitudinal electric field fails to counterbalance the reduction of plasma conductivity by the magnetic field. If the shock structure is dominated by viscosity, the electric-current profile is determined by the boundary conditions at the shock and the characteristic scale of viscosity. The decrease in conductivity here ought to lead to an increase in Joule heating j^2/σ, i.e., to a higher rise in the electronic temperature.

Apart from the described indirect influence on the shock structure, the separation of charges, like the inertia of electrons, may act as a dispersive mechanism, which shapes the oscillatory structure of a shock in a cold plasma. This dispersive mechanism will dominate as soon as its characteristic scale

$$\Delta_q = \Delta_q^{(1)}(u_x^i)^{\frac{1}{2}} \quad , \quad \Delta_q^{(1)} = c/\omega_{pi} = (10/3)^{\frac{1}{2}} \frac{M_1 r_{D1}}{M_{a1}} \quad , \tag{3.4.70}$$

cf. (3.2.26), is large compared with the collisionless skin depth Δ_d, i.e., when condition (3.2.25) is satisfied. To a first approximation, ignoring thermal pressure and collisions, the shock-layer equations including plasma polarization, are the Poisson equation (3.4.11), the Maxwell equation (3.4.12), the equations of conservation of momentum and energy in dimensionless variables normalized to state 1, i.e.,

$$u_x^i - 1 + \frac{B_y^2 - 1}{2M_{a1}^2} - \frac{3}{20M_1^2} E_x^2 = 0 \quad , \tag{3.4.71}$$

$$u_z^i = - \frac{r_{D1}}{M_1 \delta_1 \ell_1} E_x \quad , \tag{3.4.72}$$

$$\frac{(u_x^i)^2 - 1}{2} + \frac{B_y - 1}{M_{a1}^2} = 0 \quad , \tag{3.4.73}$$

equations of motion of electrons

$$u_z^e B_y - \frac{3\delta_1 \ell_1}{10M_1^2 r_{D1}} E_x = 0 \quad , \tag{3.4.74}$$

$$u_x^e B_y - 1 = 0 \quad . \tag{3.4.75}$$

The set (3.4.11,12,71-75) with the six variables u_x^i, u_x^e, u_z^i, u_z^e, E_x, B_y is not overdetermined, since (3.4.73), stating the law of conservation of energy, is derivable from the other equations. Taking into account that $B_y \sim 1$, and confining ourselves to the nonrelativistic case $c_{a1} \ll c$, we find from (3.4.72,74) that u_z^i in (3.4.12) can be neglected as small compared with u_z^e. Then, with the aid of (3.4.12,74,75) we get

$$E_x = -\frac{10M_1^2}{3M_{a1}^2} r_{D1} \frac{dB_y}{dx} . \tag{3.4.76}$$

Substituting this expression for E_x into (3.4.11,71), we obtain equations which describe the stationary structure of the shock in the adopted approximation. It is most convenient to write them by representing B_y in terms of u_x^i via (3.4.73) and introducing the Lagrangian coordinate $u_x^i \Delta_q^{(1)}(d/dx) = d/d\xi$:

$$\frac{1}{2}\left(\frac{du_x^i}{d\xi}\right)^2 + \frac{1}{8}(1-u_x^i)^2\left[\frac{4}{M_{a1}^2} - (u_x^i+1)^2\right] = 0 , \tag{3.4.77}$$

$$\frac{d^2u_x^i}{d\xi^2} + \frac{1}{2}(1-u_x^i)\left[u_x^i(u_x^i+1) - \frac{2}{M_{a1}^2}\right] = 0 . \tag{3.4.78}$$

It is clear that (3.4.77) is the integral of (3.4.78). Like (3.4.43,44), these equations have an obvious mechanical analogy: they describe the motion of a unit-mass point, the time and the coordinate being represented by ξ and u_x^i, respectively. The "potential"

$$U(u_x^i) = \frac{1}{8}(1-u_x^i)^2\left[\frac{4}{M_{a1}^2} - (u_x^i+1)^2\right]$$

has two stationary points, corresponding to $u_x^i = 1$ and $u_x^i = [(8+M_{a1}^2)^{1/2} - M_{a1}]/2M_{a1}$. The minimum of velocity u_x^i is $u_{min} = 2/M_{a1} - 1$. From the condition $u_{min} > 0$ we once again obtain the inequalities (3.4.46). The coincidence is, of course, not casual. These restrictions arise from the laws of conservation of momentum and energy, and do not depend on the actual dispersion mechanism, just as the Rankine-Hugoniot conditions do not depend on the dissipation mechanism.

The solution of (3.4.77) with the boundary condition $u_x^i(-\infty) = 1$ is a soliton of the form (3.4.45). Its damping in the present extreme case is due to viscosity, since the Joule losses in magnetized plasma are characterized by a smaller scale than the dispersion due to the inertia of electrons. Actually, soliton damping in cold plasma, resulting in an oscillatory profile, is due mainly to collisionless effects: reflection of incident thermal ions from the potential hump, which takes place for $M_{a1} < 2$ as well; excitation of various instabilities, etc. [3.42-45]. Approaching this problem in the Navier-

Stokes approximation, as in Sect.3.4.3, we conclude that the criterion for the appearance of an oscillatory profile bounds the strength of the shock from below.

3.4.5 Experimental Investigations of Transverse Shock Waves in Plasma

Detailed investigations of strong transverse shocks in a plasma were carried out with the shock tube at the Columbia University [3.47-53] with Mach numbers up to $(2\text{-}3)\cdot 10^3$. This shock tube is a coaxial electromagnetic shock tube, in which the transverse magnetic field is produced by passing an electric current via the axially located electrode (Sect.5.1). When comparing experimental results with theory one must bear in mind that the experimental shocks are often not unidimensional; in the longitudinal cross section they usually have a parabolic shape, concave towards the axial conductor (Sect.5.1.3).

For strong shock waves we can use the formulas obtained earlier for magnetized plasma. The Prandtl shock width is defined by

$$\Delta_{Pr} = (u_{x1} - u_{x2})/|du/dx|_{max} \quad .$$

If $M_1 \gg 1$, we can take advantage of (3.4.33) and obtain, with an accuracy to $O(1/M_1^2)$, the following estimate

$$\Delta_{Pr} = 2\sqrt{5}\ell_1 \quad .$$

Actually, what is measured experimentally is not the "maximum steepness" width, but the quantity

$$\Delta_{exp} = \int_{u_{min}}^{u_{max}} du_x / \left|\frac{du_x}{dx}\right| \tag{3.4.79}$$

where the deviation of u_{max} and u_{min} from the equilibrium velocities in the boundary points 1 and 2 depends on the accuracy of the experimental measurements. Formula (3.4.79) is a refinement of the estimate (1.7.17) by up to one order of magnitude.

Making use of (3.4.33), we obtain the following estimate of the shock width (3.4.79) with an accuracy to $1/M_1^2$ [3.21]:

$$\Delta_{exp} = M_1^4\ell_1 \left(\frac{1}{32} \ln \frac{3}{4(u_{x2}/u_{x1})_{min} - 1} - 0.046\right) \quad . \tag{3.4.80}$$

Figure 3.14 illustrates the curve of transverse shock width versus velocity, calculated with (3.4.80) together with the experimental results [3.50];

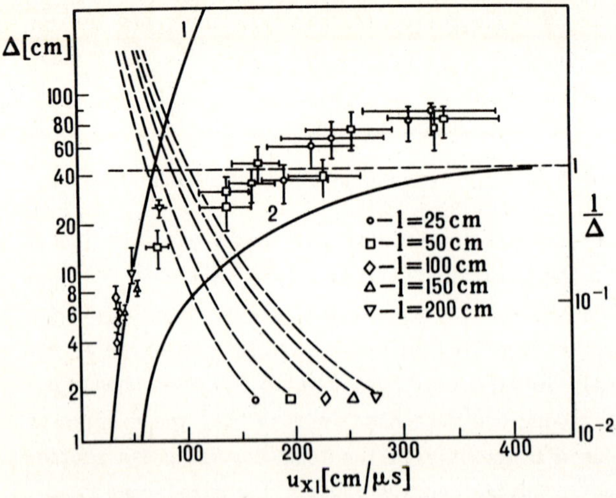

Fig.3.14. Shock widths of strong transverse shock waves in magnetized plasma: (1) calculated with (3.4.80); (2) according to unsteady-state numeric calculation [3.54]; experimental points [3.50]. (----) represent the parameter ℓ/Δ, which controls the applicability of stationary theory, as a function of shock velocity u_{x1}

the experimental accuracy is assumed to be 1%. Since the profile of the shock becomes stationary at the distance ℓ from the origin of the shock, greater than the shock width Δ, it would be natural to judge the applicability of the stationary theory by the parameter ℓ/Δ. In Fig.3.14 ℓ is the distance from the end of the shock tube to the point of observation [3.50], and Δ is the estimated stationary shock width (3.4.80). As follows from Fig.3.14, there is a fair fit between the experimental and calculated curves. The discrepancy starts with all shocks for $u_1 \geqslant 10^8$cm/s, because then $\ell/\Delta < 1$ under the experimental conditions of [3.50]. Dashed curves in Fig.3.14 indicate the limits of applicability of the stationary theory. The solid line 2 represents the results of numerical calculations for a nonstationary model [3.54]. The agreement of these calculations with experimental results verifies the applicability of a Navier-Stokes approach.

The density jump, the temperature of electrons and the compression of magnetic field in transverse shock waves with velocities of $9 \cdot 10^6$cm/2 $< u_{x1} < 2 \cdot 10^7$cm/s and the initial values of the magnetic field strength $B_{y1} = 0.1$ T and 0.36 T were measured in [3.52]. These values coincide with those computed from (3.4.2-5) to an accuracy of 20%. This small discrepancy is most likely associated with the nonunidimensionality of the experimental geometry. It is noteworthy that the region of temperature relaxation was not detected, although the distance covered by the shock was much larger than Δ_r. In these experiments the electrons behind the shock assume a temperature of about 10 eV, while the ions are heated to approximately 100 eV. The temperature of electrons, however, does not rise, even though they have time to collide often with hot ions. The reason for this is presumably associated with radiative cooling by strong resonant emission in the vacuum uv region, due to heavy impurity ions (the test gas used in [3.52] was deuterium).

Independent measurements of the compression of plasma and magnetic field in transverse shock waves [3.53] in the velocity range of 10^7 cm/s $< u_{x1} < 3 \cdot 10^7$ cm/s agree with the results of calculations (3.4.2-4) within the limits of experimental error.

There are few experimental investigations of shocks in nonmagnetized plasma. This is partly due to the fact that such shocks are not too strong, so unless special measures are taken to pre-ionize the gas, they display all the characteristic features of the ionizing shock waves, which have been extensively and carefully investigated. These matters are discussed in Chaps.4,5.

Inasmuch as the temperature in the shock front increases, and $\Delta_j \propto T^{-3/2}$, the Joule shock-front width can be estimated as

$$\Delta_{j1} = \frac{2}{u_{x1} T_{e1}^{3/2}} \left(1 + 0.23 \log \frac{T_{e1}^3}{n_{e1}}\right) \quad , \qquad (3.4.81)$$

where u_{x1} is expressed in cm/μs, T in eV, n_{e1} in units of 10^{16}cm^{-3}.

Under typical experimental conditions ($n_{e1} = 10^{16}$cm^{-3}, $T_{e1} = 1$ eV, $u_{x1} = 10$ cm/μs), we obtain from (3.4.81) $\Delta_{j1} = 0.5$ cm, which is a good fit with experiment. The shock width measured in [3.55] constitutes (0.3-0.8)cm at $n_1 = 10^{16}$cm^{-3}, $T_{e1} = 1.2$-2.1 eV, $u_{x1} = 2$ cm/μs. The value of Δ_{j1}, calculated by (3.4.81), is (0.3-0.6)cm. However, as reported in [3.55], the magnitude of Δ_{exp} does not drop as u_{x1} increases, as would follow from (3.4.81), but rather exhibits a linear increase. *Sommer* and *Barach* [3.55] suggested appropriate correction by introducing phenomenological parameters of experimental imperfection. The point is that the deviations from unidimensionality of the flow are not negligible; they account for the incomplete compression of the magnetic field in the shock.

It must also be noted that knowing the structure of shocks in a dense collision-dominated plasma is useful for estimating the amount of heating of a rarefied plasma by collisionless shocks, and for classifying collisionless shocks. For illustration, let us consider the experiments with shocks in rarefied plasma [3.56].

The authors of [3.56] studied the transverse shocks in hydrogen plasma at $n_1 = 7 \cdot 10^{14} cm^{-3}$, $T_{e1} = 1$ eV, $B_{y1} = 0.12$ T, $u_{x1} = 2.5$ cm/μs, $M_{a1} = 2.5$. The degree of compression of the magnetic field in such a shock, in accordance with (3.4.4), is $B_{y2}/B_{y1} = 2.5$, which agrees nicely with the experimental results. The shock-width estimate (3.4.81) also works quite well: the measured time of the magnetic-field build-up at a given point, i.e., the time of passage of the shock measured by the observer, is about 10 ns, while $\Delta_{j1}/u_{x1} = 12$ ns from (3.4.81). Finally, the calculated values of electronic and ionic temperatures (under the assumption of Joule heating and the absence of electron-ion heat exchange [3.57]) are $T_{e2} = 46$ eV, $T_{i2} = 2$ eV and check with experimental results [3.58]. However, the general shock pattern in these experiments indicates that the Joule heating, which dominates at the head of the shock, is then replaced by different kinds of dissipations. As low as $M_{a1} = 2.5$ the steepness of the shock in the head is greater than in the tail, whereas the reverse ought to be expected in the collision-dominated shock wave. At $M_{a1} = 3.7$, i.e., $M_2 < 1$, the shock structure exhibits an isomagnetic jump, located in the head portion of the shock.

3.5 Structures of Switch-On Shock Waves

3.5.1 Boundary Conditions and the Shock Adiabat

We write the boundary conditions for the set of two-fluid transfer equations in dimensionless variables (3.3.17). The switch-on shocks exist for $B_x \neq 0$, so we define the dimensionless magnetic field and the Alfvén Mach number as in (3.3.18), differently to the procedure in Sect.3.4.1. As indicated in Sect.3.3.3, the switch-on shock waves represent the extreme case of fast shock transitions $1 \to 2$, so the subscripts 1 and 2 pertain to the upstream and downstream states, respectively. We choose a coordinate system in which the transverse component of plasma velocity ahead of the shock is zero. We confine ourselves to a simple plasma. Then for the upstream state ($x = -\infty$) we get

$$
\begin{aligned}
&u_x^i = u_x^e = n_i = n_e = T_i = T_e = 1 \quad ; \\
&u_y^i = u_y^e = B_y = E_x = \phi = 0 \quad .
\end{aligned}
\tag{3.5.1}
$$

For the downstream state ($x = +\infty$), in accordance with (3.3.33),

$$u_x^i = u_x^e = 1/M_{a1}^2 \equiv u_2 \quad . \tag{3.5.2}$$

From (3.3.10-14) we obtain

$$n_i = n_e = 1/u_2 \quad , \tag{3.5.3}$$

$$B_y = \left[\frac{2}{3}\left(M_{a1}^2 - 1\right)\left(4 - \frac{M_1^2 + 3}{M_1^2} M_{a1}^2\right)\right]^{\frac{1}{2}} \equiv B_2 \quad , \tag{3.5.4}$$

$$u_y^i = u_y^e = B_2/M_{a1}^2 \quad , \tag{3.5.5}$$

$$T_i = T_e = 1 + \frac{5M_1^2}{9}\left(1 - \frac{1}{M_{a1}^2}\right)\left(1 + \frac{6}{5M_1^2} - \frac{1}{M_{a1}^2}\right) \equiv T_2 \quad . \tag{3.5.6}$$

Downstream $E_x = 0$, and the potential jump ϕ across the shock is determined by the shock structure. Using (3.5.3,6), we find the shock-adiabat equation in terms of pressure versus relative volume

$$\frac{\bar{p}_2}{\bar{p}_1} = \frac{1}{3\beta_1 u_2^2}\left[(5\beta_1 - 1)u_2^2 - 2\beta_1 u_2 + 1\right] \quad , \tag{3.5.7}$$

where $\beta_1 = 2\mu_0\bar{p}_1/\bar{B}_x^2$.

The range of the parameters M_1, M_{a1} pertaining to the switch-on shock waves is determined by the evolutionarity conditions. The singular point, corresponding to the state of plasma in the incident flow, is point 1. Hence, according to (3.2.14,3.27) it follows that $1 < M_1^2$, $1 < M_{a1}^2$. We derive from (3.3.20,21,5.2.6), downstream,

$$M_2^2 = 9M_1^2\left[15M_{a1}^4 - 6M_{a1}^2 + 5M_1^2(M_{a1}^2 - 1)^2\right]^{-1} \quad , \tag{3.5.8}$$

$$M_{a2}^2 = 1 \quad . \tag{3.5.9}$$

As indicated in Sect.3.3.3, in the present degenerate case the singular points 2 and 3 are indistinguishable from one another; the actual singular point downstream represents the state $2 \leftrightarrow 3$, similar to the state $1 \leftrightarrow 2$ downstream of the Chapman-Jouguet detonation wave (Sect.1.9.2; Fig.1.8d). The physical reasons for this degeneracy are easily understood by noting that in the chosen coordinate system the transverse electric field is zero, i.e., there is no way to single out a definite direction of the switched-on magnetic field among the others. Accordingly, in place of the two singular points

2 and 3 we are dealing with a singular curve, the so-called Alfvén circle in the (u_x, B_y, B_z) phase space. For the sake of compactness we retain the subscript 2 to denote the singular downstream point, viewing the on-switching shock wave as an extreme case of a fast shock transition $1 \to 2$ as the angle θ between the normal to the shock and the initial magnetic field tends to zero. From the condition $B_2^2 > 0$, in accordance with (3.5.4), we obtain

$$1 < M_{a1}^2 < \frac{4M_1^2}{M_1^2 + 3} \equiv u_{GD}^{-1} \quad . \tag{3.5.10}$$

According to (3.5.8,10), the magnitude of M_2^2 in the switch-on shock waves may range from 1/5 to infinity.

If the speed of a normal shock wave (traveling at zero angle to the magnetic field) falls outside the range defined by (3.5.10), the boundary conditions preclude the switch on of the magnetic field's transverse component. In that case, in place of (3.5.2-6), we have a solution in which $u_2 = u_{GD}$, $B_2 = 0$, i.e., the normal shock waves turn out to be the common gasdynamic waves and do not interact with the magnetic field. For $M_{a1}^2 < 1$ and $M_{a1}^2 > u_{GD}^{-1}$ they are classified as the slow and fast MHD shocks, respectively. The boundary conditions allow the existence of normal gasdynamic shocks in the range (3.5.10); these shocks, however, correspond to the transitions $1 \to 4$ and are nonevolutionary.

According to (3.5.10), the normal shocks in a plasma can be of the switch-on type only if

$$\beta_1 < 6/5 \quad , \tag{3.5.11}$$

i.e., the longitudinal magnetic field must be strong enough for the given plasma temperature. At a present value of β_1 the maximum magnitude of the switched-on transverse magnetic field is

$$B_{2max} = \left(3 - \frac{5}{2}\beta_1\right) \cdot 6^{-\frac{1}{2}} \quad . \tag{3.5.12}$$

Accordingly, the magnetic field's rotation angle and the change in its absolute magnitude in the switch-on shock do not exceed $\arctan(3/2)^{\frac{1}{2}}$ and $(5/2)^{\frac{1}{2}}$, respectively (this limit is attained as $\beta_1 \to 0$, $M_{a1}^2 = 5/2$). For switch-on shocks, similarly to the transverse shocks, see (3.4.7), the density of plasma, the temperature, the pressure, the mean-free-path length and time increase. However, in deviation from (3.4.7), the increase is restricted from above by (3.5.10) for any given initial set of parameters.

3.5.2 Switch-On Shock-Wave Structures in Nonmagnetized Plasma

The structures of switch-on shocks in a plasma were studied in [3.59-72]. We start our discussion with the methodologically simple case of a nonmagnetized plasma, $\Omega_i\tau_i \ll \Omega_e\tau_e \ll 1$ [3.65]. In contrast to the geometry of the transverse shocks, Sect.3.4.2,3, the problem of the shock structure is here nonplanar: within the confines of the shock front both transverse components of either the velocity or the magnetic field are nonzero. Accordingly, in addition to (3.4.12) we shall also need the z component of Maxwell equation (3.1.31); in dimensionless variables it has the form

$$\frac{dB_z}{dz} = -\frac{M_{a1}^2}{M_1\delta_1\ell_1}\left(\frac{u_y^i}{u_x^i} - \frac{u_y^e}{u_x^e}\right) \quad . \tag{3.5.13}$$

The notation has been defined in Sect.3.4.2; note, however, that $\delta_1 = (\Omega_i\tau_i)^{-1} \gg 1$ is a large parameter. Equations (3.4.12,5.13) define the spatial scale which, considering the fact that in the switch-on shocks $M_{a1} \sim 1$, can be presented in the form

$$\Delta_h = \Delta_h^{(1)} u_x^{1/2} \quad , \quad \Delta_h^{(1)} = \frac{M_1\delta_1\ell_1}{M_{a1}} = c_a/\Omega_i \quad . \tag{3.5.14}$$

In other words, Δ_h coincides with the characteristic scale of the Hall dispersion mechanism. All the other characteristic scales are the same as in Sect.3.4.3. Therefore, the conclusion regarding the predominance of the Joule scale under condition (3.4.50) remains valid; following (3.4.49,5.14) the Joule scale is also large in comparison with Δ_h, their ratio being $(\Omega_e\tau_e)^{-1} \gg 1$. By comparing (3.4.21,70) with (3.5.14) we verify that the dispersion scales Δ_d and Δ_q are small compared to Δ_h, whatever the magnetization of the plasma. It follows that we may write the shock-layer equations disregarding all the dissipative mechanisms except the Joule mechanism and all the dispersive mechanisms but the Hall mechanism. As in Sect.3.4.2,3, for a first approximation we assume the plasma to be quasi-neutral, cf. (3.4.20); (3.4.51) is valid by virtue of $\Delta_j \gg \Delta_r$. In dimensionless variables the equations should be written so as to retain their form, whether the normalization is carried out to the initial or to the final state of the plasma (as opposed to the case of transverse shock waves, this does not occur automatically). The equations of conservation of momentum and energy of the plasma and the equation of motion of electrons have the form

$$u_x - 1 + \frac{3}{5M_k^2}\left(\frac{T}{u_x} - 1\right) + \frac{B_y^2 + B_z^2 - B_k^2}{2M_{ak}^2} = 0 \quad , \tag{3.5.15}$$

$$u_y^i - u_{yk} - \frac{B_y - B_k}{M_{ak}^2} = 0 \quad , \tag{3.5.16}$$

$$u_z^i - \frac{B_z}{M_{ak}^2} = 0 \quad , \tag{3.5.17}$$

$$\frac{1}{2}\left(u_x^2 + u_y^{i\,2} - 1 - u_{yk}^2\right) + \frac{3}{2M_k^2}(T - 1) + \frac{B_y^2 + B_z^2 - B_k^2}{M_{ak}^4} = 0 \quad , \tag{3.5.18}$$

$$\Delta_j^{(k)} T^{-3/2} \frac{dB_y}{dx} + \frac{\Delta_h^{(k)}}{M_{ak}} u_x \frac{dB_z}{dx} + B_y\left(\frac{1}{M_{ak}^2} - u_x\right) = 0 \quad , \tag{3.5.19}$$

$$\Delta_j^{(k)} T^{-3/2} \frac{dB_z}{dx} - \frac{\Delta_h^{(k)}}{M_{ak}} u_x \frac{dB_y}{dx} + B_z\left(\frac{1}{M_{ak}^2} - u_x\right) = 0 \quad . \tag{3.5.20}$$

Subscript k here takes on the values 1 or 2, depending on the normalization procedure. At $k=1$ we set $B_k = u_{yk} = 0$, whereas at $k=2$ we set $B_k = u_{yk} = B_2$, and make the transcription $1 \to 2$ in (3.4.11,12 and 3.5.13,14). From (3.5.15-18) we find ($B^2 \equiv B_y^2 + B_z^2$):

$$u_x(B^2) = \frac{1}{8}\left\{5 + \frac{3}{M_k^2} - \frac{5(B^2 - B_k^2)}{2M_{ak}^2}\right.$$

$$\left. \pm \left[\left(5 + \frac{3}{M_k^2} - \frac{5(B^2 - B_k^2)}{2M_{ak}^2}\right)^2 - 16\left(1 + \frac{3}{M_k^2} - \frac{B^2 - B_k^2}{M_{ak}^2}\right)\right]^{\frac{1}{2}}\right\} \quad , \tag{3.5.21}$$

$$T(B^2) = \frac{M_k^2}{3}\left[1 - u^2(B^2) - \frac{B^2 - B_k^2}{M_{ak}^2}\right] \quad . \tag{3.5.22}$$

In (3.5.19,20) we replace B_y, B_z by the new variables B^2, φ; $B_y = B \sin\varphi$, $B_z = B \cos\varphi$, and obtain

$$\frac{1}{2}\Delta_j^{(k)} T^{-3/2}\left[1 + \left(\frac{u_x \Delta_h^{(k)}}{M_{a1}\Delta_j^{(k)}}\right)^2 T^3\right]\frac{dB^2}{dx} = \left(u_x - \frac{1}{M_{a1}^2}\right)B^2 \quad , \tag{3.5.23}$$

$$\Delta_j^{(k)} \frac{d\varphi}{dx} = -\frac{u_x \Delta_h^{(k)}}{M_{a1}\Delta_j^{(k)}} T^3\left(u_x - \frac{1}{M_{a1}^2}\right) \quad . \tag{3.5.24}$$

Substituting (3.5.21) into (3.5.23) and ignoring the second term in brackets on the left-hand side of (3.5.23) as being small (of the order of $(\Omega_e \tau_e)^2$), we obtain a solution for the profile of the absolute magnitude of transverse magnetic field in the switch-on shock in quadrature form

$$\frac{x}{\Delta_j^{(k)}} = \frac{1}{2} \int_0^{B^2} \frac{dB'^2 T^{-3/2}(B'^2)}{B'^2[u_x(B'^2) - M_{a1}^{-2}]} \quad . \tag{3.5.25}$$

Substituting (3.5.25) into (3.5.21,23) allows us to find the profile of φ, i.e., the direction of the transverse magnetic field. By virtue of the smallness of the ratio Δ_h/Δ_j, this direction varies little (by the amount of the order of $\Omega_e\tau_e$) within the limits of the shock, and the vector **B** remains close to the xy plane.

At the singular point $k = 1,2$ we obtain from (3.5.21)

$$u_x = 1 = \frac{1}{8}\left(5 + \frac{3}{M_k^2} \pm 3\left|1 - \frac{1}{M_k^2}\right|\right) \quad . \tag{3.5.26}$$

Inasmuch as $M_1 > 1$ by virtue of (3.5.10), to satisfy the boundary condition (3.5.26) at point 1 we must choose the plus sign before the bracket in (3.5.21). The solution in the form of a quadrature describes the entire shock structure as long as the integral curve, which enters point 2, belongs to the same branch of the curve (3.5.21) (Fig.3.15a). For (3.5.26) to be satisfied at point 2, it is necessary that $M_2 > 1$. Otherwise point 2 is entered by the branch of the curve (3.5.21) with the minus sign in front of the bracket, and the shock structure contains a discontinuity corresponding to the passage from one branch to the other (Fig.3.15b). The values of u_x on the two branches at the same value of B^2 being linked by the conservation laws, the transition from one branch to the other is possible via a gasdynamic

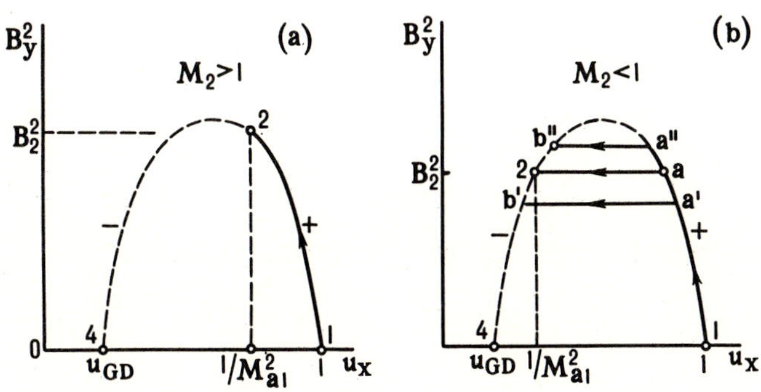

Fig.3.15a,b. Phase curves representing the structures of the switch-on shock waves in nonmagnetized plasma in the (u_x, B_y^2) plane: (**a**) purely resistive structure; (**b**) the structure containing isomagnetic subshock a → 2

jump. The situation here is entirely similar to that for transverse shocks; the internal discontinuity here is also isomagnetic due to $Pm \ll 1$, see (3.4.50). Going once again through the calculations done in Sect.3.4.3, we verify that, as long as $M_2 > 1$, the positive branch of (3.5.21) describes a monotonously decreasing function of B^2, and the quadrature (3.5.25) gives a correct description of the shock structure.

Linearizing (3.5.23) near the singular point 2 gives

$$\frac{d}{dx}(B^2 - B_2^2) = \mathrm{const}(1 - M_2^2)(B^2 - B_2^2) \quad , \tag{3.5.27}$$

where const > 0. Since for $M_2 < 1$ the limiting cycle $B^2 = B_2^2$ in the (B_y, B_z) plane is unstable as $x \to +\infty$, the integral curve of the set (3.5.23,24) does not approach it. Therefore, the solutions corresponding to the transitions $1 \to a' \to b' \to 2$, or $1 \to a'' \to b'' \to 2$ are not possible. Similarly to the transverse shock, a transition to point 2 is possible via the isomagnetic jump only, which, accordingly, is located in the tail portion of the shock structure $1 \to a \to 2$. As indicated in Sect.3.3.5, this appearance is typical of fast MHD shocks.

The structure of the isomagnetic jump is the same as the structure of the shock in nonmagnetized plasma (Sect.2.1). Similarly to Sect.3.4.3, we find that for $4/5 < M_2^2 < 1$ this jump is weak, its structure being shaped by electronic heat conduction. For $M_2^2 < 4/5$ the isomagnetic jump itself contains an electron-isothermal viscous jump.

Substituting $M_2^2 = 1$ and $M_2^2 = 4/5$ into (3.5.8), we obtain equations for the critical Mach-number lines M_{a1}^B and M_{a1}^T, corresponding to the strong and weak isomagnetic jumps in the Joule-shock structure. With $\beta_1 \to 0$, $M_{a1}^B \to (1 + 3 \cdot 5^{-1/2})^{1/2} = 1.530$, $M_{a1}^T \to (5/2)^{1/2} = 1.581$.

The profile of the longitudinal electric field is found from the equation of motion of electrons in the x coordinate and is similar in form to (3.4.58):

$$E_x = -r_{D1}\left(1.71\,\frac{dT}{du_x} - \frac{T}{u_x} + \frac{5M_1^2}{3M_{a1}^2}\,u_x\,\frac{dB^2}{du_x}\right)\frac{du_x}{dx} \tag{3.5.28}$$

in the variables normalized to state 1. Integrating (3.5.28), we find that the potential jumps across the shock (providing that the shock does not contain an isomagnetic jump), or across the Joule portion of the shock. The potential jump across the isomagnetic discontinuity is defined by the formulas of Sect.2.1.4.

3.5.3 Switch-On Shock-Wave Structure in Magnetized Plasma

It is much more difficult to account for the anisotropy of the transfer coefficients in the switch-on shock wave than to do the same for the transverse shock, since this time the plasma moves both crosswise and lengthwise to the magnetic field. If we define a unit vector collinear with the magnetic field as

$$\mathbf{b} = (1 + B^2)^{-\frac{1}{2}}\{1, B_y, B_z\} \quad , \tag{3.5.29}$$

then the longitudinal and transverse components of a vector may be represented as

$$\mathbf{a}_{\|} = \frac{\mathbf{a}\cdot\mathbf{b}}{(1 + B^2)^{\frac{1}{2}}}\{1, B_y, B_z\} \quad , \quad \mathbf{a}_{\perp} = \mathbf{a} - \mathbf{a}_{\|} \quad ; \tag{3.5.30}$$

in particular,

$$\nabla_{\|} = \frac{1}{1 + B^2}\left\{\frac{d}{dx} \, , \; B_y \frac{d}{dx} \; , \; B_z \frac{d}{dx}\right\} \quad . \tag{3.5.31}$$

For instance, eliminating the currents with the aid of Maxwell's equations, we find for the force of friction between electrons and ions

$$\begin{aligned} R_{uy} &= \frac{\varepsilon_0 m_e \bar{B}_x}{e^2 \tau_e}\left[- \frac{dB_z}{dx} - 0.49 \frac{B_y}{1 + B^2}\left(B_z \frac{dB_y}{dx} - B_y \frac{dB_z}{dx}\right)\right] \quad , \\ R_{uz} &= \frac{\varepsilon_0 m_e \bar{B}_x}{e^2 \tau_e}\left[\frac{dB_y}{dx} - 0.49 \frac{B_z}{1 + B^2}\left(B_z \frac{dB_y}{dx} - B_y \frac{dB_z}{dz}\right)\right] \quad . \end{aligned} \tag{3.5.32}$$

Similarly, for the thermal force, correct to terms small in ε and δ (note that here δ_1, δ_2 are small parameters, Sect.3.4.2), we obtain

$$R_{Ty} = - 0.71 \, \bar{n}_e \frac{B_y}{1 + B^2} \frac{d\bar{T}_e}{dx} \quad ; \quad R_{Tz} = - 0.71 \, \bar{n}_e \frac{B_z}{1 + B^2} \frac{d\bar{T}_e}{dx} \quad . \tag{3.5.33}$$

In the dimensionless shock-layer equations the terms accounting for the electron heat conduction and thermal force enter with the coefficient $(1 + B^2)^{-1}$. Insofar as for switch-on shocks $1 + B^2 \leqslant 5/2$ (Sect.2.5.1), the hierarchy of the characteristic scales at present remains the same as for shocks in nonmagnetized plasma, see (2.1.20). The greatest are the scales of electronic heat conduction Δ_{Te} and of electron-ion temperature relaxation Δ_r. The scales of ion viscosity Δ_{vi}, heat conduction Δ_{Ti}, and of Joule dissipations Δ_j are small as ε and $\varepsilon^2\delta^2$, respectively. The main dispersive mechanism is represented by the Hall mechanism with the characteristic scale

(3.5.14). For a first approximation we consider the plasma to be quasi-neutral.

Ignoring the ion viscosity and heat-conduction effects, and eliminating the transverse components of velocity with the aid of (3.5.16,17) gives

$$u_x - 1 + \frac{3}{10M_k^2}\left(\frac{T_e + T_i}{u_x} - 2\right) + \frac{B^2 - B_k^2}{2M_{ak}^2} = 0 \quad , \tag{3.5.34}$$

$$u_x^2 - 1 + \frac{3}{2M_k^2}(T_e + T_i - 2) + \Delta_{Te}\frac{dT_e}{dx} + \frac{B^2 - B_k^2}{M_{ak}^4} = 0 \quad , \tag{3.5.35}$$

$$\frac{3}{2}\frac{dT_i}{dx} + \frac{T_i}{u_x}\frac{du_x}{dx} = \frac{1}{\Delta_r}(T_e - T_i) \quad , \tag{3.5.36}$$

$$\Delta_j\frac{dB_z}{dx} - \frac{\Delta_h}{M_{ak}}\frac{dB_y}{dx} + 0.49\Delta_j\frac{B_y}{1 + B^2}\left(B_z\frac{dB_y}{dx} - B_y\frac{dB_z}{dx}\right)$$
$$+ \Delta_f B_y\frac{dT_e}{dx} = B_z\left(1 - \frac{1}{u_x M_{ak}^2}\right) \quad , \tag{3.5.37}$$

$$\Delta_j\frac{dB_y}{dx} + \frac{\Delta_h}{M_{ak}}\frac{dB_z}{dx} - 0.49\Delta_j\frac{B_z}{1 + B^2}\left(B_z\frac{dB_y}{dx} - B_y\frac{dB_z}{dx}\right)$$
$$- \Delta_f B_z\frac{dT_e}{dx} = B_y\left(1 - \frac{1}{u_x M_{ak}^2}\right) \quad . \tag{3.5.38}$$

Here Δ_r is defined by (2.1.20), Δ_j by (3.4.22), Δ_{Te} differs from the definition (2.1.20) by the factor $(1 + B^2)^{-1}$, and

$$\Delta_f = \frac{\Delta_f^{(k)}}{u_x(1 + B^2)} \quad , \quad \Delta_f^{(k)} = 0.213\frac{\delta_k \ell_k}{M_k} \quad . \tag{3.5.39}$$

Now considering (3.5.37,38) let us change from the variables B_y, B_z to the new variables B^2, φ:

$$\frac{dB^2}{dx} = 2B^2 M_{ak}\frac{\Delta_f}{\Delta_h}\frac{dT_e}{dx} + 2B^2 M_{ak}^2\frac{\Delta_j}{\Delta_h^2}\frac{1 + 0.51B^2}{1 + B^2}\left(1 - \frac{1}{u_x M_{ak}^2}\right) \quad , \tag{3.5.40}$$

$$\frac{d\varphi}{dx} = \frac{M_{ak}}{\Delta_h}\left(\frac{1}{u_x M_{ak}^2} - 1\right) \quad . \tag{3.5.41}$$

As follows from (3.5.40), since the dimensionless temperature $T_e \sim M^2$, the characteristic scale of the magnetic field variations is represented by the scale of electronic heat conduction Δ_{Te}. Accordingly, the width of the switch-on shock in magnetized plasma is of the same order as the characteristic length of electronic heat conduction. From (3.5.41) it follows that in the shock front the magnetic field vector is rotated by $\Delta\varphi$ of the order of $\Delta_{Te}/\Delta_h \sim \Omega_e \tau_e \gg 1$. To construct the solution we eliminate the dimensionless temperature of ions, T_i, from (3.5.34-36)

$$\Delta_{Te} \frac{dT_e}{dx} = (1 - u_x)\left(4u_x - 1 - \frac{1}{M_k^2}\right) - \frac{B^2 - B_k^2}{M_{ak}^2}\left(\frac{5u_x}{2} - \frac{1}{M_{ak}^2}\right) \quad , \qquad (3.5.42)$$

$$\frac{3}{2}\frac{dT_e}{dx} + \frac{5M_k^2}{2M_{ak}^2} u_x \frac{dB^2}{dx} + \left[\frac{T_e}{u_x} + \frac{5M_k^2}{3}\left(8u_x - 5 - \frac{3}{M_k^2} + \frac{5(B^2 - B_k^2)}{2M_{ak}^2}\right)\right]\frac{du_x}{dx}$$

$$+ \frac{2}{\Delta_r}\left[T_e + \frac{5M_k^2}{3} u_x^2 - u_x\left(\frac{5M_k^2}{3} + 1\right) + \frac{5M_k^2}{6M_{ak}^2} u_x(B^2 - B_k^2)\right] = 0 \quad . \qquad (3.5.43)$$

Equations (3.5.40,42,43) in B^2, T_e, u_x furnish a complete solution of the problem of shock structure. In the three-dimensional (u_x, T_e, B^2) phase space these equations have three singular points. Point 1 corresponds to the incident gas flow,

$$u_x = 1 \quad , \quad T_e = 1 \quad , \quad B^2 = 0$$

(the normalization is carried out with respect to point 1). Point 2 corresponds to the switch-on shock

$$u_x = \frac{1}{M_{a1}^2} \quad , \quad T_e = T_2 \quad , \quad B^2 = B_2^2 \quad .$$

Consistent with the MHD classification the equilibrium downstream state of the normal gasdynamic shock with Mach number M_1 is described by the singular point 4, the corresponding values of the dimensionless velocity, temperature and magnetic field being

$$u_x = u_{GD} = \frac{M_1^2 + 3}{4M_1^2} \quad ,$$

$$T_e = T_{GD} = (5M_1^2 - 1)(M_1^2 + 3)/16M_1^2 \quad , \quad B^2 = 0 \quad .$$

The desired solution for the switch-on shock is given by the integral curve of (3.5.40,42,43) in the (u_x, T_e, B^2) phase space, which comes out of point 1 at $x=-\infty$ and enters point 2 at $x=+\infty$.

Let us examine the general features of the structure. For this purpose we linearize (3.5.40,42,43) in the vicinity of the singular points 1,2,4 and consider the relevant characteristic equations. At point 1 we get

$$\left[\frac{\Delta_h^{(2)^2}}{M_{a1}^2\Delta_j^{(1)}} k - 2\left(1 - \frac{1}{M_{a1}^2}\right)\right]\left\{k^2\Delta_{Te}^{(1)}\Delta_r^{(1)}(5M_1^2 - 4) - k\left[\frac{9}{2}\Delta_r^{(1)}\left(1 - \frac{1}{M_1^2}\right) - 2\Delta_{Te}^{(1)}\left(\frac{5M_1^2}{3} - 1\right)\right] - 6\left(1 - \frac{1}{M_1^2}\right)\right\} = 0 \quad . \tag{3.5.44}$$

The first cofactor in (3.5.44) defines a positive root, since $M_{a1} > 1$; of the remaining two roots one is positive and one negative. The two positive roots of (3.5.44) define a two-dimensional surface $S_2^-(1)$ of integral curves originating at point 1. As point 1 in the plane $B^2=0$ is a saddle point, a unique integral curve exists that comes out of point 1 in the sense of increasing T_e and decreasing u_x.

The characteristic equation at the singular point 3 is

$$\left[\frac{\Delta_h^{(2)^2}}{M_{a1}^2\Delta_j^{(1)}} k - 2\left(1 - \frac{4M_1^2}{M_{a1}^2(M_1^2 + 3)}\right)\right]\left\{k^2\Delta_{Te}^{(4)}\Delta_r^{(4)}(5M_4^2 - 4) - k\left[\frac{9}{2}\Delta_r^{(4)}\left(1 - \frac{1}{M_4^2}\right) - 2\Delta_{Te}^{(4)}\left(\frac{5M_4^2}{3} - 1\right)\right] - 6\left(1 - \frac{1}{M_4^2}\right)\right\} = 0 \quad . \tag{3.5.45}$$

Here the variables in brackets are normalized to state 1, and those in braces to state 4. Mach number M_4 in the downstream flow of the gasdynamic shock wave is expressed via M_1 by (2.1.23) with the transcription $0 \to 1$, $2 \to 4$; by virtue of (3.5.10) $M_4 < 1$.

The root of (3.5.45), pertaining to the first cofactor, is negative for $M_{a1}^2 < 4M_1^2/(M_1^2+3)$, which corresponds to the switch-on shock wave. The other two roots are both negative for $M_4^2 > 4/5$ $(M_1^2 < 19/15)$ and have opposite signs for $M_4^2 < 4/5$ $(M_1^2 > 19/15)$. Accordingly, the integral curves which enter point 4 occupy either the three-dimensional space $S_3^+(4)$ for $M_1^2 < 19/15$, or the two-dimensional surface $S_2^+(4)$ for $M_1^2 > 19/15$.

The characteristic equation at point 2 (normalization to state 2) has the form

$$k^3(5M_2^2 - 4)\frac{\Delta_h^{(2)^2}\Delta_r^{(2)}\Delta_{Te}^{(2)}}{\Delta_j^{(2)}} + k^2\Big[\frac{5M_2^2}{6}B_2^2\Delta_r^{(2)}\Delta_{Te}^{(2)}\frac{1+0.51B_2^2}{1+B_2^2}$$

$$+\frac{1}{2}B_2^2\frac{\Delta_r^{(2)}\Delta_f^{(2)}\Delta_h^{(2)}}{\Delta_j^{(2)}} + \frac{\Delta_{Te}^{(2)}(\Delta_h^{(2)})^2}{3\Delta_j^{(2)}}\left(\frac{5M_2^2}{3} - 1\right) - \frac{3}{4}\left(1 - \frac{1}{M_2^2}\right)\frac{\Delta_r^{(2)}(\Delta_h^{(2)})^2}{\Delta_j^{(2)}}\Big]$$

$$+ k\Big[B_2^2\frac{1+0.51B_2^2}{1+B_2^2}\left(\frac{5M_2^2}{9}\Delta_{Te}^{(2)} - \frac{3}{4}\Delta_r^{(2)}\right) + \frac{(\Delta_h^{(2)})^2}{\Delta_j^{(2)}}\left(\frac{1}{M_2^2} - 1\right)$$

$$+ 2B_2^2\frac{\Delta_h^{(2)}\Delta_f^{(2)}}{\Delta_j^{(2)}}\Big] - B_2^2\frac{1+0.51B_2^2}{1+B_2^2} = 0 \quad . \qquad (3.5.46)$$

For $M_2^2 > 4/5$ (3.5.46) has one positive root and two negative ones; for $M_2^2 < 4/5$ two roots of (3.5.46) are positive and one negative. Accordingly, for $M_2^2 > 4/5$ the integral curves entering point 2 fill up the two-dimensional surface $S_2^+(2)$, whereas for $M_2^2 < 4/5$ point 2 is entered by only one integral curve $S_1^+(2)$ in the sense of increasing B^2 and decreasing u_x.

The general pattern of the integral curves is shown schematically in Fig. 3.16 for $M_1^2 < 19/15$, i.e., for $M_2^2 > 4/5$ and $M_{GD}^2 > 4/5$. The two-dimensional surfaces $S_2^-(1)$ and $S_2^+(2)$ intersect along the unique integral curve $1 \to 2$, which defines the structure of the switch-on shock wave. Here the shock structure is entirely shaped by the electron heat conduction, the electron-ion temperature relaxation, the Hall effect and the Joule dissipations, i.e,, friction between electrons and ions.

The variables B^2, T_e, u_x change monotonously across the shock, and the magnetic-field vector (the transverse component of the magnetic-field strength) rotates axially in the plane of the shock, completing a large number of revolutions across the shock layer. The tip of the magnetic-field vector describes a conical helix (diverging in the downstream direction) with the lead of the order of Δ_h. The intersection of the three-dimensional space $S_3^+(4)$ with the two-dimensional surface $S_2^-(1)$ defines a portion of the latter, which is filled up by integral curves, coming out of point 1 as $x \to -\infty$ and going into point 4 as $x \to +\infty$. Curve $1 \to 4$ in the plane $B^2 = 0$ represents a solution of the problem of the shock structure in nonmagnetized plasma for $M_2^2 > 4/5$, i.e., the structure of a weak isomagnetic jump. It is evident from Fig.3.16 that such an integral curve is not unique; an infinitely large number of suitable integral curves exist, which connect points 1 and 4 and

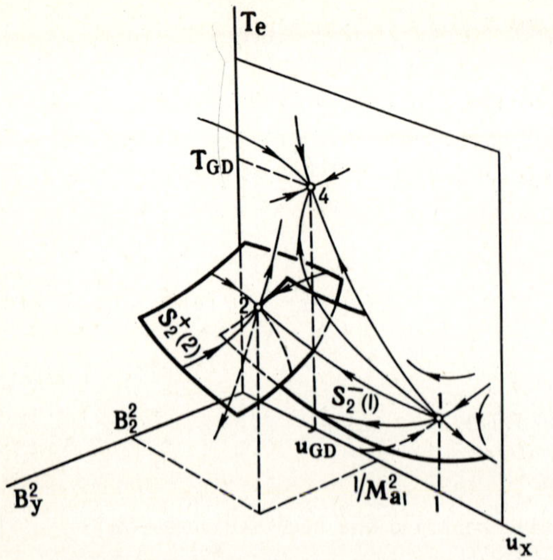

Fig.3.16. Integral curves of shock-layer equations in the (u_x, B_y^2, T_e) phase space for a switch-on shock wave in magnetized plasma dominated by electron heat conduction and Joule dissipations

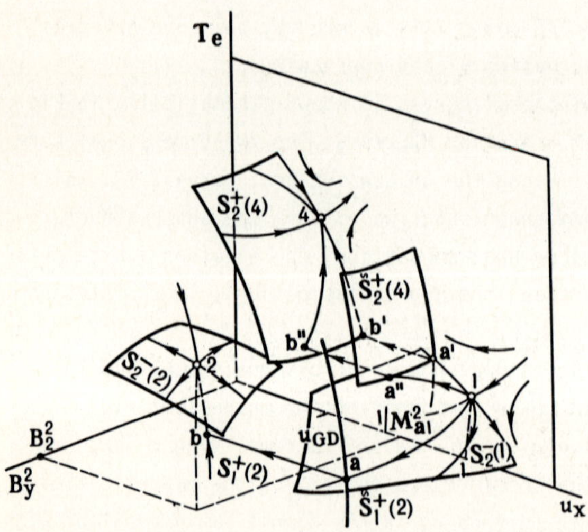

Fig.3.17. Integral curves of shock-layer equations in the (u_x, B_y^2, T_e) for a switch-on shock-wave in magnetized plasma, including isomagnetic electron-isothermal subshock a →b

occupy the intersection $S_2^-(1) \,_\Omega S_3^+(4)$. The fact that the solution for the structure of the gasdynamic shock transition 1→4 is not unique implies instability of the shock.

For $M_2^2 < 4/5$ the singular point 2 is entered by the unique integral curve of the set $S_1^+(2)$, as indicated in Fig.3.17. This curve does not coincide with $S_2^-(1)$. Substituting (3.5.40,42) into (3.5.43), we find that the sign of the term in braces before du_x/dx is reversed in going from 1 to 2 if

$M_2^2 < 4/5$. Accordingly, on the corresponding integral curve of the set (3.5.40,42,43), going from point 1 to point 2, the velocity becomes a double-valued function of coordinate x, which is physically unacceptable. This can be helped by introducing the appropriate internal discontinuity, the ion shock, to accomplish the transition from $S_2^-(1)$ to $S_1^+(2)$. The restriction $M_2^2 < 4/5$ is equivalent to the evolutionarity condition for the ion shock (Sect.2.1.3), and coincides with the condition of solvability of (3.5.58), which defines the relevant boundary condition.

As the scales of ion viscosity and ion heat conduction are small in comparison with the scale of electronic heat conduction, this discontinuity, being isomagnetic, is evidently electron-isothermal as well, i.e., within the ion shock $T_e = \text{const}$, being correct to small ε.

The components of the ion viscous stress tensor in dimensionless variables normalized to state 1 may be represented as

$$\frac{\pi^i_{xx}}{m_i\bar{n}_1\bar{u}^2_{x1}} = -\Delta_{vi}\frac{(2-B^2)}{3}G \quad ; \qquad \frac{\pi^i_{xy}}{m_i\bar{n}_1\bar{u}^2_{x1}} = \Delta_{vi}B_yG \quad ;$$

$$\frac{\pi^i_{xz}}{m_i\bar{n}_1\bar{u}^2_{x1}} = -\Delta_{vi}B_zG \quad , \tag{3.5.47}$$

where

$$G = \left[(2-B^2)\frac{du_x}{dx} + 3\left(B_y\frac{du^i_y}{dx} + B_z\frac{du_z}{dx}\right)\right]/(1+B^2)^2 \quad ,$$

and Δ_{vi} is defined in (3.1.20).

The equations of conservation of momentum of plasma in the transverse direction may then be written in the form

$$\Delta_{vi}B_yG = u^i_y - B_y/M^2_{a1} \quad , \tag{3.5.48}$$

$$\Delta_{vi}B_zG = u^i_z - B_z/M^2_{a1} \quad , \tag{3.5.49}$$

whence

$$u^i_y/B_y = u^i_z/B_z = \Delta_{vi}G + 1/M^2_{a1} \quad . \tag{3.5.50}$$

Using the equality, inferred from (3.5.50) correct to $O(\varepsilon)$

$$B_y\frac{du^i_y}{dx} + B_z\frac{du^i_z}{dx} = \Delta_{vi}B^2\frac{dG}{dx} \quad ,$$

we obtain the following equations for the structure of a viscous isomagnetic jump

$$(2 - B^2)\frac{du_x}{dx} + 3B^2\frac{d(\Delta_{vi}G)}{dx} = G(1 + B^2)^2 \quad , \tag{3.5.51}$$

$$u_x - 1 + \frac{3}{10M_1^2}\left(\frac{T_e + T_i}{u_x} - 2\right) + \frac{B^2}{2M_{a1}} = \frac{\Delta_{vi}G}{3}(2 - B^2) \quad , \tag{3.5.52}$$

$$\frac{3}{2}\frac{dT_i}{dx} + \frac{T_i}{u_x}\frac{du_x}{dx} = \frac{10M_1^2}{9}\Delta_{vi}G\left[(2 - B^2)\frac{du_x}{dx} + 3B_y\frac{du_y^i}{dx} + 3B_z\frac{du_z^i}{dz}\right]$$

$$+ \frac{10M_1^2}{3}\frac{d}{dx}\left(\Delta_{Ti}\frac{dT_i}{dx}\right) = \frac{10M_1^2}{9}G^2(1 + B^2)^2$$

$$+ \frac{10M_1^2}{3}\frac{d}{dx}\left(\Delta_{Ti}\frac{dT_i}{dx}\right) \quad . \tag{3.5.53}$$

Here (3.5.51) follows from (3.5.50); (3.5.52) follows from the equation of conservation of momentum of plasma; (3.5.53) is the equation of ion heat conduction, where the scale of ion heat conduction Δ_{Ti} differs from the definition (3.1.20) by a factor of $(1 + B^2)^{-1}$. Integrating (3.5.53) with the aid of (3.5.51,52) gives

$$\frac{10M_1^2}{3}\Delta_{Ti}\frac{dT_i}{dx} = \frac{3}{2}T_i - \frac{5M_1^2}{3}u_x^2 + u_x\left(\frac{10M_1^2}{3} + 2 - \frac{5M_1^2B^2}{3M_{a1}^2}\right)$$

$$- \frac{5M_1^2}{3}B^2\Delta_{vi}^2G^2 - T_e \ln u_x + F \quad , \tag{3.5.54}$$

where F is the integration constant. Here is a complete analogy with (2.1.46), which describes the structure of an internal ion shock in unmagnetized plasma.

The boundary conditions for the ion shock are set by reducing all the derivatives in (3.5.51,52,54) to zero, thus getting

$$G = 0 \quad , \tag{3.5.55}$$

$$u_x - 1 + \frac{3}{10M_1^2}\left(\frac{T_e + T_i}{u_x} - 2\right) + \frac{B^2}{2M_{a1}^2} = 0 \quad , \tag{3.5.56}$$

$$\frac{3}{2}T_i - \frac{5M_1^2}{3}u_x^2 + u_x\left(\frac{10M_1^2}{3} + 2 - \frac{5M_1^2}{3M_{a1}^2}B^2\right) - T_e \ln u_x + F = 0 \quad . \tag{3.5.57}$$

Eliminating T_i from (3.5.56,57), we obtain an equation which couples the velocity values in the points a and b (the terminal points of the ion shock):

$$T_e \ln \frac{u_x}{u_b} = \frac{5M_1^2}{3}(u_x - u_b)\left[5 + \frac{3}{M_1^2} - \frac{5B^2}{2M_{a1}^2} - 4(u_x + u_b)\right] . \quad (3.5.58)$$

For every point b in the (B^2, T_e, u_x) phase space (3.5.58) enables a point to be found a so that the passage from a to b via the ion shock is possible.

Thus, for $M_2^2 < 4/5$ the solution is set up in the following way. With the aid of (3.5.58) for each point of the integral curve $S_1^+(2)$ we find the corresponding point a. The locus of these points is some curve $\tilde{S}_1^+(2)$ (Fig.3.17). Let a be the intercept of $\tilde{S}_1^+(2)$ and $S_2^-(1)$, and b the corresponding point on $S_1^+(2)$. Then the structure of the switch-on shock will be represented by the integral curve $1 \to a \to b \to 2$, which contains an ion shock $a \to b$. In a strong shock (large M_1) the temperature of ions in the internal discontinuity rises above the equilibrium temperature in point 2. Then point b is followed by the relaxation layer (cf. similar solutions for the strong shock structures in Sect.2.1.3).

As previously, the transition 1-4, pertaining to the normal gasdynamic shock, corresponds to a multitude of possible structures. Following (3.5.58), for the surface $S_2^+(4)$ we can construct another surface $\tilde{S}_2^+(4)$ so that the transition from $\tilde{S}_2^+(4)$ to $S_2^+(4)$ occurs via the ion shock. The intersection of $\tilde{S}_2^+(4)$ with $S_2^-(1)$ is a locus of all points a", the onset of the ion shock in the structure $1 \to a'' \to b'' \to 4$. In particular, the transition $1 \to a' \to b' \to 4$ is always possible in the plane $B^2 = 0$, which represents a shock in an unmagnetized plasma for $M_2^2 < 4/5$.

The profile of longitudinal electric field on the scale Δ_{Te} has the form

$$E_x = -r_{D1}\left(\frac{d}{du_x}\frac{T_e}{u_x} + \frac{0.71}{u_x(1 + B^2)}\frac{dT_e}{du_x} + \frac{5M_1^2}{3M_{a1}^2}u_x\frac{dB^2}{du_x}\right)\frac{du_x}{dx} . \quad (3.5.59)$$

Integrating (3.5.59), we find the jump of potential across the shock front if $4/5 < M_2^2$, or the change in the potential outside. The electric field in the ion shock and the corresponding potential jump can readily be found along the lines of Sect.2.1.4.

Experiments with switch-on shocks [3.66-59,70,72] generally confirm the conclusions about the shock structure; in particular, the existence of magnetic-field oscillations at the head of the shock. However, attempts to assess these experiments quantitatively must account for the departure of plasma ionization from unity [3.71]; therefore, we shall postpone this discussion till Chap.4.

3.6 Structures of Switch-Off Shock Waves

Switch-off shocks may be viewed as the extreme case of slow MHD shocks, when their velocity tends to the highest possible velocity of a slow shock, the Alfvén velocity. Accordingly, we label the upstream singularity with index 3, although it would be more consistently denoted by the index $2 \equiv 3$, like the degenerate downstream singularity of the switch-on shock. In the present case the transverse electric field also vanishes, which causes degeneration. According to the MHD classification, the downstream singularity is represented by point 4. The set of boundary conditions for the switch-off shock coincides formally with (3.5.1-6,8,9), as long as we use the normalization to the final state 4, i.e., (3.5.1) now pertains to $x = +\infty$ and (3.5.2-6) to $x = -\infty$, with the transcription $1 \to 4$, $2 \to 3$. The appropriate expressions, derived from (3.5.4,8,9), have the form

$$B_3^2 = \frac{2}{3}\left(M_{a4}^2 - 1\right)\left(4 - \frac{M_4^2 + 3}{M_4^2} M_{a4}^2\right) \quad , \tag{3.6.1}$$

$$M_3^2 = 9M_4^2\left[15M_{a4}^4 - 6M_{a4}^2 + 5M_4^2(M_{a4}^2 - 1)^2\right]^{-1} \quad , \tag{3.6.2}$$

$$M_{a3}^2 = 1 \quad . \tag{3.6.3}$$

Reversing (3.6.1,2), we can express the downstream Mach numbers M_4, M_{a4}, and thus the entire set of boundary conditions, in terms of the parameters of the initial state 3.

Note that despite the formal analogy between the boundary conditions for the switch-on and switch-off shocks, an essential dissimilarity exists, deriving physically from the fact that the zero transverse magnetic field state corresponds to the minimum entropy in one case and to the maximum entropy in the other. This fact is reflected in the different ranges of the corresponding Mach numbers. The upstream sonic Mach number in the switch-off shocks varies from zero to infinity, whereas downstream (by virtue of $B_3^2 > 0$, $M_3^2 > 0$)

$$\frac{4M_4^2}{M_4^2 + 3} < M_{a4}^2 < 1 \quad , \quad \frac{3M_{a4}^2(2 - 5M_{a4}^2)}{5(M_{a4}^2 - 1)^2} < M_4^2 < 1 \quad . \tag{3.6.4}$$

Let it be emphasized that for a given initial upstream state of the plasma a unique shock wave exists capable of switching off the existing transverse magnetic field, whose Mach numbers are given by (3.6.3) and

$$M_3 = (6/5\beta_3)^{\frac{1}{2}} \cos\theta \quad , \tag{3.6.5}$$

where $\beta_3 = 2\mu_0\bar{p}_3/\bar{B}_3^2$, and θ is the angle between the initial magnetic field and the normal to the shock front. Each point of the two-dimensional space defined by the inequalities (3.6.4) corresponds to a definite initial state. Here we are dealing with the manifestation of one of the differences between switch-on and switch-off shocks: the switched-on component of the magnetic field does not exceed $(3/2)^{\frac{1}{2}}\bar{B}_x$, while there are no restrictions on the magnitude of the switched-off components.

The structures of the switch-off shock waves are described by the same equations, as used in Sects.3.5.2,3, and their analysis is similar.

The principal process which shapes the shock structure in nonmagnetized plasma is represented by Joule dissipations. In this case the shock-layer equations in dimensionless variables normalized to state 3 are reduced to (3.5.15-18,23,24), where we set $k=3$ and $M_{a3}=1$. Eventually, we get

$$\Delta_j^{(3)} \cdot \frac{dB^2}{dx} = T^{3/2}B(u_x - 1) \quad , \tag{3.6.6}$$

$$\frac{\Delta_j^2}{\Delta_h^{(3)}} \frac{d\varphi}{dx} = u_x T^3 (1 - u_x) \quad , \tag{3.6.7}$$

$$u_x - 1 + \frac{3}{5M_3^2}\left(\frac{T}{u_x} - 1\right) + \frac{B^2 - B_3^2}{2} = 0 \quad , \tag{3.6.8}$$

$$u_x^2 - 1 + \frac{3}{M_3^2}(T - 1) + B^2 - B_3^2 = 0 \quad . \tag{3.6.9}$$

From (3.6.8,9) we obtain

$$F(B,u_x) = 4u_x^2 - \left[5 + \frac{3}{M_3^2} - \frac{5}{2}(B^2 - B_3^2)\right]u_x + 1 + \frac{3}{M_3^2} - (B^2 - B_3^2) = 0 \quad . \tag{3.6.10}$$

Let us illustrate the structure of the switch-off shock in the (B,u_x) phase plane. In Fig.3.18 the transition from point 3 to point 4 is represented by the portion of curve (3.6.10) originating in point 1 with coordinates $u_x = 1$, $B = B_1$, and terminating at the intercept of this curve with the zero-field hyperbola, which presently degenerates into two straight lines $u_x = 1$ and $B = 0$ in accordance with the boundary conditions for the switch-off shock. Arrows in the diagram indicate the direction of motion along the curve (3.6.10) in the sense of increasing compression and increasing entropy downstream of the shock. As follows from Fig.3.18, a smooth transition from point 3 to point 4 (including only the Joule dissipations)

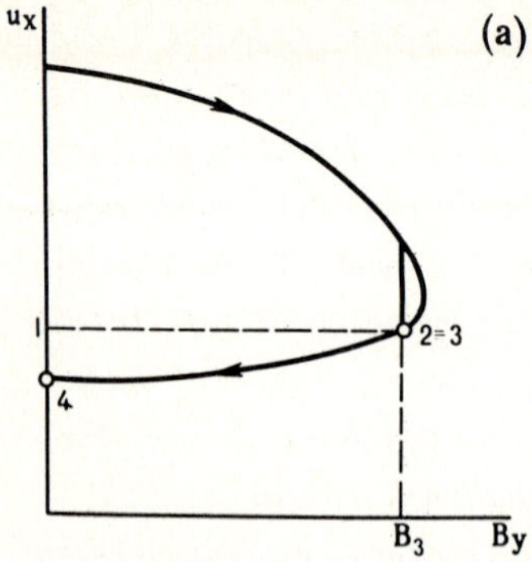

Fig.3.18a-c. Phase curves representing the structures of switch-off shock waves in non-magnetized plasma in the (u_x, B_y) plane: (**a**) purely resistive structure; (**b**,**c**) the structures containing isomagnetic subshocks $2 \equiv 3 \rightarrow b$

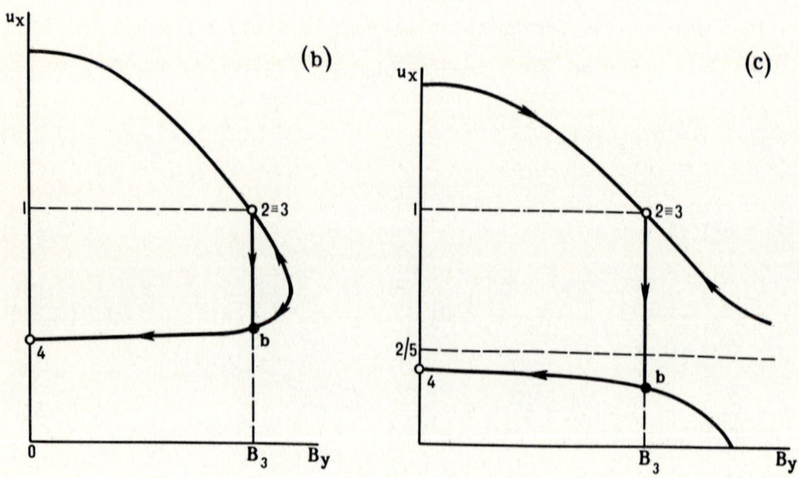

is possible only for the subsonic flow for $M_3 < 1$ (Fig.3.18a). With the supersonic flow ($M_3 > 1$) the structure of the switch-off shock, as follows from Fig.3.18b,c, starts with an isomagnetic jump, whose internal structure is determined by heat conduction and viscosity. The velocity and temperature in the isomagnetic jump change at constant $B = B_3$ from $u_x = 1$, $T = 1$ to $u_x = u_{GD} = (M_3^2 + 3)/4M_3^2$, $T = (5M_3^4 + 14M_3^2 - 3)/16M_3^2$.

The isomagnetic jump is followed by the Joule region, associated with the compression of the plasma (if $1 < M_3^2 < 5$) or expansion (if $M_3^2 > 5$), as shown in Fig.3.18b,c. The shock structure on the Joule scale Δ_j is readily found from (3.6.6-9); for instance, for B^2 we obtain

$$\frac{x}{\Delta_j^{(3)}} = \frac{1}{2} \int_{B_3^2}^{0} \frac{T^{-3/2}(B^2)\,dB^2}{B^2[u_x(B^2) - 1]} \quad . \tag{3.6.11}$$

Note that in the particular case $M_3^2 = 5$ the dimensionless velocity behind the isomagnetic jump $u_x(B^2) = 2/5 = \text{const}$, and the integral (3.6.11) is easily taken in elementary functions.

In the slow subsonic shocks ($M_3 \ll 1$), an important role is assumed by the electron heat conduction, which heats the plasma far ahead of the shock compression region. The scale of electron heat conduction reaches maximum for $M_3^2 < \Omega_{e3}\tau_{e3} \ll 1$. The compression is then negligibly small, i.e., we may set $u_x = 1$. The temperature change in the shock is described by ($T_e = T_i = T$)

$$\frac{3}{5M_3^2}(T - 1) + \frac{B^2 - B_3^2}{2} = 0 \quad , \tag{3.6.12}$$

$$\frac{3}{M_3^2}(T - 1) + B^2 - B_3^2 = \Delta_{Te}^{(3)} T^{3/2} \frac{dT}{dx} \quad , \tag{3.6.13}$$

where Δ_{Te} is defined in (2.1.20) with the transcription $0 \to 3$. From these equations we deduce that the temperature exhibits a smooth change from its equilibrium value $T = T_3$ at $x = -\infty$ to $T = T_4$ at $x = 0$:

$$\begin{aligned} \frac{x}{M_3^2 \Delta_{Te}^{(3)}} &= -\frac{2}{9}(T_4^{5/2} - T^{5/2}) - \frac{10}{27}(T_4^{3/2} - T^{3/2}) \\ &\quad - \frac{10}{9}(T_4^{1/2} - T^{1/2}) - \frac{5}{9}\ln\frac{(T_4^{\frac{1}{2}} - 1)(T^{\frac{1}{2}} + 1)}{(T_4^{\frac{1}{2}} + 1)(T^{\frac{1}{2}} - 1)} \quad . \end{aligned} \tag{3.6.14}$$

For $x > 0$ the solution is once again given by (3.6.6-9), where we set $T = T_4$.

Consider now the structure of the switch-off shock in magnetized plasma. This time the shock width depends on the processes of electron heat conduction and electron-ion heat exchange, and the shock-layer equations coincide with (3.5.34-36,40,41) with the transcription $1 \to 3$, $2 \to 4$, i.e., $k = 3,4$.

As follows from (3.5.40), the characteristic scale of the magnetic-field variations B^2 is now the scale of electron heat conduction. From (3.5.41) it is clear that the transverse component of the magnetic field completes a large number of revolutions about the axis (the total angle of rotation $\Delta\varphi \sim \Omega_e\tau_e \gg 1$).

In the three-dimensional (u_x, B^2, T_e) phase space the shock structure is represented by the integral curve of the set (3.5.40,42,43), which comes out of point 3 with the coordinates (1, B_3^2, 1) at $x = -\infty$, and enters the singular point 4 with the coordinates (u_4, 0, T_4) at $x = +\infty$. This set of equations has yet another singular point 1, which corresponds to the initial state from which the passage to point 4 is accomplished via a normal gasdynamic shock. Such a transition is physically acceptable as long as $M_1^2 > 1$, which is equivalent to $M_2^2 \geqslant 1/5$.

The characteristic equation of the set (3.5.40,42,43) at point 3 coincides with (3.5.44) with the transcription $1 \leftrightarrow 3$. For $M_3^2 > 4/5$ this equation has one root positive and two negative; for $M_3^2 < 4/5$ two of its roots are positive, and one negative. Hence it follows that for $M_3^2 > 4/5$ a unique integral curve $S_1^-(3)$ exists, which goes to point 3 as $x \to -\infty$ in the sense of increasing B^2, whereas for $M_3^2 < 4/5$ the integral curves form a two-dimensional surface $S_2^-(3)$. At point 4 the equation has one positive root for $M_4^2 < 4/5$, while for $4/5 < M_4^2 < 1$ all three roots are negative. Accordingly, the integral curves, which enter point 4 at $x \to +\infty$, fill up a two-dimensional surface $S_2^+(4)$ for $M_4^2 < 4/5$, and a three-dimensional region $S_3^+(4)$ for $4/5 < M_4^2 < 1$. It follows that if $M_3^2 > 4/5$ and $M_4^2 > 4/5$, then the unique integral curve, which comes out of point 3 in the sense of decreasing B^2, arrives at point 4. It is exactly this integral curve that represents the solution for the shock structure in the (u_x, B^2, T_e) phase space.

For $M_3^2 < 4/5$ and $M_4^2 < 4/5$, the sought-for structure is represented by the intersection of the two-dimensional surfaces $S_2^-(3) \cap S_2^+(4)$, as shown in Fig. 3.19.

In either case the shock front is shaped by the processes of electron heat conduction, electron-ion temperature relaxation, the Hall effect, thermal force, and friction between electrons and ions. The variables u_x, B^2, T_e all change monotonously across the shock layer, and the tip of vector **B** describes a tapered helix, converging downstream.

For $M_3^2 > 4/5$ and $M_4^2 > 4/5$, the set of equations in u_x, B^2, T_e has no solution for the unambiguous and smooth transition from point 3 to point 4, since the term in braces in front of du_x/dx in (3.5.42) has opposite signs at points 3 and 4. This calls for introducing an internal discontinuity in the shock structure, the ion shock transition from $S_1^-(3)$ to $S_2^+(4)$, just as in Sect.3.5.3:

$$(2 - B^2)\,\Delta_{vi}\,\frac{du_x}{dx} + 3B^2\Delta_{vi}\,\frac{dG}{dx} = G(1 + B^2)^2 \quad , \tag{3.6.15}$$

$$u_x - 1 + \frac{3}{10M_3^2}\left(\frac{T_e + T_i}{u_x} - 2\right) + \frac{B^2 - B_3^2}{2} = \frac{G(2 - B^2)}{3} \quad . \tag{3.6.16}$$

Integrating the equation of ion heat conduction once, we obtain

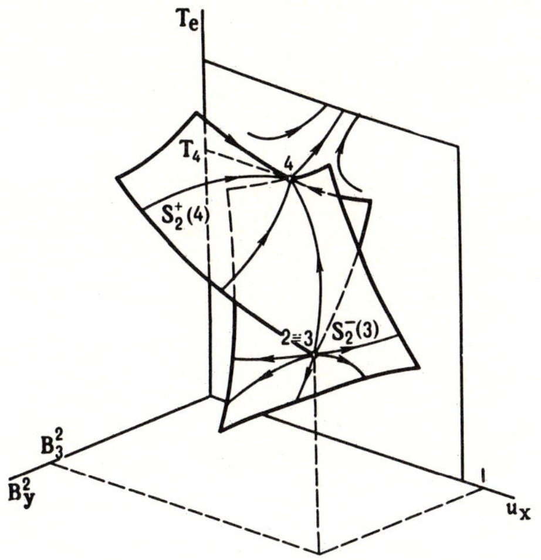

Fig.3.19. Integral curves of shock-layer equations in the (u_x, B_y^2, T_e) phase space for a switch-off shock wave in magnetized plasma, dominated by electron heat conduction and Joule dissipations

$$\frac{10M_1^2}{3}\Delta_{Ti}\frac{dT_i}{dx} = \frac{3}{2}T_i - \frac{5M_3^2}{3}u_x^2 + u_x\left(\frac{10M_3^2}{3} + 2 - \frac{5M_3^2(B^2 - B_3^2)}{3}\right)$$
$$- \frac{5M_3^2}{3}B^2G^2 - T_e \ln u_x + F \quad . \qquad (3.6.17)$$

The boundary conditions for the ion shock are imposed by bringing to zero the value of G, the left-hand side of (3.6.16) and the right-hand side of (3.6.17). Eliminating T_i, we obtain

$$T_e \ln\frac{u_x}{u_b} = \frac{5M_3^2}{3}(u_x - u_b)\left[5 + \frac{3}{M_3^2} - \frac{5(B^2 - B_3^2)}{2} - 4(u_x + u_b)\right] \quad . \qquad (3.6.18)$$

Equation (3.6.18) allows us to find for each point b of the surface $S_4^+(4)$ the corresponding point a, from which transition to point b via the ion shock is possible. The locus of these points is a surface $\tilde{S}_2^+(4)$ (Fig.3.20). Let a be the intersection of $S_1^-(3)$ and $S_2^+(4)$, and b the corresponding point of $S_2^+(4)$. Then the structure of the switch-off shock has the form $3 \to a \to b \to 4$ and contains the internal ion shock transition a-b, as indicated in Fig.3.20. If $M_3 \gg 1$, then the temperature jump is concentrated in the ion shock (portion $a \to b$), where the ion temperature may rise well above the equilibrium downstream value T_4. Relaxation of the ion temperature takes place in the region $b \to 4$.

The meaning of the restrictions $M_3^2 > 4/5$, $M_4^2 < 4/5$ was discussed in Sect. 2.1.3: they represent the evolutionarity conditions of the ion shock, coin-

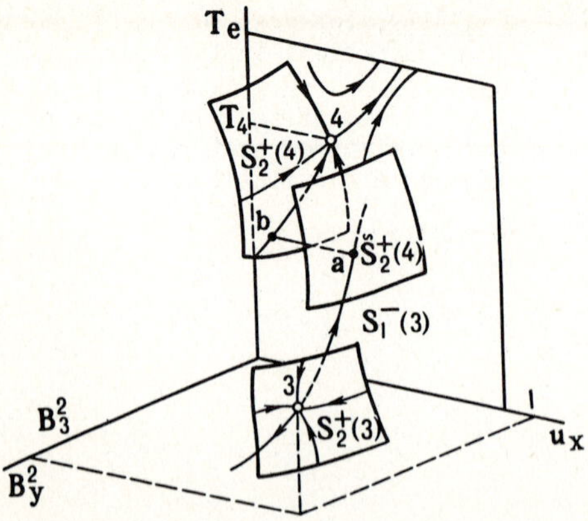

Fig.3.20. Integral curves of shock-layer equations in the (u_x, B_y^2, T_e) phase space for a switch-off shock wave in magnetized plasma, including isomagnetic electron-isothermal sub-shock $a \to b$

ciding with the condition of solvability of (3.6.18). In the present case of the slow shock, which can be subsonic, the former of these conditions is not satisfied automatically unlike the cases in Sects.2.1.3 and 3.5.3.

For $B_3 \gg 1$ the structure of the off-switching shock bears no analogy to the switch-on shock structure. For $B_3 \gtrsim \varepsilon^{-1/2}$ the terms including the electron and ion heat conductivity contain the small coefficient $1/B_3^2$ and are negligibly small at the head of the shock. From (3.5.42) follows that presently B and u_x are linked by (3.6.10), and the shock structure for $M_3 > 1$, as indicated in Fig.3.18b, should start with the viscous isomagnetic jump. The electron heat conductivity being small, the electrons in the viscous jump are heated by the adiabatic compression,

$$T_e = u_x^{-2/3} \quad . \tag{3.6.19}$$

By virtue of (3.6.19), we obtain in place of (3.6.17)

$$T_i = \frac{10M_3^2}{9}(1 - u_x)^2 + u_x - u_x^{-2/3} + \frac{2}{3} + \frac{5M_3^2}{3} B_3^2 G^2 \quad . \tag{3.6.20}$$

Eliminating T_e and T_i from (3.6.16), we obtain for a first approximation (correct down to terms small in $1/B_3^2$)

$$G = -\frac{(u_x - 1)(4u_x - 1 - 3/M_3^2)}{B_3^2 - 2} \quad . \tag{3.6.21}$$

Substituting (3.6.21) into (3.6.15), we obtain the solution for the ion shock structure in the form of a quadrature

$$\frac{x}{\Delta_{vi}^{(3)}} = \frac{1}{4} \int_{(1+u_4)/2}^{u_x} \frac{T_i^{5/2}(u)du}{(u - 1)[u - (M_3^2 + 3)/4M_3^2]} . \tag{3.6.22}$$

For $M_3 \gg 1$ the right-hand side of (3.6.20) is dominated by the first term, while the last term is small, being of the order of $1/B_3^2$. To this accuracy the integral (3.6.22) can be taken in elementary functions:

$$\frac{x - x_0}{0.074 M_1^4 \ell_1} = -\frac{u_x^4}{4} + \frac{5}{4} u_x^3 - \frac{81}{32} u_x^2 + \frac{175}{64} u_x - \frac{81}{265}\left(u_x - \frac{1}{4}\right) . \tag{3.6.23}$$

The solutions obtained are similar to the solutions for the transverse shock structure in magnetized plasma.

In both cases the solution in quadratures becomes feasible, because the ion heat conduction is suppressed by the transverse magnetic field. Behind the viscous discontinuity the magnetic field strength drops following (3.6.13), and the electrons are heated by collisions with ions. Therefore, the electron heat conductivity increases, and the further structure is defined by the same relations as those discussed above for $M_3^2 > 4/5$, $M_4^2 < 4/5$.

In conclusion, let us observe that the peculiar feature of slow shocks – the downstream fall-off of magnetization – can be manifest in the structure of the switch-off shock. Indeed, $\Omega\tau \sim \bar{B}\bar{T}^{3/2}/\bar{n}$, and the decrease in $\bar{B} = \bar{B}_x(1 + B^2)^{\frac{1}{2}}$ may happen to overcompensate the increase in the characteristic time of Coulomb interactions, $\tau \propto \bar{T}^{3/2}/\bar{n}$, so that the plasma may be magnetized ahead of the switch-off shock, and nonmagnetized behind the shock. Such a situation may be encountered with $B_3 \gg 1$, $M_3 \ll 1$, $M_4 \ll 1$. Then the head of the shock is dominated by the electron heat conduction, and the shock structure is the same as in unmagnetized plasma. Switch-off shocks have been observed experimentally [3.73]; unfortunately, the poor reproducibility of the results has so far prevented comparison with theory.

4. Ionizing Shock Waves in Magnetic Fields: Structures and Stability

Ionizing shock waves convert the initially nonconducting neutral gas into a plasma. If the gas is confined to the region of a magnetic field, then the incipient conduction in the shock front switches on the MHD interactions between the flow and the external magnetic field. In this respect the ionizing shocks differ from the shocks in a plasma with magnetic field, where the MHD interaction is observed throughout the entire flow.

Extensive experimental and theoretical investigations of ionizing shock waves in magnetic fields were started in the fifties, after the construction of powerful electromagnetic shock tubes enabled studying such shocks in the laboratory [4.1,2]. It was also then that a peculiar feature of ionizing shocks in a magnetic field was discovered: the basic boundary conditions at the shock front, derived from the conservation laws and Maxwell equations, are not sufficient for unambiguously determining the downstream state in terms of the shock velocity and the initial parameters of the flow [4.3,4]. With varying types of ionizing shocks in magnetic fields two-parameter, one-parameter and zero-parameter families of final states and corresponding shock structures may exist, described by zero, one and two additional boundary conditions (ABC) respectively, as required by the evolutionarity criteria (Sect.1.7).

Discussion of the form of these additional boundary conditions continued into the early eighties. They were shown to depend mainly on the profiles of ionization build-up, and thus to be determined by the mechanisms of precursor ionization ahead of the shock front.

In Sect.4.1 we specify the basic boundary conditions for ionizing shocks in magnetic field (Sect.4.1.1) and classify the shocks on the basis of the evolutionarity conditions (Sect.4.1.2). In Sect.4.2.1 we discuss the case of $Pm \ll 1$, common for typical experiments with ionizing shocks. In Sect.4.2.2 we find how small the initial conductivity of the gas should be for the distinction between the ionizing shock waves and the MHD shock waves to be manifest. Section 4.2.3 is dedicated to precursor ionization in ionizing shocks in a magnetic field. This case differs from that discussed in Sect. 2.3.2 by the existence of the induced electric field ahead of the shock in neutral gas, which may strongly affect the profile of precursor ionization. In Sect.4.2.4 we derive the ABC and consider the general features of the shock structures for all types of ionizing shocks in a magnetic field. The ultimate transition from the ionizing shocks to MHD shocks, as their intensity increases, is traced in Sect.4.2.5.

In Sects.4.3-5 we consider the most interesting configurations of flow in the magnetic field. In addition, we discuss some phenomena, the most interesting of which is the Joule heating of the heavy plasma component in normal ionizing shocks, derived from the tensor nature of conductivity in the precursor region [4.5,6]. We also discuss the results of analytic and numerical computations of structures of the transverse (Sect.4.3.3) and the normal (Sect.4.4.2) ionizing shocks in the MHD approximation, and discuss the experimental data available.

4.1 Classification and the Problem of Boundary Conditions

4.1.1 The Basic Boundary Conditions

The ideal flow on either side of the ionizing shock front in a magnetic field is described by 9 variables: density ρ, temperature T (or entropy S), 3 components of velocity $\mathbf{u}$ and the transverse components of the electric and the magnetic fields E_y, E_z, B_y, B_z; $B_x = \text{const}$ by virtue of (3.1.1). The basic boundary conditions state the continuity of the transverse components of the electric field and the mass fluxes, and of each of the three components of momentum and energy; thus, the number of the boundary conditions is 7, as in the MHD model. However, for a MHD shock wave, both upstream and downstream,

$$\mathbf{E} + \mathbf{u} \times \mathbf{B} = 0 \quad , \tag{4.1.1}$$

i.e., E_y, E_z can be expressed via the components of $\mathbf{u}$, $\mathbf{B}$, and thus the number of independent variables is reduced to 7, equaling the number of basic boundary conditions. For an ionizing shock in a magnetic field, (4.1.1) holds only behind the shock, where the conductivity is finite. Ahead of the ionizing shock the left-hand side of (4.1.1) may be nonzero. Zero current in the equilibrium upstream state, cf. (3.3.23), is ensured by zero conductivity of neutral gas. The requirement of zero upstream conductivity is essential. Obviously, with finite (however small) conductivity ahead of the shock the boundary conditions become the same as in the MHD approximation: the left-hand side of (4.1.1) is zero ahead of the shock as well. The transition from MHD to ionizing shocks as $\sigma_0 \to 0$ is discussed in Sect.4.2.2.

Consequently, the number of flow variables is two more than the number of basic boundary conditions. Therefore, the basic boundary conditions do not predetermine the final state of the flow, given the parameters of the initial state and the velocity of the shock front. An ambiguity still remains, which can be removed by fixing the values of two additional parameters, e.g., the two components of electric field ahead of the shock.

In accordance with (2.1.2), the transverse electric field in the shock layer is constant. Let the z axis of the reference frame run along the electric field:

$$E_y = \text{const} = 0 \quad . \tag{4.1.2}$$

Now we write the set of basic boundary conditions in dimensionless variables, normalized to state 0, see (3.3.17); the dimensionless electric field is defined as

$$E_s = \bar{E}_z/\bar{u}_{x0}\bar{B}_x \quad . \tag{4.1.3}$$

The subscript k pertains to the downstream state. In the adopted frame, by virtue of (4.1.1) we may write for the state behind the shock

$$\mathbf{u}_k = \{u_{xk}, u_{yk}, 0\} \quad , \qquad \mathbf{B}_k = \{1, B_{yk}, 0\} \quad . \tag{4.1.4}$$

It is convenient to fix the reference frame by setting the y component of the upstream velocity equal to zero:

$$\mathbf{u}_0 = \{1, 0, u_{z0}\} \quad . \tag{4.1.5}$$

[As follows from (4.1.4), this condition is compatible with (4.1.2)]. The continuity equation in dimensionless variables has the form (3.3.19). Eliminating the dimensionless density ρ, we obtain equations of continuity of fluxes of momentum and energy:

$$u_{xk} - 1 + \frac{1}{\gamma M_0^2}\left(\frac{T_k}{u_{xk}} - 1\right) + \frac{\mathbf{B}_{\perp k}^2 - \mathbf{B}_{\perp 0}^2}{2M_{a0}^2} = 0 \quad , \tag{4.1.6}$$

$$u_{yk} - \frac{B_{yk}}{M_{a0}^2} = - \frac{B_{y0}}{M_{a0}^2} \quad , \tag{4.1.7}$$

$$u_{zk} - \frac{B_{zk}}{M_{a0}^2} = 0 \quad , \tag{4.1.8}$$

$$\mathbf{u}_k^2 - 1 - u_{z0}^2 + \frac{1}{(\gamma - 1)M_0^2}(T_k - 1) + \frac{2E_s(B_{y0} - B_y)}{M_{a0}^2} = 0 \quad , \tag{4.1.9}$$

plus two equations expressing the continuity of the electric field:

$$E_s + u_{xk}B_{yk} - u_{yk} = 0 \quad , \tag{4.1.10}$$

$$u_{xk}B_{zk} - u_{zk} = 0 \quad . \tag{4.1.11}$$

Here $\mathbf{B}_\perp^2 = B_y^2 + B_z^2$; and the sonic and the Alfvén Mach numbers are defined in Sect.3.3. In order not to burden our discussion with irrelevant details, we do not include in (4.1.9) the term accounting for the energy expended on the ionization of the gas, although these expenditures ought to be included in the calculations.

According to (4.1.4,8), $u_{zk} = B_{zk} = 0$; (4.1.11) is also satisfied. The remaining equations do not contain u_{zk}, B_{zk}, and are essentially the same as the equations determining MHD shock adiabats, see (3.3.22,23), the difference being only due to the fact that the normalization is here carried out with respect to the point 0, where (4.1.10) is not valid. Accordingly, we obtain four roots for u_{xk} from the set (4.1.6,7,9,10), corresponding to the MHD singular points $k = 1, 2, 3, 4$. The parameters of the corresponding states should, evidently, comply with the inequalities (3.3.27,32) [4.7-9].

The arguments developed above allow ionizing shock waves in a magnetic field to be classified according to the type of the final state, i.e., ionizing shocks of the types 1, 2, 3, 4. The basic boundary conditions do not determine the compression, heating and variation of magnetic field in the shocks of each of these types. These quantities depend on the values of two additional parameters, such as, for instance, the magnitude and the direction of the electric induction field ahead of the shock. For our present purposes it is more convenient, however, to use E_s and B_{z0}, which, of course, makes no difference whatever.

4.1.2 Evolutionarity Conditions

To formulate the evolutionarity conditions, we have to assess the number of linear running waves going away from the ionizing shock front in a magnetic field. Assuming the nonconducting medium ahead of the ionizing shock to stretch unrestrictedly, the disturbances of the transverse components of electric and magnetic field are carried away from the shock by electromagnetic waves having either of the two possible polarizations, and the disturbances of gasdynamic variables are carried by only the sonic waves. (Recall that the entropy and the vortex waves ahead of the shock cannot be outgoing.) The number of basic boundary conditions being equal to 7, by virtue of (1.7.8) and (3.3.37) the number r' or additional boundary conditions at the evolutionary ionizing shock front of type k is

$$r' = \begin{cases} 4 - k, & M_0 > 1 \;; \\ 5 - k, & M_0 < 1 \;. \end{cases} \tag{4.1.12}$$

Inasmuch as the flow is characterized by 9 variables, we are left with only two free parameters which can be fixed by the ABC; therefore,

$$0 \leqslant r' \leqslant 2 \;. \tag{4.1.13}$$

As follows from (4.1.12,13), the existence of evolutionary ionizing shocks of type 1 is precluded. Supersonic ionizing shock waves ($M_0 > 1$) may pertain to types 2,3,4, whereas the subsonic waves ($M_0 < 1$) may pertain only to types 3 and 4.

For $M_0 > 1$, $k = 4$ the number of ABC $r' = 0$. Accordingly, the supersonic ionizing shocks of type 4 are evolutionary if there are no additional relations connecting the values of variables on either side of the shock. For a given shock velocity a two-parameter family of different downstream states and corresponding structures exists, which is represented by a two-dimensional region on the (E_s, B_{y0}) plane.

A single additional boundary condition ($r' = 1$) corresponds to supersonic ionizing shocks of type 4 and subsonic ionizing shocks of type 3. In this case for a given shock velocity a one-parameter family of different downstream states and corresponding shock structures exists, represented by a certain curve in the (E_s, B_{y0}) phase plane.

Finally, two additional boundary conditions ($r' = 2$) correspond to supersonic ionizing shocks of type 2 and subsonic ionizing shocks of type 3. Then for a given shock velocity the shock structure and the jumps of all the variables across the shock front are unambiguously determined by the set of basic and additional boundary conditions. The possible types of evolutionary ionizing shocks and the outgoing waves are shown schematically in Fig.4.1.

In accordance with the results of Sect.1.7.1, we also have to make sure that the matrix of coefficients for the amplitudes of outgoing waves does

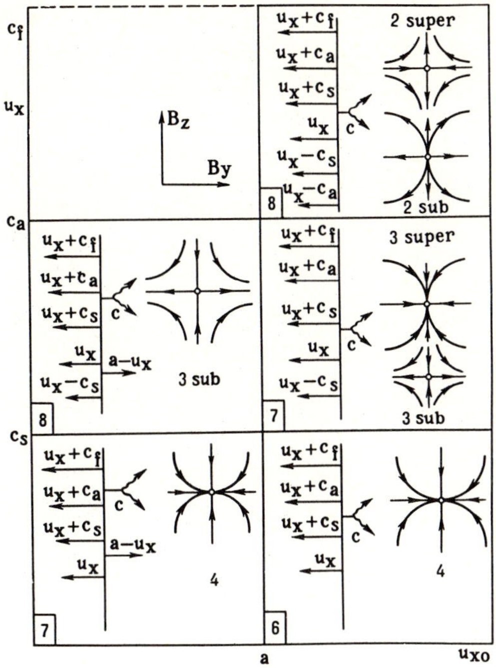

Fig.4.1. The number and types of waves diverging from the front of an evolutionary ionizing shock wave of indicated type, and the behavior of integral curves of (4.2.7,8) in the neighborhood of the corresponding downstream singular points in the (B_y, B_z) plane

not exhibit a block structure, i.e., the perturbations of the flow variables do not fall apart into separate groups, transmitted exclusively by certain kinds of waves (as, for instance, is the case with magnetohydrodynamics, where the perturbations of u_z and B_z are carried only by the Alfvén waves). Here we confine ourselves to supersonic ionizing shocks; a similar treatment can be applied to subsonic shocks.

Ahead of the shock two outgoing electromagnetic waves exist, pertaining to two independent polarizations. Designating the amplitudes of perturbations of E_z in one of the waves and of E_y in the other by E_1 and $-E_2$, respectively, we may write for the amplitudes of perturbations of the magnetic field in the electromagnetic waves

$$\delta B_{y0} = E_1 \quad , \quad \delta B_{z0} = E_2 \quad . \tag{4.1.14}$$

For the amplitudes of perturbations of all variables in the downstream MHD wave we retain the notation of Sect.3.3.5, dropping the subscript ℓ. Now we linearize the relations at the discontinuity, bearing in mind that disturbed ahead of the shock are only the four electromagnetic variables (we leave out the amplitudes of perturbations which obviously pertain to the incident waves), whereas in the downstream region generally all 7 variables describing the MHD flow are perturbed. The amplitudes of F^+, A^+, S^+ and S^0 always pertain to the outgoing waves; which of the amplitudes of F^-, A^-, S^- pertain to the outgoing waves depends on the type of shock. After some straightforward algebra in dimensional variables we get from the equation of conservation of the x component of momentum

$$B_{y0}E_1 + B_{z0}E_2 + L_1(F^+, F^-, S^+, S^-, S^0) = 0 \quad ; \tag{4.1.15}$$

from the equation of conservation of the y component of momentum

$$- B_xE_1 + L_2(F^+, F^-, S^+, S^-) = 0 \quad ; \tag{4.1.16}$$

from the equation of conservation of the z component of momentum

$$- B_xE_2 + L_3(A^+, A^-) = 0 \quad ; \tag{4.1.17}$$

from the equation of conservation of energy

$$u_{x0}B_{y0}E_1 + (u_{x0}B_{z0} - u_{z0}B_x)E_2 + L_4(F^+, F^-, S^+, S^-, S^0) = 0 \quad ; \tag{4.1.18}$$

and from the condition of continuity of the tangential component of the electric field

$$- E_2 + L_5(A^+, A^-) = 0 \quad ; \tag{4.1.19}$$

$$E_1 + L_6(F^+, F^-, S^+, S^-, S^0) = 0 \quad . \tag{4.1.20}$$

Here, as in Sect.3.3.5, $L_1 \ldots L_6$ are linear quantics, whose exact form does not concern us.

We see that the set of (4.1.15-20) generally does not separate into subsets for separate subgroups of variables. The same is true for the subsonic ionizing shocks. Here the Alfvên waves are not singled out similarly to magnetohydrodynamics: the perturbations of u_z, B_z downstream are expressible via E_z by the strength of (4.1.17,19), i.e., they are related to the disturbances of other variables by (4.1.15). Accordingly, the ionizing shocks in a magnetic field, differently to MHD shocks, may be trans-Alfvên ($M_{a0} > 1$, $M_{a3} < 1$ or $M_{a4} < 1$) [4.7,8]. Note that the degenerate case of merged singular points 3 and 4, which is of little interest in the study of MHD shocks as it describes simply the limit of zero intensity of the slow MHD shock, may here correspond to the ionizing shock, pertaining to the transition $0 \to 3 \equiv 4$. Then the downstream velocity coincides with the local velocity of the slow magnetosonic waves. By analogy with the theory of detonation (Sect.1.9), these shocks are called the Chapman-Jouguet ionizing shocks; in accordance with the definitions of Sects.1.7.1, 3.3.6 they are classified as singular shock waves. As we shall see below, these shocks, like the Chapman-Jouguet detonation waves in detonation theory, play an important role in the theory of ionizing shocks in magnetic field.

The experimental reality imposes certain restrictions on the flow of plasma and on the electromagnetic fields; this may call for the modification of the obtained results. The simplest way to assess these restrictions is to consider a flow sandwiched between two conducting planes (a rough model of the walls of a coaxial electromagnetic shock tube with $R_{out} - R_{in} \ll R_{out}$). Let the normal to these planes be directed along the z axis. Confining ourselves to the quasi-unidimensional case, where all the variables depend only on x, we get: $u_z \equiv 0$, $B_z \equiv 0$, $E_y \equiv 0$. We see that the number of flow variables is reduced to 6, and the number of boundary conditions is reduced to 5. Accordingly, the number of outgoing waves is also reduced: the Alfvên wave is precluded downstream, and one of the electromagnetic waves upstream. Going once again through the arguments developed above, we verify that the evolutionary ones are the supersonic shocks of types 2 and 3 and the subsonic shocks of type 4 (all of them requiring one additional boundary condition), and the one-parameter families of supersonic shocks of type 4 [4.9,10]. We shall not give this case an individual treatment, since all the relevant results are readily obtained within the general context of the problem if we state one of the additional boundary conditions in the form $B_{z0} = 0$.

It ought to be emphasized that the conclusions regarding the evolutionarity of ionizing shocks in a magnetic field hold only if the neutral gas ahead of the shock may be considered a dissipative medium, in accordance with the requirements of Sect.1.7.2. In the present context this means that the perturbations of the degree of ionization (or, which is the same, the perturbations of conductivity), which is strictly zero in the equilibrium upstream state, are either absent or strongly damped.

Observe finally that the evolutionarity criteria enable one to determine the number of ABC required by particular ionizing shocks, but say nothing about the exact form of ABC.

4.2 Shock Structures and Additional Boundary Conditions

4.2.1 Magnetic Structures of Ionizing Shocks as Pm → 0

Our qualitative analysis of the structures of ionizing shocks in magnetic fields is based on several simplifying assumptions. First, let us note that the experiments with ionizing shocks are characterized by temperatures of several electronvolts and densities of the order of $10^{16}cm^{-3}$. According to (3.3.60), $Pm \ll 1$ in this range. Below we shall confine our discussion to the zero-order approximation for a small Pm. Accordingly, here we consider the plasma to be an ideal (inviscid, thermally nonconducting) medium with finite conductivity. The action of viscosity, heat conduction and other collisional mechanisms, whose scales are small in comparison with the diffusion length of magnetic field Δ_j, is manifest in the shock structures only as internal subshocks, which in this approximation are represented by isomagnetic zero-width discontinuities.

Secondly, here, as in Sect.4.1.1, we ignore energy expenditure for the ionization of the gas.

Thirdly, we use the scalar form of Ohm's law (3.1.3). This is the strongest of our assumptions. Being concerned mainly with the qualitative analysis of the problem, we use this model for its convenience.

In the adopted approximation from Ohm's law (3.1.3) and the Maxwell equation (3.1.2) we derive

$$\Delta_j \frac{dB_y}{dx} = E_s + u_x B_y - u_y \quad , \tag{4.2.1}$$

$$\Delta_j \frac{dB_z}{dx} = u_x B_z - u_z \quad , \quad \text{where} \tag{4.2.2}$$

$$\Delta_j = \nu_m/\bar{u}_{x0} = m_e \nu_e/\mu_0 e^2 n_e \bar{u}_{x0} \quad , \tag{4.2.3}$$

ν_m being the magnetic viscosity, see (3.1.7), and ν_e the rate of elastic collisions of electrons. With low ionization, as follows from (4.2.3),

$$\Delta_j \propto \alpha^{-1} \quad . \tag{4.2.4}$$

Apart from (4.2.1,2), the set of shock-layer equations includes the equations of continuity and conservation of momentum and energy, and those of ionization kinetics and thermal balance of electrons. Incidentally, in the accepted approximation the qualitative analysis of the magnetic structures of ionizing shocks in the (u_x, B_y, B_z) phase space can be carried out regardless of the actual form of the last two equations.

Indeed, the dimensionless density can be eliminated from the continuity equation: $\rho = u_x^{-1}$. The equations of conservation have the form (4.1.6-9), where we change the notation by dropping the subscript k of the variables u_x, u_y, u_z, B_y, B_z and replacing T_k by $\alpha T_e + T$ (see the notations in Sect. 2.2.1). The equations of conservation can be used to express the transverse components of velocity on the right-hand sides of (4.2.1,2) via the magnetic-field components. The relationship between u_x, B_y, B_z is given by

$$F(u_x,B_y,B_z) \equiv 4u_x^2 - u_x\left[5 + \frac{3}{M_0^2} - \frac{5}{2M_{a0}^2}(B_y^2 + B_z^2 - B_{y0}^2 - B_{z0}^2)\right]$$
$$+ 1 + \frac{3}{M_0^2} - \frac{B_z^2 - B_{z0}^2}{M_{a0}^4} - \frac{(B_y - B_{y0})^2}{M_{a0}^4} + 2E_s\frac{B_y - B_{y0}}{M_{a0}^2} = 0 \tag{4.2.5}$$

(for simplicity henceforth we assume $\gamma_i = \gamma_e = 5/3$). Changing the variable $x \to \xi$, where

$$d\xi = dx/\Delta_j \quad , \tag{4.2.6}$$

we obtain equations which together with (4.2.5) describe the magnetic structures of ionizing shocks in the phase space:

$$\frac{dB_y}{d\xi} = E_s + \frac{B_{y0}}{M_{a0}^2} + B_y\left(u_x - \frac{1}{M_{a0}^2}\right) \quad , \tag{4.2.7}$$

$$\frac{dB_z}{d\xi} = B_z\left(u_x - \frac{1}{M_{a0}^2}\right) \quad . \tag{4.2.8}$$

Of course, to obtain the profiles of all the variables as functions of x we have to integrate (4.2.6) with due regard to the explicit form of the kinetic equation of ionization and the equation of electron heat balance. However, the pattern of integral curves in the (u_x,B_y,B_z) phase space does not depend on these equations, as long as $\alpha \neq 0$, $\Delta_j < \infty$ [4.7,8].

It is convenient to rewrite (4.2.5) in an equivalent form [4.8]

$$[B_y - B_y^*(u_x)]^2 + B_z^2 = [B_{y0} - B_y^*(u_x)]^2 + B_{z0}^2$$
$$+ \frac{4M_{a0}^4(u_x - 1)(u_x - u_{GD})}{1 - (5/2)u_x M_{a0}^2} \quad , \tag{4.2.9}$$

where

$$B_y^*(u_x) = \frac{E_s M_{a0}^2 + B_{y0}}{1 - (5/2)u_x M_{a0}^2} \quad , \qquad u_{GD} = \frac{M_0^2 + 3}{4M_0^2} \quad . \tag{4.2.10}$$

The ionizing shock's magnetic structure is represented in the phase space by a curve on the surface (4.2.5). In particular, this surface contains points 0 and k, pertaining to the initial ($x = -\infty$, $u_x = 1$, $B_y = B_{y0}$, $B_z = B_{z0}$), and the final ($x = +\infty$) states of the gas, as well as the point pertaining to the transition from point 0 via the gasdynamic shock, its coordinates being

$$u_x = u_{GD},\ B_y = B_{y0},\ B_z = B_{z0} \quad . \tag{4.2.11}$$

From (4.2.9) it follows that the sections of the surface (4.2.5) cut by planes $u_x = \text{const}$ are circles, whose centers lie on the hyperbola

$$B_z = 0,\ B_y = B_y^*(u_x) \quad .$$

The sections of the same surface cut by the planes $B_y = \text{const}$ and $B_z = \text{const}$ are third-order curves, which usually consist of several separate branches. The surface (4.2.5) itself may be composed of a number of sheets. Note also that at $E_s + B_{y0}/M_{a0}^2 = 0$, i.e., at $B_y^*(u_x) \equiv 0$, the variables B_y and B_z enter (4.2.9) only as a combination $(B_y^2 + B_z^2)$, i.e., the surface (4.2.5) is invariant under rotations about the u_x axis.

The expressions for the local Mach numbers M and M_a, defined with respect to the characteristic velocites at the given point of the flow, have the form (3.3.20,21). From (3.3.21) follows that the circles $u_x = \text{const}$ on the surface (4.2.5) correspond also to constant values of the local Alfvén Mach numbers. Expressing the temperature of the gas T in terms of u_x, B_y, B_z via the equations of conservation and making use of (3.3.20), we obtain

$$\frac{\partial F}{\partial u_x} = 8u_x - 5 - \frac{3}{M_0^2} - \frac{5}{2M_{a0}^2}\left(B_y^2 + B_z^2 - B_{y0}^2 - B_{z0}^2\right) = 3u_x\left(1 - \frac{1}{M^2}\right) \quad . \tag{4.2.12}$$

A fixed value of the local Mach number corresponds to a paraboloid of revolution in the phase space, its equation being readily obtainable from (3.3.20):

$$\frac{5}{3u_x}\left(1 + \frac{3}{5M_0^2} - u_x - \frac{B_y^2 + B_z^2 - B_{y0}^2 - B_{z0}^2}{2M_{a0}^2}\right) = \frac{1}{M^2} \quad . \tag{4.2.13}$$

Obviously, only the values $M^2 > 0$ are physically acceptable. Therefore, the curves representing the magnetic structures must fall within the paraboloid defined by (4.2.13), where the right-hand side is set equal to zero, i.e., within the region $M^2 = \infty$. The portion of the surface (4.2.5), which transcends the indicated paraboloid, represents a nonphysical region.

Certain general properties of the surface (4.2.5) can be established from analyzing the neighborhood of point 0. Indeed, we may examine the same surface by using the variables normalized with respect to any given point. Equation (4.2.5) then retains its form; what is changed is only the set of parameters $(M_0, M_{a0}, B_{y0}, B_{z0}, E_s)$.

In particular, for given values of B_y and B_z the surface (4.2.5) contains point 0 and a point with coordinates (4.2.11). Using (3.3.20), we find the local sonic Mach number M' at the point (4.2.11):

$$M'^2 = \frac{M_0^2 + 3}{5M_0^2 - 1} \quad . \tag{4.2.14}$$

From (4.2.14) it follows that

$$\begin{aligned} 1/5 < M'^2 < 1 \quad , &\quad 1/4 < u_{GD} < 1 \quad \text{for} \quad M_0 > 1 \\ M'^2 > 1 \quad , &\quad 1 < u_{GD} < 4 \quad \text{for} \quad 1/5 < M_0 < 1 \\ M'^2 < 0 \quad , &\quad u_{GD} > 4 \quad \text{for} \quad M_0 < 1/5 \quad . \end{aligned} \tag{4.2.15}$$

We see that the surface (4.2.5) is naturally divided into two regions, in one of which the local sonic Mach number M is always greater than unity (supersonic sheet of the surface), and in the other always less than unity (subsonic sheet). Given B_y, B_z, the point corresponding to the larger of the two values of u_x belongs to the supersonic sheet, and the point corresponding to the smaller of the two lies on the subsonic sheet. As follows from (4.2.15), the nonphysical portion of the surface (4.2.5) is part of its supersonic sheet. The supersonic and subsonic sheets may either be unconnected, or join along the line $M=1$, the so-called sonic line, represented by the intersection of the surface (4.2.5) with the paraboloid (4.2.15) at $M=1$. If at the given point of the sonic line the surface (4.2.5) is nondegenerate, i.e., one of the derivatives $\partial F/\partial B_y$, $\partial F/\partial B_z$ is nonzero, then the normal to the surface at this point is parallel to the plane (B_y, B_z). If both derivatives vanish, the regularity of the surface (4.2.5) at this point of the sonic line is violated and the normal to the surface is indefinite. The regularity may be violated at some point of the sonic line, and along the entire sonic line with the appropriate adjustment of the parameters, e.g., with $E_s + B_{y0}/M_{a0} = 0$ and $M_{a0}^2 = 2/5$ or $8/5$, as $M_0 \to \infty$. This situation is shown schematically in Fig. 4.2 below. Thus, the sonic line represents the locus of critical points when the surface (4.2.5) is projected onto the (B_y, B_z) plane, the projection of the sonic line being the discriminant curve of (4.2.5).

Obviously, in the vicinity of any point of the projection of the surface (4.2.5) onto the (B_y, B_z) plane—as long as this point does not belong to the discriminant curve—the velocities u_{x+} and u_{x-} pertaining to the supersonic and the subsonic sheets of the surface can be unambiguously expressed via B_y and B_z by virtue of (4.2.5), viewed as a quadratic equation in u_x. With the aid of these expressions the set (4.2.7,8) is converted into two sets of equations in B_y and B_z describing the magnetic structure of ionizing shock on the (B_y,B_z) phase plane: one for the supersonic sheet of the surface (4.2.5), $u_x = u_{x+}(B_y,B_z)$, and the other for the subsonic sheet, $u_x = u_{x-}(B_y,B_z)$. Both sets of equations coincide with one another on the discriminant curve in the (B_y,B_z) plane.

Downstream of the ionizing shock $\sigma \neq 0$, and the scale Δ_j takes on a certain finite value. Integrating (4.2.6) we find that $\xi \to \infty$ as $x \to \infty$. Therefore, the singular point k of the original set of shock-layer equations is simultaneously the singularity of the set (4.2.7,8), where the right-hand sides of these equations become zero. The integral curve, which represents the magnetic structure of the shock, must enter this point as $\xi \to +\infty$. On the contrary, as $x \to -\infty$ the right-hand sides of (4.2.1,2) are finite; the derivatives dB_y/dx and dB_z/dx go to zero together with $1/\Delta_j$. If $1/\Delta_j$ [and thus, in accordance with (4.2.4), the degree of ionization α] decreases rapidly enough with $x \to -\infty$, so that the integral

$$\xi(-\infty) = \int^{-\infty} \frac{dx'}{\Delta_j(x')} \qquad (4.2.16)$$

converges, then the state of the gas ahead of the ionizing shock is represented by a nonsingular point of the set (4.2.7,8). It is this case that we are presently concerned with, since the divergent integral (4.2.16) pertains to magnetic structures of MHD shocks. In other words, the peculiarities of ionizing shocks in a magnetic field are not manifest if the upstream ionization drops at such a slow rate. In Sect.4.2.3 we demonstrate that this integral is finite for typical experimental parameters of ionizing shocks.

The singular points, which represent the downstream state of the gas in the (u_x,B_y,B_z) phase space, are the intercepts of the surface (4.2.5) with the zero-field hyperbola, see (3.3.23), defined by

$$B_z = 0 \ , \quad E_s + B_{y0}/M_{a0}^2 + B_y(u_x - 1/M_{a0}^2) = 0 \ . \qquad (4.2.17)$$

As indicated in Sect.4.4.1, there are no more than four such points, which comply with the conventional MHD classification (Sect.3.3.2); in particular, the inequalities (3.3.27-32) still hold. As in magnetohydrodynamics, the equality signs are restricted to the degenerate cases, corresponding to the merging of two or more points. As follows from (3.3.27-30), this implies the equality of the velocity of the downstream flow to one of the characteristic velocities. In particular, the coincidence of points 2 and 3 is possible only when the downstream velocity equals the Alfvén velocity, i.e., $M_a = 1$.

From (3.3.21 and 4.2.17) it is clear that this also requires that $E_s + B_y/M_{a0}^2 = 0$. Consequently, in the present degenerate case instead of the isolated singularities 2 and 3 we are dealing with one singularity on the circumference $u_x M_{a0}^2 = 1$, both on the surface (4.2.5) and in the (B_y, B_z) plane. As already indicated, the integral curves in the (u_x, B_y, B_z) phase space are then invariant under rotation about the u_x axis (Fig.4.2), and the integral curves in the (B_y, B_z) plane are straight-line segments, passing through the origin of coordinates.

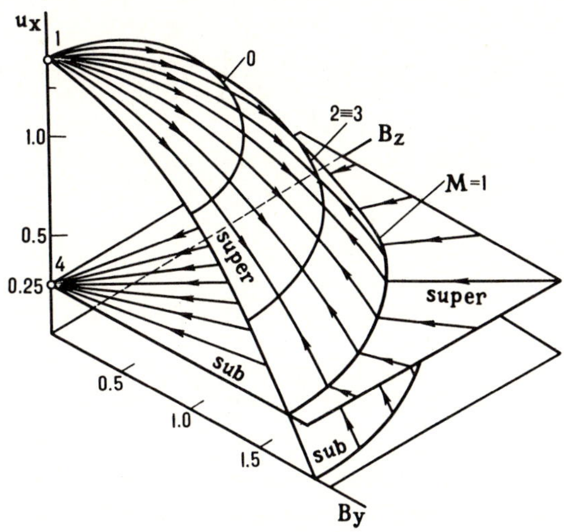

Fig.4.2. Integral curves of (4.2.7,8) on the surface (4.2.5) in the (u_x, B_y, B_z) space in the degenerate case. The curves are symmetrical with respect to rotations about the u_x axis because the relation $E_s + B_{y0}/M_{a0}^2 = 0$ holds. On the sonic line $M = 1$ the regularity of the surface is violated due to a special choice of the parameters: $M_0 \to \infty$, $M_{a0} = 8/5$

The merging of points 3 and 4 corresponds to the Chapman-Jouguet shock (Sect.4.1.2). Usually the Chapman-Jouguet shock is represented by the point where the section of the surface (4.2.5) by the plane $B_z = 0$ touches the zero-field hyperbola. This is quite natural, as the two intercepts of these curves ultimately come together and become a tangent point. An exception is the degenerate case when the tangent to the curve at the meeting point with the hyperbola cannot be defined. Such a situation arises only when $B_y = 0$ at this point, i.e., with the switch-off shock. From (4.2.17) it follows that then $E_s + B_y/M_{a0}^2 = 0$. The singularity, similar to that occurring on the sonic line in Fig.4.2, exists here at a single point on the u_x axis.

The magnetic structures of ionizing shocks are represented by the integral curves of (4.2.5,7,8) which, as $\xi \to +\infty$, tend to one of the singular points $k = 1,2,3,4$. The families of these integral curves can be plotted either on

◄ Fig.4.3

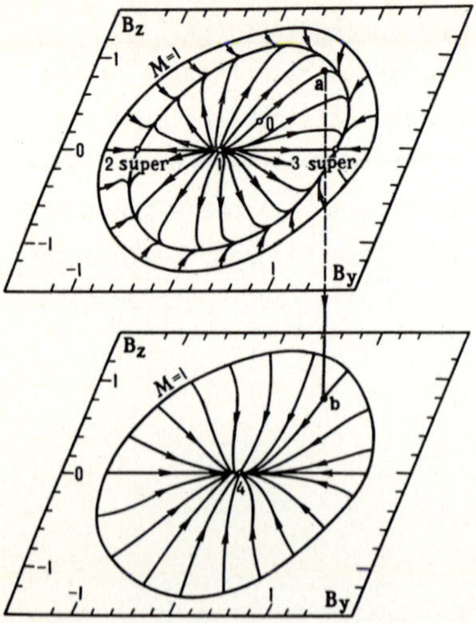

Fig.4.4 ▼

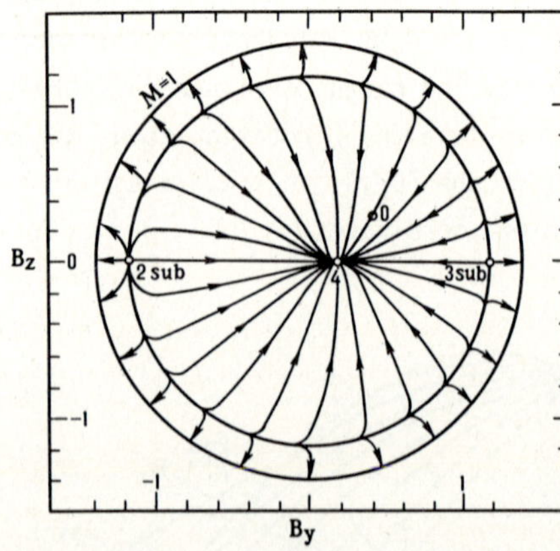

Fig.4.3. Integral curves of (4.2.7,8) in the (B_y, B_z) plane, corresponding to the supersonic sheet of the surface (4.2.5) (*top*) and to the subsonic sheet (*bottom*) for a supersonic ionizing shock wave ($M_0 \to \infty$). The vertical arrow indicates the isomagnetic jump from the supersonic sheet to the subsonic one

Fig.4.4. Integral curves of (4.2.7,8) in the (B_y, B_z) plane, corresponding to the subsonic sheet of the surface (4.2.5) for a subsonic ionizing shock wave ($M_0 < 1$)

the (B_y, B_z) plane (the supersonic and the subsonic sets separately), or on the surface (4.2.5). It would be natural to directionalize these curves in the sense of increasing ξ (and increasing entropy S). The direction of increasing entropy is indicated by arrows in Figs.4.2-4. At fixed B_y, B_z, the laws of conservation allow the jump from the supersonic to the subsonic sheet. In the ionizing-shock structure such a jump corresponds to an internal subshock, similar to the common gasdynamic shock in partly ionized plasma. In a zero approximation with respect to the small parameter Pm, the magnetic field within such a subshock remains constant and the entropy increases.

The transition from the supersonic to the subsonic sheet of the surface (4.2.5) depicts physically realizable isomagnetic subshocks only if the supersonic point does not fall into the nonphysical region, in compliance with (4.2.13-15). Accordingly, the magnetic structure of the supersonic ionizing shock ($M_0 > 1$) can be pictured in the following way. The phase point, representing the flow in the space of variables (u_x, B_y, B_z), starts moving from the nonsingular point 0 on the supersonic sheet of the surface (4.2.5) along the integral curve of the set (4.2.7,8) in the sense of increasing ξ (as indicated by arrows in Figs.4.2-4). The phase path may either terminate on the supersonic sheet by entering one of the singular points as $\xi \to +\infty$, or switch at some point to the subsonic sheet via an isomagnetic jump. Generally, this brings us to a nonsingular point on the subsonic sheet, and the phase point then keeps moving along the new integral curve, eventually entering one of the singular points on the subsonic sheet as $\xi \to +\infty$.[1] It is also possible that the isomagnetic jump starts right at point 0, or hits one of the singular points on the subsonic sheet, terminating the shock structure.

The magnetic structure of a subsonic ionizing shock is represented by a portion of the integral curve on the subsonic sheet, starting from a nonsingular point 0 at $\xi = \xi(-\infty)$ and ending in one of the singular points as $\xi \to +\infty$.

Let us point out an obvious property of the set (4.2.7,8). If $B_z = 0$ for one of the points of the integral curve, then the entire integral curve corresponds to $B_z = 0$, i.e., the segments of the straight line $B_z = 0$, connecting the singular points on the (B_y, B_z) plane with the point where the same straight line crosses the discriminant curve, are the integral curves of the given set of equations. In particular, if $B_{z0} = 0$, then $B_z = 0$ throughout the entire shock structure. The condition $B_{z0} = 0$ thus singles out from the entire class of skew ionizing shocks a subdivision of oblique ionizing shocks. The MHD shocks, considered as the extreme case of ionizing shocks, as their intensity is raised indefinitely, belong to the oblique shock waves by virtue of being plane-polarized. The oblique shocks can be discriminated by the boundary conditions, which should prevent the production of other types of ionizing shocks in the experimental setup (Sect.4.1.2).

Let us now consider the behavior of the integral curves of the set (4.2.7, 8) near the singular points. For this purpose we linearize the equations in the vicinity of a k-type singularity and try a solution of the form

$$B_y = B_{yk} + ae^{qx} \quad , \qquad B_z = be^{qx} \quad .$$

The eigenvalues $q_{1,2}$ and the corresponding eigenvectors $\mathbf{Q}_{1,2}$ in the (B_y, B_z) plane are defined by

1 It should be noted that by moving along the integral curve indicated on the supersonic sheet, we cannot arrive at the nonphysical region of the surface (4.2.5). To prove this point, it suffices to demonstrate that, as $M^2 \to \infty$, $\lim d(1/M^2)/d\xi > 0$, and in accordance with the above-developed arguments the derivative can be computed at point 0. This inequality is proved with the aid of (2.5.7,8,13)

$$q_1 = \frac{u_{xk}}{M_{ak}^2}\left(M_{ak}^2 - 1 - \frac{M_k^2 B_{yk}^2}{M_k^2 - 1}\right) \quad , \qquad Q_1 = \{1,0\} \quad , \tag{4.2.18}$$

$$q_2 = u_{xk}(1 - 1/M_{ak}^2) \quad , \qquad Q_2 = \{0,1\} \quad . \tag{4.2.19}$$

The type of the singularity k depends on the signs of the eigenvalues q_1 and q_2. As follows from (3.3.21,27-30), $q_2 > 0$ for the points 1 and 2; $q_2 < 0$ for the points 3 and 4. Using (3.3.25), the condition $q_1 > 0$ can be rewritten in the form

$$\begin{cases} M_{fk} > 1 \quad \text{at} \quad M_k > 1 \quad , \\ M_{sk} > 1 \quad \text{at} \quad M_k < 1 \quad , \end{cases} \tag{4.2.20}$$

where M_{fk} and M_{sk} are Mach numbers at point k in relation to the velocities of the fast and slow magnetosonic waves, respectively.

Recall (Sect.3.3.2) that point 1 is always supersonic ($M_1 > 1$), point 4 is always subsonic ($M_4 < 1$), points 2 and 3 may be subsonic or supersonic; we shall distinguish the latter by the subscripts super and sub. From arguments developed above it follows that as $\xi \to +\infty$ the singular points 1 and 2_{sub} are unstable nodes in the (B_y, B_z) plane, points 2_{super} and 3_{sub} are saddle points, and points 3_{super} and 4 are stable nodes. The behavior of integral curves in the neighborhoods of these singularities is illustrated in Fig.4.1. Insofar as the motion along the integral curves takes place in the sense of increasing entropy, we see that the singularities 1 and 2_{sub} correspond to the local maxima of entropy, points 3_{super} and 4 correspond to the local minima of entropy, and points 2_{super} and 3_{sub} are the minimax points. As $\xi \to \infty$, none of the integral curves enters an unstable node; to the contrary, a stable node is a center of attraction for all integral curves which come close enough to it. A saddle point is approached as $\xi \to +\infty$ by only two integral curves (the two separatrices of the saddle). The degenerate case of coinciding points 2 and 3 is associated with the appearance of a singular curve (circle $M_a = 1$); this case, however, does not require special attention, since due to its symmetry the problem can always be reduced to one dimension (Fig.4.1). The Chapman-Jouguet shock corresponds to a degenerate singularity (saddle node), arising from the merging of points 3_{sub} and 4. The two separatrices of this singularity divide the regions, in one of which all the integral curves come into the degenerate singularity as $\xi \to \infty$ (similarly to a node), and in the other only one integral curve (the separatrix) enters the singularity as $\xi \to \infty$, whereas all the other depart from it (similarly to a saddle).

Let us now analyze the behavior of integral curves near the singular points 1,2,3,4 in the downstream flow. The vicinity of point 1 is not shown in Fig.4.1, since this point cannot represent a downstream state. The phase point cannot approach the singular point 1 as $\xi \to \infty$, moving from point 0 on the supersonic sheet along any one of the integral curves. Consequently, the ionizing shocks of type 1 have no structure and hence do not exist. This complies with the conclusion of Sect.4.1.2 regarding the nonevolutionarity.

With the shocks of type 2 we have the following options. Point 2_{super}, as $\xi \to \infty$, is encountered by two integral curves, corresponding to the eigenvector $\mathbf{Q}_1$ (Fig.4.1), i.e., to $B_z = 0$. Accordingly, the relevant magnetic structure is represented in the (B_y, B_z) plane by a length of the straight line $B_z = 0$ from point 0 on the supersonic sheet (note that $B_{z0} = 0$) to point 2_{super}. We cannot arrive at point 2_{sub} as $\xi \to \infty$ along the integral curves pertaining to the subsonic sheet, therefore, subsonic ionizing shocks of type 2 do not exist, just as predicted on the basis of their nonevolutionarity (Sect.4.1.2). Point 2_{sub} can be approached by moving along the integral curve on the supersonic sheet from point 0 to the point with coordinates $B_z = 0$, $B_y = B_y(2_{sub})$, from which the transition to point 2_{sub} is allowed via an isomagnetic jump (then also $B_{z0} = 0$). Here the magnetic structure of the ionizing supersonic shock is represented by a straight line segment $B_z = 0$ on the supersonic sheet, followed by the isomagnetic jump. All the change in the magnetic field across the shock takes place before the isomagnetic jump. It follows that the condition of existence of the structure imposes the following restrictions on the ionizing shocks of type 2: the shock should be supersonic and belong to the class of oblique shock waves, i.e., $B_{z0} \equiv 0$. Ionizing shocks of this kind include, in particular, transverse ionizing shocks in which $B_x = 0$. As in Sect.3.3.2, proceeding to the limit $B_x \to 0$ (and hence $c_a \to 0$) we find that $M_a \to \infty$ downstream, i.e., $M_a > 1$, and the downstream state is represented by the singular point 2.

From (4.2.5,17) it is clear that as $M_{a0} \to \infty$ the downstream state can be represented only by the singular point 2, i.e, that the ionizing shocks of types 3 and 4 are possible only in a limited range of values of M_{a0}.

Now let us turn to ionizing shocks of type 3. Point 3_{super} as $\xi \to \infty$ is entered by all integral curves as soon as they come close enough to this point. The corresponding magnetic structures are represented by portions of the integral curves on the supersonic sheet, going from point 0 to point 3_{super}. Such shocks are generally skew, although they classify as oblique if $B_{z0} = 0$.

Point 3_{sub} as $\xi \to \infty$ is entered by two integral curves, corresponding to the eigenvector $\mathbf{Q}_2$ (Fig.4.1), allowing thus the existence of subsonic ionizing shocks of type 3_{sub}. For the structure to exist, point 0 must belong to one of the said two integral curves; then the magnetic structure is represented by the portion of this curve from point 0 to point 3_{sub}. The structure of the supersonic ionizing shock of type 3_{sub} is depicted by a portion of the integral curve on the supersonic sheet, an isomagnetic jump from the supersonic sheet to one of the integral curves entering point 3_{sub} on the subsonic sheet, and a portion of the same curve from the isomagnetic jump to point 3_{sub}. From Fig.4.3 it is clear that $B_z \neq 0$ for both curves, so an oblique supersonic ionizing shock of type 3_{sub} must contain the isomagnetic transition from the supersonic sheet directly to the point 3_{sub}, as in the case of a supersonic ionizing shock of type 2_{sub}.

Finally, point 4 as $\xi \to \infty$ is encountered by all the nearby integral curves. Ionizing shocks of type 4 may be subsonic and supersonic, skew and oblique.

The integral curves in the (B_y, B_z) plane are plotted in Fig.4.3 for $M_0 \to \infty$, $M_{a0}^2 = 5/3$, $B_{y0} = 0.2$, $B_{z0} = 0.3$, $E_s + B_{y0}/M_{a0}^2 = 0.04$ and in Fig.4.4 for $M_0^2 = 0.2$, $M_{a0}^2 = 0.8$, $B_{y0} = 0.4$, $B_{z0} = 0.3$, $E_s + B_{y0}/M_{a0}^2 = 0.05$. Figure 4.3 shows the supersonic (upper) and the subsonic (lower) sheets of the surface (4.2.5), projected onto the (B_y, B_z) plane, since the starting point 0 $(x = -\infty)$ lies on the supersonic sheet, while the terminal point (at $x = +\infty$) may pertain either to the supersonic sheet, e.g., 3_{super}, or to the subsonic sheet. Figure 4.4 shows the projection of only the subsonic sheet onto the (B_y, B_z) plane, because the initial point 0 belongs to this sheet, and the only structure possible is $0 \to 4$. Observe that the neighborhood of the singular point 1 on the supersonic sheet (Fig.4.3) falls into the nonphysical region of the surface (4.2.5), because point 0 lies on the border of this region as $M_0 \to \infty$, and in the direction of the arrows the value of M^{-2} increases as M tends to infinity.

The magnetic structures of MHD shocks are represented by integral curves of a similar set of equations, which as $\xi \to -\infty$ enter one of the singular points. If we picture these curves in the (u_x, B_y, B_z) phase space rather than on the (u_x, B_y) phase plane, we clearly see (in the adopted zero-order approximation in small Pm) the concatenation of the evolutionarity criteria and the conditions of existence of the stationary structure, pointed out in Sect.3.3.6. From Figs.4.3,4 it is evident that the magnetic structures pertaining to nonevolutionary shock transitions are not unique: they form a two-parameter family for the transitions $1 \to 4$, one-parameter families for the transitions $1 \to 3$ and $2 \to 4$. Two different structures also exist corresponding to the nonevolutionary trans-Alfvén shock transition $2 \to 3$.

As follows from Figs.4.3,4, the context of the present model, in principle, allows the existence of structures $0 \to 3_{super}$ and $0 \to a \to b \to 4$, represented in Fig.4.3, and only one structure $0 \to 4$ in the cases shown in Fig.4.4. The theory then should be able to answer the question, which of the singularities (3_{super} or 4) depicts the downstream state in the context of Fig.4.3. Should it be point 4, then the structure of the evolutionary ionizing shock of type 4 must be defined unambiguously (the parameters of the initial and the final states being fixed), which requires pinpointing the location of point a exactly (the starting point of the isomagnetic subshock) on the integral curve $0 \to 3_{super}$. [We become aware of the fact that even with the supersonic ionizing shock of type 4, for which the existence of additional boundary conditions is not required by the evolutionarity criteria (in fact, what is required is the absence of ABC), the structure is still somewhat ambiguous, a snag that the theory should be capable of dealing with successfully.] Yet another question is whether the shock transition $0 \to 4$ is actually possible under the circumstances of Fig.4.4.

The relations which evolve from the laws of conservation and the Maxwell equations are not sufficient to answer these questions.

The treatment must straightforwardly include the process of ionization in the shock; thus, we are confronted with the choice of a physically correct model of ionization. The important point here is that, as a rule, it is nonequilibrium ionization that plays the essential role in the shock waves. In other words, it will not suffice to state that the gas becomes ionized in passing via the shock layer: we have to determine exactly the processes responsible for the ionization of the gas, locate the limit where the ionization becomes significant, etc. As follows from (4.2.16), this amounts to assessing the asymptotic behavior of the decline of ionization as we move away from the shock. In this connection we have to deal with the question of how remote an asymptote is still physically sensible, how small a conductivity cannot yet be ignored. Let us consider this problem.

4.2.2 The Criterion for Distinguishing Between Ionizing and MHD Shock Propagation Regimes

As we know from the foregoing discussion, the peculiar features of ionizing shock waves in a magnetic field are manifest only when the upstream conductivity converges to zero rapidly enough: perfect nonconduction enables the electromagnetic waves to carry ahead the electric field induced by the shock. In reality, however, any gas exhibits a certain degree (if perhaps quite small) of conduction ($\sigma_0 \neq 0$); then the integral in (4.2.16) diverges and the boundary conditions for the shock wave are reduced to the MHD set, so that apparently no problem of additional boundary conditions arises. In this connection, let us look more closely at the conditions under which the upstream-gas conductivity may be safely neglected.

Consider an ionizing shock wave, driven by a moving piston at constant velocity u_{x0}. In the shock frame the transverse electric field $\mathbf{E}_\perp$, carried ahead by the electromagnetic wave propagating in the gas with conductivity σ_0, is found from the exact wave equation

$$\frac{\partial^2 \mathbf{E}_\perp}{\partial x^2} - \frac{1}{c^2}\frac{\partial^2 \mathbf{E}_\perp}{\partial t^2} - \mu_0\sigma_0 \frac{\partial \mathbf{E}_\perp}{\partial t} = 0 \quad . \tag{4.2.21}$$

Here we discount nonlinear effects, assuming that $\mathbf{E}_\perp$ is not large enough to modify the ionization and conduction ahead of the shock wave.

The asymptotic solution of (4.2.21), with the initial condition $\mathbf{E}_\perp(x,0) = 0$ at $t = 0$ (the instant when the piston starts moving and the shock is produced) and the boundary condition $\mathbf{E}_\perp(0,t) = \mathbf{E}_{\perp 0}$ at the shock front or on the piston surface, is given by

$$\mathbf{E}_\perp(x,t) = \begin{cases} \mathbf{E}_{\perp 0} \exp(-\mu_0\sigma_0 cx/2)\theta(ct - x) \ , & t \ll t_0 \\ \mathbf{E}_{\perp 0} \operatorname{Erfc}[x(\mu_0\sigma_0/4t)^{\frac{1}{2}}] \ , & t \gg t_0 \ , \end{cases} \tag{4.2.22}$$

where $\theta(x)$ is step function, Erfc(x) is the error function, and $t_0 = 1/\mu_0\sigma_0 c^2$.

For $t \ll t_0$ the exponent in (4.2.22) is small for all $0 < x < ct$, and the solution describes an undamped electromagnetic wave. For $t \gg t_0$ the displacement currents are small in comparison with the conduction currents, and the solution describes electric-field diffusion.

If σ_0 is very small, then certainly $ct_0 \gg L$ (L being the length of the shock tube, $\sim 10^2$ cm), and we are dealing with the wave regime. In this case the establishment time of the electric field, determined by the boundary condition on the shock front, is of the order of L/c and much smaller than the time of formation of the steady-shock structure. If the upstream conductivity, however, is nonzero, then it is possible that $t_0 \ll L/c$. In this case the stationary (quasi-stationary in reference with electromagnetic waves) solution requires that the time of diffusion of the electric field, $\mu_0\sigma_0 L^2$, be small compared with the characteristic time scale of the GD flow, L/u_{x0}. This requirement supplies the upper restriction on the conductivity of the gas upstream of the ionizing shock wave in the form

$$Rm_0 = \mu_0\sigma_0 u_{x0} L \ll 1 \quad . \tag{4.2.23}$$

In other words, the reconstruction time of the shock front (whose thickness for $Pm \ll 1$ is characterized by the scale $\Delta_j^{(0)} = 1/\mu_0\sigma_0 u_{x0}$), defined as $\tau = \Delta_j^{(0)}/u_{x0}$, should be large in comparison with L/u_{x0}, which is the meaning of inequality (4.2.23). It is also obvious that over a sufficiently long time, or over a sufficiently long path, the ionizing shock structure will ultimately become the steady MHD structure.

For a weakly ionized gas the condition (4.2.23) is expressed in terms of a restriction on the initial ionization α, so that for $\alpha \ll \tilde{\alpha}$ the experiment produces ionizing shock waves, and MHD shock waves for $\alpha \gg \tilde{\alpha}$. The threshold value $\tilde{\alpha}$ is given by

$$\tilde{\alpha} = \frac{\mu_0 m_e v_{Te}\sigma_{ea}}{e^2 u_{x0} L} \cong 2\cdot 10^{-5}(\varepsilon_e[\mathrm{eV}])^{1/2}\left(\frac{\sigma_{ea}}{10^{-15}\,\mathrm{cm}^2}\right)\left(\frac{u_{x0}}{10\ \mathrm{cm}/\mu\mathrm{s}}\right)^{-1}(L[\mathrm{m}])^{-1} \quad , \tag{4.2.24}$$

where v_{Te} is the mean electron velocity, ε_e is the mean upstream energy of electrons, and σ_{ea} is the cross section of electron-neutral collisions.

Thus, the theory of ionizing shock waves in a magnetic field is valid as long as the initial conductivity is below $1/\mu_0 u_{x0} L$. From (4.2.24) it is clear that α crosses the threshold in the region of precursor ionization (Sect.2.3.2), which indicates the importance of this region in this theory. Note that (4.2.21) describes also propagation of the constant transverse magnetic field, imposed by the boundary conditions on the shock front, and perturbations of the transverse components of electric and magnetic fields ahead of the front. Therefore, in determining the number of diverging waves required by the evolutionarity conditions one should include the electromagnetic waves also when the electromagnetic field in the flow is quasi-stationary.

4.2.3 Precursor Ionization in a Magnetic Field. Conditions for the Ionization Stability of the Upstream Flow

The connection between the kinetics of precursor ionization and the additional boundary conditions for the ionizing shock waves in a magnetic field was evolved in [4.10-16]; we shall stick to the results of [4.10]. As follows from (4.2.24) and the results of Sect.2.3.2, the critical degree of ionization is reached in the near precursor region, associated with direct photoionization of atoms (molecules) of neutral gas by radiation coming from behind the shock. In the far precursor region $\alpha \ll \tilde{\alpha}$ and can be ignored.

As in Sect.2.3.2, we assume the radiating shock surface to be represented by the plane $x=0$, whereas the precursor region corresponds to $x<0$. In the presence of a magnetic field the boundary between the precursor region and the region of shock compression and heating is not always defined as clearly as in the case discussed in Sect.2.3.2, where the interface is represented by the narrow gasdynamic subshock. Nevertheless, let us assume that the shock can be divided into the regions indicated. Then the analysis of the precursor region, where $\alpha \ll 1$, is facilitated by the fact that the shock-layer equations can be linearized for small α, and the entire set of equations can thus be reduced to one linear equation in α, the values of velocity, magnetic and electric fields, etc., being fixed. Thus we can obtain the profiles of ionization (and conductivity) in the precursor region for preset values of shock variables, considered at this stage as free parameters. In the next approximation the obtained profile of conductivity is used for assessing the compression of magnetic field and the changes in the gasdynamic variables, which allows us to solve the problem self-consistently.

The linearized equation of ionization kinetics in the near precursor region ahead of the ionizing shock in a magnetic field has the form

$$\frac{d\alpha}{dx} = j(x) + \alpha K \quad . \tag{4.2.25}$$

Here $j(x)$ is defined as in (2.3.31); $K=(\nu_i-\nu_\ell)/u_{x0}$ is the coefficient of electron multiplication, where ν_i is the rate of ionization, ν_ℓ is the rate of electron loss (linear in α), not accounted for in (2.3.31), such as diffusive escape to the tube walls, attachment to neutral atoms, etc. An important feature of precursor ionization in a magnetic field is the possible multiplication of electrons which experience heating due to the electric induction field E_0^*. Coefficient K depends on the temperature of precursor electrons (in the region of interest the degree of ionization is sufficiently high for the distribution of the electrons in an electric field to be Maxwellian). The temperature, in turn, is determined by the balance between the Joule heating in the electric field E_0^* and both elastic and inelastic energy losses, e.g., due to ionization and excitation of atoms. The threshold value $E_0^*=E_d$, above which $K(E_0^*)>0$, corresponds to the electric-field strength capable of maintaining nonequilibrium ionization in the gas. The values of E_d are known from experiments on glow discharge. For most gases the ratio E_d/p varies between 0.1 and 30 V/cm·Torr, when $p\Lambda$ is in the range of 0.4 - 4 cm·Torr (p being the pressure, and Λ the characteristic length of diffusion) [4.17,18], i.e., E_d is not high. With hydrogen, for instance, $E_d=7.5$ V/cm at $p=0.25$

Torr, $\Lambda = 10$ cm. The electric induction field ahead of the ionizing shock may have a strength of the order of 100 V/cm, and the inclusion of the last term in (4.2.25) is necessary. It ought to be mentioned that the electric breakdown of argon by the electric induction field ahead of the transverse ionizing shock has been observed experimentally [4.19].

In the expression for j(x) let us ignore the correction for finiteness of the angular dimension of radiating plasma, and introduce a phenomenological factor R to account for the optical thickness of the radiating layer and for the deviation of the density of ionizing radiation from the blackbody law. A more detailed account of the finiteness of optical thickness would certainly modify the expression for j(x) more drastically; such an extension, however, is pointless in the context of the stationary problem. For a shock driven by a piston, for instance, the thickness of the radiating layer, i.e., the region between the piston and the shock front, increases with time. Introducing the phenomenological constant R to account for the finite optical thickness is strictly justified as long as the radiating layer over a reasonable time span remains optically thick ($R \approx 1$), or optically thin ($R \to 0$, its actual value being inessential, Sect.4.2.5), or when its thickness is maintained at a constant level, e.g., by the action of wall-boundary effects (Sect.2.4.1). A strong source of ionizing radiation, traveling with the shock, e.g., the current sheet of the magnetic piston, can also be included in this model through the factor R; then, possibly, $R > 1$. Let us denote the term in front of the left-hand brace in (2.3.32) by j_0. Then the solution of (4.2.25) is

$$\alpha(x) = C\,e^{Kx} + \frac{j_0 R}{K^2}\left\{e^{Kx}\,E_1\!\left[-\frac{x}{\ell_{ph}}(1 - K\ell_{ph})\right] - E_1\!\left(-\frac{x}{\ell_{pH}}\right) - K\ell_{ph}E_2\!\left(-\frac{x}{\ell_{pH}}\right)\right\} \quad , \qquad (4.2.26)$$

where C is an arbitrary integration constant, the other designations being the same as in (2.3.32).

The boundary condition $\alpha(-\infty) = 0$ is satisifed by setting $C = 0$, if $K < 0$. If $K > 0$, the value of C is arbitrary; however, it is necessary that

$$K\ell_{ph} < 1 \quad . \qquad (4.2.27)$$

Otherwise the stationary solution of the kinetic equation of ionization, decreasing as $x \to -\infty$, does not exist. Then the solution of the linearized nonstationary equation of ionization kinetics with the initial condition $\alpha(x,0) = 0$ describes exponential build-up of ionization with time at every point, which is naturally interpreted as the upstream breakdown of neutral gas [4.14]. Accordingly, below we shall call the strength of the electric field $E_0^* = E_*$, such that $K(E_0^*)\ell_{ph} = 1$, the breakdown threshold. To avoid possible misunderstanding let it be emphasized that the breakdown in question occurs in the domain of action of a strong external source of ionizing radiation, the shock front. The breakdown threshold is found from the condition

$$\nu_i = \nu_\ell + \ell_{ph}/u_{x0} \quad ; \tag{4.2.28}$$

usually it is the first term that dominates on the right-hand side of (4.2.28). Accordingly, E_* is of the order of E_d and thus below the threshold of ordinary (Paschen) breakdown of neutral gas by a static electric field E_b [typically, $E_b/p \sim (3\text{-}5) \cdot 10^2$ V/cm·Torr].

Strictly speaking, the feasibility of calculating stationary shock structures is restricted to the range of electric induction fields where $K(E_0^*) < 0$, since it is only then that the conditions of the theorem stating the relationship between evolutionarity and existence of stationary structures are satisfied (Sect.1.7.2). The neutral gas ahead of the shock is a dissipative medium, the upstream disturbances of α (as well as the perturbations of all other variables) are damped. Then, on account of $C = 0$, the profile of precursor ionization is unambiguously determined by the flux of radiation coming from the downstream region.

However, in ionizing-shock experiments in magnetic fields the induced upstream electric fields are often such that $K(E_0^*) > 0$. Then the conditions of the theorem are not satisfied, and the profiles of precursor ionization (and thus the entire shock structures), the value of C being arbitrary, are ambiguous even when all the parameters of the shock are fixed, including B_{y0} and E_s [4.15]. The investigation of stationary structures might seem to come to a deadlook: to determine which structure is actually realized one has to deal with a time domain problem of the shaping and development of the shock structure. Nevertheless, the theory of stationary structures of ionizing shocks in a magnetic field is capable of describing the situation reasonably, if we set $C = 0$ even if $K > 0$. Physically, this means that the shock front is taken to be the source of ionization. For $K < 0$ the partial solution with $C \neq 0$, increasing as $x \to -\infty$, describes the subsidence of nonequilibrium ionization, produced by an external, i.e., different from the ionizing shock front, source. By setting $C = 0$ we ensure the absence of such a source. For $K > 0$ it can be formally represented by any small perturbation of α. However, if the characteristic time of development of the electron avalanche in the electric field is larger than the time of travel of the shock through the tube, then the electric breakdown of gas far from the shock does not have time enough to occur. Then we are dealing essentially with the ionization wave, induced by and traveling along with the shock; its phase velocity depends on the profile of upstream disturbance of ionization, caused by photoionization. As a rule, in the experiments there are no alternative sources of ionization capable of producing a comparably large disturbance of α (observe that this "small" perturbation is actually rather considerable and may be as large as 10^{-5} - 10^{-3}; such ionization does not arise spontaneously).

Here we may draw an analogy with the similar problem of ignition of a steady combustion wave at $x = -\infty$ in the theory of combustion (the so-called cold boundary problem). If we assume the combustion rate in the cold inflammable mixture to take on an arbitrarily small though finite value (which might seem a quite natural assumption), we shall obtain indefinitely long structures of combustion waves, with the reaction going as slow as desired, while the reaction rate is still indeterminate. Obviously, such solutions are meaningless, since they do not comply with the initial assumptions, above all, the assumption of unidimensionality of the problem. The model which gives a physically acceptable result is based on the assumption that the reaction rate is strictly zero up to a certain temperature T^*, characterizing the combustion. Then the rate of combustion is readily computable and fits in with the experimental results. This model is essentially based on the concept that the only source of heat for maintaining the combustion is the combustion wave itself.

Rewriting (4.2.27) as an inequality for E_0^*, we obtain the restrictions on the parameters of stationary structures of ionizing shocks in a magnetic field [4.10]

$$E_0^* < E_* \quad , \tag{4.2.29}$$

$$C = 0 \quad . \tag{4.2.30}$$

Condition (4.2.29) ensures the existence of a solution for the shock structure, while (4.2.30) guarantees its being unique with a fixed set of the parameters of the shock and of the initial state. Investigating stationary structures, searching for additional boundary conditions and the like make sense only if (4.2.29,30) are satisfied simultaneously. We shall call these restrictions the conditions of ionization stability of the upstream flow in the ionizing shocks in a magnetic field. Insofar as these conditions are deduced from the analysis of asymptotic behavior of precursor ionization, the conditions of ionization stability are exact in the sense that any modification of the theoretical model does not change these criteria (4.2.29,30) essentially. In other words, they retain validity beyond the context of our present model with the precursor region defined explicitly in the shock structure.

4.2.4 Additional Boundary Conditions and the Magnetic Structures of Ionizing Shocks

Any possible way to define where in the ionizing shock structure the region of precursor ionization ends and the main shock transition begins, is to a reasonable extent based on convention. The region of precursor ionization may be considered the entire head portion of the shock, where $\alpha(x) \ll 1$. The profile of ionization in this region is then given by the solution of the linearized kinetic equation (4.2.25). For ionizing shocks in a magnetic field it is described by the second term on the right-hand side of (4.2.26), which accounts for both the photoionization and the shock ionization due to the Joule heating of electrons. This definition of the region of precursor ionization greatly facilitates the task of getting a solution in closed analytic form. In the general case, when this technique cannot be used, we arrive at a highly sophisticated many-dimensional problem of eigenvalues of a set of integrodifferential equations, including the Maxwell equations, the equations of plasma dynamics and ionization kinetics, and the equation of transfer of ionizing radiation. Actually, the conditions of ionization stability, (4.2.29,30), remain valid.

Let us use the following convention for separating the shock into regions of precursor ionization and shock compression. If the shock structure includes an isomagnetic subshock, it will serve as a natural boundary between the precursor region (which includes the supersonic part of the magnetic structure) and the shock-compression region. If the isomagnetic subshock is not present, then the boundary may be drawn in just any physically sensible way: the region of precursor ionization may be terminated where the density or temperature gradient is largest, or where the heat exchange in the collisions of electrons and heavy particles becomes comparable with the magnitude of Joule heating, etc.

Let us assume that the shock layer has been separated into regions of shock compression and precursor ionization. On the integral curve originating from point 0 we locate point a_p so that the precursor region is described by the portion of the integral curve from 0 to a_p, where the entropy is below $S(a_p)$. For the precursor region, by virtue of $\alpha \ll 1$, we may set

$$\Delta_j = \tilde{\Delta}_j/\alpha \quad , \qquad \tilde{\Delta}_j = \text{const} \quad . \tag{4.2.31}$$

The asymptotic expression (4.2.31) is exact as $x \to -\infty$, where $\alpha \to 0$, see (2.3.4). It would not be hard to include the dependence of $\tilde{\Delta}_j$ on compression, electron temperature, etc. [4.14]; this accuracy here is unnecessary, since in the precursor region $\tilde{\Delta}_j$ varies within one order of magnitude.

Using (4.2.6-8), we may write

$$\frac{dx}{\Delta_j} = \frac{[(dB_y)^2 + (dB_z)^2]^{\frac{1}{2}}}{E^*(B_y,B_z)} \quad , \tag{4.2.32}$$

where the dimensionless electric field in the gas frame is

$$E^*(B_y,B_z) = [(E_s + B_{y0}/M_{a0}^2)^2 + 2B_y(u_x - 1/M_{a0}^2)(E_s + B_{y0}/M_{a0}^2)$$

$$+ (B_y^2 + B_z^2)(u_x - 1/M^2_{a0})^2]^{\frac{1}{2}} \quad . \tag{4.2.33}$$

Here u_x is assumed to be expressed in terms of B_y,B_z via (4.2.5) on the sheet of the surface (4.2.5) where point 0 belongs to. Now we substitute (4.2.31) into (4.2.32) and integrate between the limits of the precursor region, yielding

$$\int_0^p \frac{[(dB_y)^2 + (dB_z)^2]^{\frac{1}{2}}}{E^*(B_y,B_z)} = \frac{1}{\tilde{\Delta}_j}\int_{-\infty}^{0} \alpha(x)dx \quad . \tag{4.2.34}$$

The integral on the left-hand side is taken along the integral curve of the set (4.2.7,8) from point 0 to point p. Point p, marking the end of the precursor region, lies on the integral curve between 0 and a_p if the shock structure contains an isomagnetic subshock, or coincides with a_p if there is no isomagnetic jump. Note that B_y and B_z along the integral curve are linked by a functional dependence, and the variable B_y (or B_z) can be chosen as the integration variable in (4.2.34) as long as $dB_y/dx \neq 0$ (or $dB_z/dx \neq 0$).

The integral on the right-hand side of (4.2.34) is computed with the aid of (4.2.26) and with due account of (4.2.29,30):

$$\frac{1}{\tilde{\Delta}_j}\int_{-\infty}^{0} \alpha(x)dx = \frac{j_0 R \ell_{ph}}{\tilde{\Delta}_j}\, g(K\ell_{ph}) \equiv \phi_K \quad , \tag{4.2.35}$$

where

$$g(\xi) = -\frac{1}{2\xi} - \frac{1}{\xi^2} - \frac{\ln(1-\xi)}{\xi^3} \quad .$$

The quantity ϕ_K is a function of the parameters of the initial state and the shock wave N_0, T_0, M_0, M_{a0}, B_{z0}, B_{y0}, E_s. These parameters being fixed, ϕ_K depends only on the downstream temperature of the plasma, i.e., on the type of the ionizing shock ($k = 2,3,4$). As the strength of ionizing radiation increases with the downstream plasma temperature, from (3.3.32) it follows that

$$\phi_2 < \phi_3 < \phi_4 \quad . \tag{4.2.36}$$

Now let us demonstrate how (4.2.34) can be used for solving the questions formulated in Sect.4.2.1. With the ionizing shock of type 2, the evolutionarity conditions call for the existence of two additional boundary conditions. As demonstrated in Sect.4.2.1, such shocks are oblique, and thus the first of the two required conditions has the form $B_{z0} = 0$ (Figs.4.1,3,4). The second ABC is given by (4.2.34) with $k = 2$ and $\mathbf{B}_\perp(p) = \mathbf{B}_\perp(2)$. For brevity, let us denote the left-hand side of (4.2.34) by Ψ. This quantity depends on the same set of arguments as ϕ_k; henceforth they will be omitted. The equation of E_s for the shock of type 2 has the form

$$\Psi(p) = \phi_2 \quad . \tag{4.2.37}$$

One additional boundary condition is needed for the evolutionarity of the supersonic shock of type 3. If it is a shock of type 3_{sub}, then the downstream singularity is a saddle point (Fig.4.1). Accordingly, this point is entered by a single integral curve as long as the sign of B_y is fixed. Then point a is the origin of the isomagnetic subshock, the unique point of the integral curve coming out of point 0 on the upper (supersonic) sheet of (B_y, B_z), from which the isomagnetic jump is possible onto the integral curve on the lower (subsonic) sheet, which enters point 3_{sub}. The only additional boundary condition that is required is given by (4.2.34) with $k = 3$ and $p = a$. Similarly, the only additional boundary condition for the supersonic shock of type 3_{super} is given by (4.2.34) with $k = 3$ and $p = a_p$.

The evolutionarity of the subsonic shock of type 3 requires two additional boundary conditions. The singular point 3_{sub} being a saddle point (Fig.4.1), the first of the ABC fixes point 0 to the unique integral curve that enters point 3_{sub}. Such a shock is plane-polarized (oblique): $B_{z0} = 0$. The other ABC is given by (4.2.36) with $k = 3$, $p = a_p$.

The condition of evolutionarity of the supersonic ionizing shock of type 4 does not require any additional boundary conditions, except those arising from conservation laws and the Maxwell equations. The possible restrictions in this case may come from the following reasons. First, (4.2.34) with $k = 4$ must have a solution which determines the shock structure by fixing point a (the onset of the isomagnetic subshock) on the integral curve coming out of point 0. This solution does not exist if, for instance, with the given set of parameters the electric field transcends the breakdown threshold. Secondly, the isomagnetic subshock must bring point a into the two-dimensional domain of attraction of the singular point 4 (stable node) on the subsonic sheet.

The one additional boundary condition, required for the evolutionarity of the subsonic shock of type 4 is given by (4.2.34) with $k = 4$, $p = a_p$.

Let us now demonstrate how (4.2.34) can be used to determine the type of shock transition realized with the given set of parameters, e.g., in the case shown in Fig.4.3. Observe that the integrand on the left-hand side of (4.2.34) is always positive, and thus $\Psi(p)$ increases steadily as point p recedes from the starting point 0. Consequently, the solution for the supersonic shock of type 4 exists if inequality

$$\Psi(p') > \phi_4 \tag{4.2.38}$$

is satisfied at the furthermost point p' of the portion of the integral curve, originating at point 0, which is brought by the isomagnetic subshock into the domain of attraction of point 4. Then at some point between 0 and p' the function $\Psi(p)$ may be equal to ϕ_4 (this equality is bound to take place if the isomagnetic subshock brings the entire portion 0-p' into the domain of attraction of point 4).

Let us consider the integral curves plotted in Fig.4.3). Presumably, the solutions here may correspond to both the structure $0 \to 3_{super}$ and the structure $0 \to a \to b \to 4$. Let a_p be the region of shock compression on the integral curve $0 \to 3_{super}$, i.e., $p' = a_p$. A shock of type 4 is possible if $\Psi(a_p) > \phi_4$, and the solution of type 3_{super} requires that $\Psi(a_p) = \phi_3$. Insofar as $\phi_4 > \phi_3$, in the case of Fig.4.3 the solution corresponds to the shock of type 4 if $\Psi(a_p) > \phi_4$, or type 3_{super} if $\Psi(a_p) = \phi_3 < \phi_4$, or there is no solution if $\phi_3 \neq \Psi(a_p) < \phi_4$.

Similarly, the type of shock wave for which the solution exists is determined for any given set of parameters.

4.2.5 Limiting Regimes

Let us now apply the theory developed above to the extreme cases of "weak" and "strong" ionizing shocks. Obviously, any ionizing shock causes strong enough heating of the gas, and is "strong" in this respect. Let us therefore emphasize that we now call an ionizing shock weak if it is not strong enough to qualify for the MHD limit, i.e., that the downstream flow in such a shock corresponds to one of the MHD singular points, whereas the upstream flow does not.

The extreme case of "weak" ionizing shocks will be called the gasdynamic (GD) limit of the theory. In corollary, the ionizing shock, whose upstream and downstream states correspond to MHD singular points, will be referred to as "strong".

To pass to the GD limit, we must assume the action of both mechanisms of precursor ionization to be negligibly small, i.e., set formally $R \to 0$ and $E_* \to \infty$. Then the right-hand side of (4.2.34) vanishes in the entire range of variation of B_{z0} and E_s, allowed by the structural limitations. The vanishing of the integral on the left-hand side of (4.2.34) implies that $p = 0$, i.e., that in this case only the trivial structures are allowed, corresponding to supersonic-supersonic or subsonic-subsonic shock transitions. Point 0 should belong to the supersonic sheet and coincide with the starting point of the isomagnetic subshock. Thus, in the GD limit the ionizing shock starts with a gasdynamic subshock, in which the gas becomes conductive, while the downstream state corresponds to one of the subsonic MHD singular points. Then the structure of ionizing shock in a magnetic field complies with the Zel'dovich-von Neumann-Döring model (Sects.1.9.2,2.3.1). This regime was analyzed in [4.7,8,20-25].

Obviously enough, in the GD limit the number of ABC complies with the evolutionarity criteria. For instance, for the supersonic ionizing shocks of type 2_{sub} the structural restrictions result in one additional relation $B_{z0} = 0$ and also require that the isomagnetic shock transition from the supersonic to the subsonic sheet should land directly at point 2_{sub} (Sect.4.2.1). Together with the relation $p = 0$, we get the other ABC in the form $B_{y2} = B_{y0}$. To put it differently, the transverse magnetic field cannot change on the upper sheet (the property of the GD limit), neither does it change on the lower sheet (by virtue of structural restrictions), thus remaining constant throughout the entire shock. Incidentally, this conclusion applies to the transverse shocks.

For the supersonic ionizing shocks of type 3_{sub} the only ABC that is necessary evolves from the requirement that the isomagnetic subshock from point 0 should fall onto the one integral curve which (at the prescribed sign of B_{z0}) enters the singular point 3_{sub} as $\xi \to \infty$ on the subsonic sheet. No additional boundary conditions are required for the supersonic ionizing shocks of type 4, as long as the isomagnetic shock transition from point 0 falls into the two-dimensional domain of attraction of point 4 on the subsonic sheet. No other types of shock transitions are possible in the GD limit.

In the context of the GD limit it is easy to answer the questions formulated at the end of Sect.4.2.1. The conditions of Fig.4.3 correspond to the shock of type 4; then point a coincides with point 0. An ionizing shock without an isomagnetic subshock (Fig.4.4) does not exist in the GD limit.

Another extreme case, pertaining to a finite-valued breakdown threshold E_* with a low level of photoionization ($R \to 0$) will be referred to as the electrostatic breakdown limit [4.10,15]. The breakdown threshold is given by (4.2.28), although this time the process of photoionization supplies only the "priming" electrons, while further multiplication of electrons in the precursor layer is due only to Joule heating. Recall that the right-hand side of (4.2.34) holds only for $E_0^* < E_*$; and, moreover, $\phi_k \to 0$ as $R \to 0$ for any $E_0^* < E_*$, while $\phi_k \to \infty$ only as $E_0^* \to E_*$. Therefore, (4.2.34) has two solutions: either $E_0^* < E_*$, $\Psi = 0$, $p = 0$ (GD limit), or $E_0^* = E_*$ (if the topmost value of the electric induction field with the given set of shock parameters is not less than E_*). In this extreme (apart from the GD limit) the theory provides solutions with $E_0^* = E_*$ for shocks of types 2 and 3. The relevant structures of supersonic shocks of types 2_{sub} and 3_{sub} start not with the isomagnetic subshock, but rather with the magnetic field compression in the precursor region. In this extreme both supersonic-supersonic and subsonic-subsonic shock transitions are allowed. Finally, the structures of supersonic ionizing shocks of type 4 do not differ from those in the GD limit, although the structural restrictions are supplemented by condition (4.2.29).

The breakdown threshold of neutral gas depends on its pressure and on the dimensions of the shock tube, but not on the parameters of the shock. It would be natural therefore to define the dimensionless breakdown threshold as $E_* = \bar{E}_*/c_a\bar{B}_x$ (the upper bar marks dimensional quantities). Then from (4.2.29) or (4.2.34) we get

$$E^*(B_{y0}, B_{z0}) \leqslant E_*/M_{a0} \quad . \tag{4.2.39}$$

Hence it follows that by formally reducing the parameter E_* we may pass on to the MHD regime. With $E_* \to 0$ the upstream electric field, which is proportional to the left-hand side of (4.2.39), also becomes zero. Expression (4.2.39) can be straightforwardly used for finding the law governing the transition to the MHD limit with increasing M_{a0} and fixed E_* for oblique shocks of type 2. Recall that only such shocks are compatible with unrestricted increase in M_{a0} (Sect.4.2.1). Then

$$|E_s + B_{y0}| = E_*/M_{a0} \quad . \tag{4.2.40}$$

Note that the transition to the MHD regime, as $M_{a0} \to \infty$, occurs slowly: the dimensionless upstream electric field decreases as $1/M_{a0}$.

Now we examine the effects of precursor ionization. As R tends to infinity, the right-hand side of (4.2.36) increases unrestrictedly. This increase can be matched by the left-hand side of this equation only if one of the limits of the range of integration is brought close to the singular point of the integrand. If the integration range remains finite after the transition to this limit, the integral on the left-hand side of (4.2.36) will not diverge towards the upper limit, since point a_p, marking the end of the precursor region, by definition cannot pertain to the downstream flow, i.e., coincide with the MHD singularity k. Then the left-hand side of (4.2.36) can be made to increase indefinitely only by bringing point 0 close to one of the MHD singular points, which means just passing to the MHD limit. The integral will diverge on the upper limit only if with the given set of parameters the MHD limit of the ionizing shock of type k in question corresponds to the trivial transition $k \to k$. Then as $R \to \infty$ point 0 and point a_p both tend to the singular point k.

If the precursor ionization is included in the consideration, the transition from the GD limit to the MHD limit occurs in the range of velocities which is narrower than that pertaining to the electrostatic breakdown limit. This is due to the strong (exponential) dependence of the flux of ionizing radiation on the downstream temperature. The larger the preexponential coefficient in the expression for ϕ_k, the narrower the velocity range that pertains to the transition to the MHD limit.

Correct to the order of magnitude, from (2.3.32) and (4.2.35) we obtain

$$\phi_k = \Gamma(T_k/I)\exp(-I/T_k) \quad , \tag{4.2.41}$$

where Γ is a dimensionless quantity which depends on the parameters of the initial state and the parameters of the shock; as a rule, $\ln\Gamma \gg 1$, and I is the dimensionless ionization energy, (Sect.2.2.1). The expression in the exponent is also greater than unity, therefore, the right-hand side of (4.2.41) exhibits a substantial change when the shock velocity changes only slightly, while the change in $\ln\Gamma$ is small.

The dimensionless temperature downstream of the ionizing shock of type k is defined by

$$T_k(M_{a0}) = \frac{2}{\beta_0'} u_{xk}(M_{a0})\left\{\left[1 - u_{xk}(M_{a0})\right]^2 M_{a0}^2 - \frac{1}{2}(B_{\perp k}^2 - B_{\perp 0}^2)\right\} \quad , \qquad (4.2.42)$$

where $\beta_0' = 2\mu_0 p_0/B_x^2$.

The transition to the MHD regime, i.e., the transition from $\phi_k \ll 1$ to $\phi_k \gg 1$, takes place mainly in the velocity range obtainable from (4.2.41,42) with double logarithmic accuracy ($\ln\Gamma \gg 1$):

$$T_{k,GD}^{-1}\left(\frac{I}{\ln\Gamma - \ln\ln\Gamma}\right) < M_{a0} < T_{k,MHD}^{-1}\left(\frac{I}{\ln\Gamma - \ln\ln\Gamma}\right) \quad . \qquad (4.2.43)$$

Here $T_{k,GD}(M_{a0})$ and $T_{k,MHD}(M_{a0})$ are the dimensionless downstream temperatures in the shock of type k in the GD and MHD limits respectively, expressed as functions of the Alfvén Mach number M_{a0} in accordance with (4.2.42), the initial state of the gas being fixed.

If we pass formally to the limit $I \to \infty$, $\ln\Gamma \to \infty$ for $I/\ln\Gamma = \text{const}$, then the transition from the GD to the MHD limit is strictly confined to the range of M_{a0} values defined in (4.2.43), the GD and MHD regimes being realized outside this range.

4.3 Transverse Ionizing Shock Waves

4.3.1 Magnetic Structures

Now we apply the general results of Sect.4.2 to transverse ionizing shocks, where the magnetic field lies in the plane of the shock front. As indicated in Sect.4.2, they may belong to type 2 only, i.e., shocks of this kind are supersonic ($M_0 > 1$) and oblique ($B_{z0} = 0$). The latter specification represents one of the two additional boundary conditions required by the criteria of evolutionarity (Sect.4.1.2). The other ABC, which is deduced from (4.2.34), together with the basic boundary conditions (Sect.4.1.1) determine the jumps of all variables in the transverse ionizing shock, and thus its structure, given the parameters of initial state and the shock velocity.

The dimensionless variables are the same as in Sect.4.2, except that B_y and M_a are renormalized as in Sects.3.3.2,4.1, since normalization with respect to $\bar{B}_x = 0$ and $c_a = 0$ is impossible. The transition to the limit $\bar{B}_x \to 0$ ($B_y \to \infty$, $M_a \to \infty$) in the formulas of Sect.4.2 is carried out by transcribing $B_y/B_{y0} \to B_y$, $M_a/B_{y0} \to M_a$, i.e., by normalizing the transverse magnetic field and the velocity to $\bar{B}_{y0}$ and c_a', respectively, see (3.2.11). Different from (4.1.3), the dimensionless electric field is defined by

$$E_s = c\bar{E}_z/\bar{u}_{x0}\bar{B}_{y0} \quad . \tag{4.3.1}$$

In the definition of the dimensionless breakdown threshold E_*, (Sect.4.2.5) we also replace c_a by c'_a and $\bar{B}_x$ by $\bar{B}_{y0}$.

The surface (4.2.5) now becomes a curve in the phase plane (u_x, B_y):

$$F(u_x, B_y) \equiv 4(u_x - 1)(u_x - u_{GD}) + \frac{5u_x}{2M_{a0}^2}(B_y^2 - 1) + \frac{2E_s}{M_{a0}^2}(B_y - 1) = 0 \quad . \tag{4.3.2}$$

Equations (4.2.7,8) reduce to

$$\frac{dB_y}{d\xi} = B_y u_x + E_s \quad . \tag{4.3.3}$$

The condition of zero electric field in the downstream flow is

$$B_{y2}u_{x2} + E_s = 0 \quad . \tag{4.3.4}$$

In the GD limit ("a weak shock") the magnetic field does not change, $B_{y2} = 1$. Hence for the electric field we obtain

$$E_s = -u_{x2} = -u_{GD} \quad . \tag{4.3.5}$$

Usually, in experiments with transverse ionizing shocks $\beta_0 \ll 1$, i.e., $a_0 \ll c'_{a0}$ and $M_0 \gg 1$ even with the "weak" supersonic ionizing shocks; thus $u_{GD} \simeq 1/4$. Accordingly, the upper bound of the dimensionless electric field in the transverse ionizing shock is $E_s = 1/4$. In the MHD limit the upstream electric field is zero. Since as $x \to -\infty$ we also have $B_y = 1$, $u_x = 1$, from (4.3.4) we find the lower bound for the electric field in the shock frame: $E_s = -1$. It follows that the values of E_s pertaining to evolutionary transverse ionizing shocks fall within the range

$$-1 \leq E_s \leq -1/4 \quad . \tag{4.3.6}$$

The value of u_{x2} for a fixed E_s is found from (4.3.2) with $B_y = -E_s/u_x$; the root necessarily exists and falls within the range $1/4 \leq u_{x2} \leq 1$ as long as (4.3.6) is satisfied.

It ought to be observed that equality $E_s = -1$ can take place only for $M_{a0} > 1$, the corresponding solution describing the MHD shock. If, however, $M_{a0} < 1$, then the equality $E_s = -1$ corresponds to a trivial or null solution: $\bar{u}_{x2} = \bar{u}_{x0}$, $\bar{T}_2 = \bar{T}_0$, etc. Obviously enough, this solution cannot by itself describe an ionizing shock, although ionizing shocks may exist which approach this solution as closely as desired, insofar as for $M_0 \gg 1$ the temperature jump across the shock front $\bar{T}_2/\bar{T}_0 \simeq M_0^2(1 - u_{x2})$ may be large enough with any $u_{x2} < 1$.

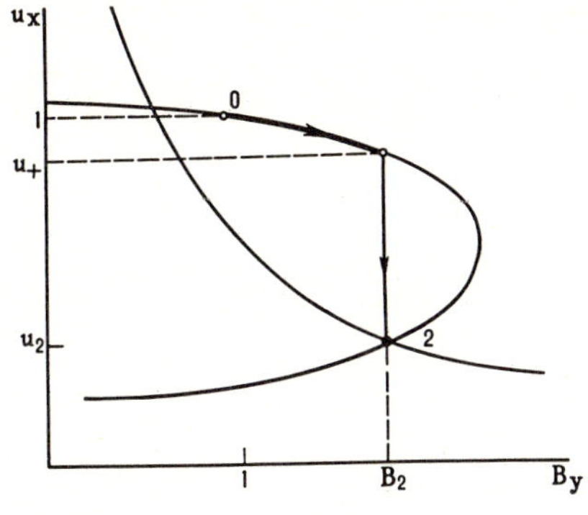

Fig.4.5

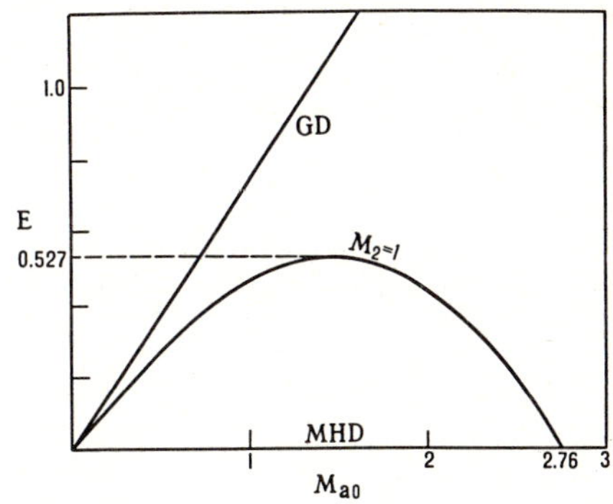

Fig.4.6

Fig.4.5. Magnetic structure of a transverse ionizing shock wave with an isomagnetic subshock ($M_2 < 1$)

Fig.4.6. The region between the rays GD and MHD ($E = 0$) in the (M_{a0},E) plane corresponds to the evolutionary transverse ionizing shock waves, chemistry being neglected. The isomagnetic subshock is absent in shock waves whose parameters lie below the sonic line $M_2 = 1$

The typical shape of the phase curve (4.3.2) and the magnetic structure of a transverse ionizing shock in the (u_x,B_y) phase plane are shown in Fig. 4.5. As follows from the diagram, if the downstream state corresponds to the singular point 2_{sub}, i.e., $M_2 < 1$, and the shock structure contains an isomagnetic subshock, then shock compression of the magnetic field occurs entirely in the precursor region, ahead of the isomagnetic subshock.

Equation (4.2.34) for the transverse shock becomes

$$\int_1^{B_p} \frac{dB_y}{B_y u_x(B_y) + E_s} = \phi_2 \quad , \tag{4.3.7}$$

where $u_x(B_y)$ is the larger of the two roots of (4.3.2). By B_p we denote the magnetic field at point p, corresponding to the end of the precursor region (Sect.4.2.2). The looseness of the division of the shock into a precursor region and a region of shock compression and heating vanishes if $M_2 < 1$ and the structure contains an isomagnetic subshock. Then $B_p = B_{y2}(M_{a0},E_s)$, and the integral in (4.3.7) converges as long as $E_s \neq -1$. The range of M_{a0} and $E = (1 + E_s)M_{a0}$, pertaining to the transverse ionizing shocks, is represented in Fig.4.6; the isomagnetic subshock is contained in the structures of shocks whose parameters fall between the curves marked GD and $M_2 = 1$. As follows from Fig.4.6, the structure of the transverse ionizing shock necessarily contains the isomagnetic subshock for $M_{a0} > 2.76$.

Let us now examine the behavior of the solutions of (4.3.7). In the GD limit, i.e., as $E_s \to -1/4$, $B_{y2} \to 1$. Designating the left-hand side of (4.3.7) by $\Psi(M_{a0}, E_s)$, we obtain

$$\Psi(M_{a0}, E_s) \sim \frac{4}{3}(B_{y2} - 1) \to 0 \text{ as } B_{y2} \to 1 \quad . \tag{4.3.8}$$

In the MHD limit, as $E_s \to -1$ and $M_{a0} > 1$, the asymptotic behavior of $\Psi(M_{a0}, E_s)$ is described by

$$\Psi(M_{a0}, E_s) \sim \frac{M_{a0}^2}{M_{a0}^2 - 1} \ln \frac{(M_{a0}^2 - 1)[B_p(-1, M_{a0}) - 1]}{M_{a0}^2(E_s + 1)} \quad . \tag{4.3.9}$$

We see that if $M_{a0} > 1$, the function $\Psi(M_{a0}, E_s)$ varies from zero to infinity when E_s varies from -1/4 to -1. The function ϕ_2 remains finite if $E_* > 3M_{a0}/4$; otherwise it tends to infinity when $E_s \to (E_*/M_{a0} - 1)$. Consequently, the solution of (4.3.7) exists and is unique when

$$-1 \leqslant E_s \leqslant \min\{-1/4, E_*/M_{a0} - 1\} \quad . \tag{4.3.10}$$

When the flux of ionizing radiation from the shock is negligibly low (the electrostatic breakdown limit), photoionization is capable of producing only the "priming" electrons, while the level of ionization in the precursor region is maintained by Joule heating. In this case we must set $R \to 0$ in (4.2.35); then $\phi_2 \to 0$ everywhere except the point $E_s = (E_*/M_{a0} - 1)$, where $\phi_2 \to \infty$. The solution (4.2.40) for E_s in the transverse ionizing shock in the electrostatic breakdown limit has the form

$$E_s(M_{a0}) = \begin{cases} -1/4 & \text{for} \quad M_{a0} < 4E_*/3 \quad ; \\ E_*/M_{a0} - 1 & \text{for} \quad M_{a0} > 4E_*/3 \quad . \end{cases} \tag{4.3.11}$$

Disregarding this mechanism of precursor ionization as well, i.e., setting $E_* \to \infty$, we come to the GD limit. In this extreme for all M_{a0} the transverse ionizing shocks can have only the trivial magnetic structure, $E_s = -1/4$, $B_{y2} = 1$ [4.7,25,26]. On the other hand, setting $E_* \to 0$ we come to the MHD limit $E_s = -1$. The case of infinitely large precursor ionization also reduces to the MHD limit. Indeed, with $R \to \infty$ the right-hand side of (4.3.7) becomes infinitely large, and therefore the solution of (4.3.7) tends to the singularity of the function $\Psi(M_{a0}, E_s)$, i.e., to $E_s = -1$.

The solution of (4.3.7) for $M_{a0} < 1$ calls for special treatment, since in this case the form of the function $\Psi(M_{a0}, E_s)$ as $E_s \to -1$ depends essentially on the choice of the point B_p. Then the range of integration becomes of the same order of smallness as the denominator of the integrand. Therefore, the singularity of type (4.3.9) will not arise [and $\Psi(M_{a0},E_s)$ will remain bounded when $E_s \to -1$] as long as the upper limit of the integration range in (4.3.7) approaches the singularity at a rate not exceeding the rate of contraction of the integration range into a point:

$$0 < |B_p - B_{y2}| < B_{y2} - 1 \to 0 \quad .$$

At the same time it is obvious (inasmuch as both routes to the MHD limit, whether $R \to \infty$ or $E_* \to 0$, should lead to the same result) that a physically correct division of the shock structure into the precursor region and the region of shock compression for $M_{a0} < 1$ should ensure the existence of a singularity in the behavior of $\Psi(M_{a0},E_s)$ when $E_s \to -1$, as required for the solution to exist.

Now let us examine how the provision for finite strength of the ionizing radiation will modify the rate at which the solution of (4.3.7) passes from the GD to the MHD limit, as the shock velocity increases [4.27]. As demonstrated in Sect.4.2.5, the range of the GD-MHD transition (in terms of shock velocity, or in dimensionless units M_{a0}) depends on the parameter in (4.2.41) and is usually rather narrow, because $\ln\Gamma \gg 1$. For instance, with typical parameters of the ionizing shock in hydrogen and assuming $N_0 = 10^{16}\,cm^{-3}$, we obtain $\Gamma = 1.7 \cdot 10^7 R$.

Figure 4.7 shows the curves of

$$\frac{B_{y2}/B_{y0} - 1}{\rho_2/\rho_0 - 1} = - \frac{E_s + u_{x2}}{1 - u_{x2}} \tag{4.3.12}$$

plotted against M_{a0} for different values of R: $R = 1$, $R = 0.1$, $R = 0.01$ (curves 1,2,3 respectively), and $R \to 0$ (curve "breakdown"). The curves were plotted on the basis of a numeric solution of (4.3.7) together with (4.2.35). As follows

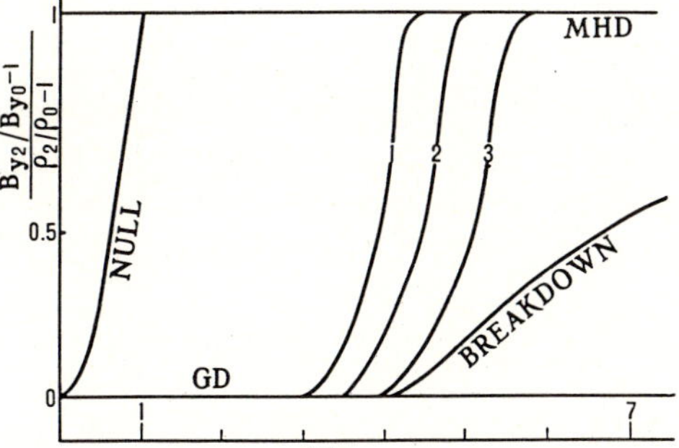

Fig.4.7. Transition from the GD to the MHD regime with increasing intensity of transverse ionizing shock waves. For the curves 1,2,3 the ratio of the density flux of plasma ionizing radiation from behind the shock front to the density flux of the radiation of a blackbody of the same temperature is taken to be 1,0.1 and 0.01, respectively. The curve "breakdown" is plotted in the electrostatic breakdown limit where this ratio tends to zero

from the diagram, the transition from the GD limit ($B_{y2} = B_{y0} = 1$) to the MHD limit in the electrostatic breakdown approximation ($R \to 0$) occurs in a slow power-law fashion in accordance with (4.3.11). The inclusion of the finite photoionization level of the upstream neutral gas restricts the transition from the GD to the MHD limit to a more or less narrow range of shock velocities; the large value of $\ln\Gamma$ makes this restriction effective even when the strength of ionizing radiation is by several orders of magnitude less than the corresponding strength of blackbody radiation. The range pertaining to the transition is defined by (4.2.43) with $k = 2$.

Substituting into (4.2.43) the numerical values of the variables, in accordance with the curves of Fig.4.7, we find the following limits for the dimensionless shock velocity:

$$\begin{aligned} 3.67 \lesssim M_{a0} \lesssim 4.36 \quad &\text{at} \quad R = 1 \;, \\ 4.00 \lesssim M_{a0} \lesssim 4.65 \quad &\text{at} \quad R = 0.1 \;, \\ 4.42 \lesssim M_{a0} \lesssim 5.03 \quad &\text{at} \quad R = 0.01 \;. \end{aligned} \tag{4.3.13}$$

From a comparison of Fig.4.7 with (4.3.13) we observe that the estimate (4.2.43) correctly describes the width and location of the transition range.

4.3.2 Additional Boundary Conditions and Structures of Transverse Ionizing Shocks

The early works on the transverse ionizing shocks [4.7,8,20-25] took no account of nonequilibrium precursor ionization, thus being confined to the GD approximation. As indicated in Sect.4.2, the additional boundary condition is equivalent to the statement that the magnetic field is not compressed in the shock. However, if we take into account the energy spent on ionization and the corresponding additional amount of compression of the plasma in the relatively wide region of ionization relaxation in the tail portion of the shock (Sect.2.3.5), the magnetic Reynolds number within the limits of this region may reach the order of unity, thus allowing a certain (if small) amount of compression of the magnetic field [4.12,26]. Note that this compression must take place after the pressure jump in the gasdynamic shock (the isomagnetic subshock in the head of the shock structure). The transition to the MHD regime in the context of this model occurs when the value of Pm increases from $Pm \ll 1$ to $Pm \sim 1$, and the magnetic field is then compressed within the gasdynamic subshock. In hydrogen ($p_0 \sim 0.1$ Torr) this transition should pertain to the range of shock velocities of $(1\text{-}3) \cdot 10^7$ cm/s.

These conclusions, however, are disputed by experiment [4.28,29]. With the transverse ionizing shock velocities in hydrogen and deuterium in the range of $(0.9\text{-}3)\cdot 10^7$ cm/s the compression of magnetic field and plasma is such as to comply with the MHD boundary conditions [4.29]. The transition from the GD to the MHD regime, i.e., the rise of the left-hand side of (4.3.12) from 0 to 1, occurs in the velocity range of $(5\text{-}7)\cdot 10^6$ cm/s, where $\mathrm{Pm} \ll 1$. The value of Pm exhibits a very sharp increase with the velocity: $\mathrm{Pm} \propto T_2^4$, $T_2 \propto M_0^2$, see (3.3.60). Moreover, the magnetic-field oscillograms obtained in [4.28] indicate that the compression of magnetic field takes place ahead of the pressure jump rather than behind.

To reconcile theory with experiment it is necessary to go further into the ionization processes in shocks in a magnetic field. The expedience of taking due account of the precursor ionization was emphasized in [4.11,12]. As demonstrated in Sect.4.3.1, the theory which includes the precursor-ionization effect is capable of explaining the qualitative aspect of the experimental results: the compression region preceding the pressure jump when $\mathrm{Pm} \ll 1$. We have to ascertain, however, that the action of the mechanisms of precursor ionization under typical experimental conditions can be responsible for the compression of the magnetic field, corresponding to the transition to the MHD regime.

Let us make order-of-magnitude estimations for the precursor phenomena in the electrostatic breakdown approximation. In accordance with the results of Sect.4.2.5 (see also Fig.4.7), the velocity range of the transition from the GD to the MHD limit so obtained will be the widest: including photoionization can only narrow it further.

A rough estimate of electron temperature in the precursor region T_{e0} can be obtained from the equation of balance between the Joule heating of electrons in the electric induction field E_0^* and the energy loss by ionization and elastic collisions which in dimensional variables has the form

$$\sigma_0 E_0^{*2} = n_e \nu_i \left(J + \frac{3}{2} T_{e0}\right) + \frac{2m_e}{m_a} n_e \nu_{ea} (T_{e0} - T_0) \quad . \tag{4.3.14}$$

where $\sigma_0 = e^2 n_e / m_e \nu_{ea}$ is the conductivity of the gas, ν_{ea} is the rate of elastic electron-atom collisions (2.2.18), J is the energy of ionization,

$$\nu_i = C_e N_0 \left(\frac{8T_{e0}}{\pi m_e}\right)^{1/2} \left(\frac{J}{T_{e0}} + 2\right) \exp(-J/T_{e0})$$

is the rate of ionization by electron impact from the ground state, see (2.2.10). Here $C_e = [d\sigma_{ion}(\varepsilon)/d\varepsilon]_{\varepsilon=J}$; N_0 and T_0 are, respectively, the concentration and temperature of atoms in the incident flow. As follows from (4.3.14), in first approximation in small α the estimate for T_{e0} does not depend on n_e.

The rate of α-linear electron loss in (4.2.28) is

$$\nu_\ell = \nu_a + \nu_\Lambda \quad , \tag{4.3.15}$$

where $\nu_a = N_0\sigma_a v_{Te}$ is the rate of attachment to neutral atoms (σ_a being the attachment cross section, v_{Te} the thermal velocity of electrons); $\nu_\Lambda = D_a/\Lambda^2$ is the rate of diffusive escape of electrons to the walls of the tube (D_a being the coefficient of ambipolar diffusion, Λ the characteristic length of diffusion). If we also consider the time required for the ionization to rise to the level $n_e = n_{e0}$, for which $Rm \sim 1$, then in place of (4.2.28) we get a somewhat more rigid breakdown criterion:

$$\nu_i = \nu_a + \nu_\Lambda + \frac{u_{x0}}{L} \ln \frac{n_{e0}}{n_{ph}} \quad . \tag{4.3.16}$$

Here n_{ph} is the order-of-magnitude initial density of photoelectrons; the last term on the right-hand side of (4.3.16) is only weakly sensitive to the chosen value of n_{ph}.

The threshold values of electric field, pertaining to each of the right-hand terms in (4.3.16), may be estimated as [4.15]

$$E_{*a} = N_0 J[8\sigma_a\sigma_{ea}/\pi e^2 \ln(C_e J/\sigma_a)]^{\frac{1}{2}} \quad , \tag{4.3.17}$$

$$E_{*\Lambda} = (J/\Lambda)[8/\pi e^2 \ln(C_e J\sigma_{ea} N_0^2 \Lambda^2)]^{\frac{1}{2}} \quad , \tag{4.3.18}$$

$$E_{*L} = J[N_0\sigma_{ea}u_{x0} \ln(n_{e0}/n_{ph})/e^2 L]^{\frac{1}{2}}(8m_e/\pi J\gamma_L)^{\frac{1}{4}} \quad , \tag{4.3.19}$$

where

$$\gamma_L = \ln A - \frac{1}{2} \ln \ln A \quad ,$$

$$A = (8/\pi m_e)^{\frac{1}{2}} C_e J^{3/2} N_0 L/u_{x0} \ln(n_{e0}/n_{ph}) \quad .$$

In (4.3.17-19) setting $J = 13.6$ eV (hydrogen), and, consistent with the experimental conditions [4.28], $N_0 = 10^{16}\,\mathrm{cm}^{-3}$, $u_{x0} = 5 \cdot 10^6$ cm/s, $L = 20$ cm, $\Lambda = 5$ cm, we find $E_{*a} = 0.4$ V/cm, $E_{*\Lambda} = 2$V/cm, $E_{*L} = 8$ V/cm; i.e., the breakdown threshold is determined by the time of ionization buildup. Let us impose the additional boundary condition, corresponding to the electrostatic breakdown limit, in the form

$$u_{x0}B_{y0} - u_{x2}B_{y2} = E_{*L} \quad . \tag{4.3.20}$$

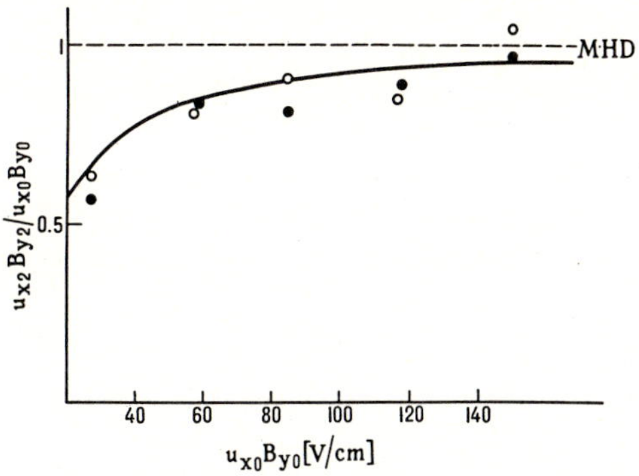

Fig.4.8. The ratio $u_{x2}B_{y2}/u_{x0}B_{y0}$ as a function of $u_{x0}B_{y0}$ for transverse ionizing shock waves. The curve is plotted using (4.3.19, 20) for the experimental conditions and results from [4.28]

The amount of compression of the magnetic field is plotted in Fig.4.8 against $u_{x0}B_{y0}$, the latter being proportional to the maximum strength of the electric induction field attained in the GD regime. Experimental points are reproduced from [4.28]; we see that the agreement between theory and experiment is quite fair. Of course, this concord should not be overrated, insomuch as our estimates are rather rough, especially the estimation of electron temperature from (4.3.14). In a molecular gas, such as hydrogen, the electron energy loss by collisions is much larger than assumed by the last term in (4.3.14), and the breakdown threshold is thus higher (Sect.4.2.3). A refined theoretical curve would fall somewhat below the experimental points in Fig.4.8.

At any rate, these considerations allow us to make some important conclusions. First, we have verified that the inclusion of precursor ionization even in the most straightforward model allows us to obtain fair order-of-magnitude estimates for the parameters of the transition to the MHD regime. Secondly, the transition to the MHD limit occurs as $u_{x0}B_{y0}$ increases, which indicates the importance of breakdown ionization, i.e., impact ionization by electrons accelerated in the electric induction field. Thirdly, the fact that this mechanism by itself is not sufficient for explaining the observed transition to the MHD regime, points to the importance of photoionization, as confirmed also by the narrow velocity range pertaining to the transition (Fig.4.7): $(5\text{-}7)\cdot 10^6$cm/s. In this way we have succeeded in assessing the physical mechanisms which underlie the shaping of the structure of transverse ionizing shocks. This knowledge could be used to determine their steady-state structures within the context of a more realistic model than the one developed in Sect.4.2, thus extending the results obtained in [4.26] to include the effects of precursor ionization. However, it seems much more expedient to consider the process of structure shaping, since under the experimental conditions the nonstationary effects may be quite considerable. At the same time, the steady-state structures may be considered as the final state of nonstationary structures as $t \to \infty$ (Sect.5.4).

4.3.3 Structure of Transverse MHD Shocks in Partially Ionized Plasma

Now we come to considering shock waves which occupy an intermediate position between ionizing shocks and shocks in completely ionized plasma. The similarity of the relevant boundary conditions (the initial ionization being nonzero) puts them closer to the latter. Consistent with Sect.3.2.2 these shocks classify as fast MHD shock waves, so for this reason we mark the initial state with subscript 1. Because the initial ionization is less than unity, we have to include yet another process in the consideration, that of ionization, which may modify the shock structure. Such shocks which are produced in the electromagnetic shock tube SUPPER VI [4.30], are a rewarding object for theoretical study.

The structures of fast shocks in completely and partially ionized plasmas were calculated in [4.31] on the basis of three-fluid Navier-Stokes equations. Considering the difference in the temperature and velocity of atoms and ions, the set of shock-layer equations is reduced to $m = 9$ ordinary first-order differential equations in dimensionless variables B_y, α, u_x^i, u_x^a, T_i, T_e, T_a, dT_e/dx, dT_a/dx (in the accepted notation). The attracting manifolds of the singular points 1 and 2 in the nine-dimensional phase space ($x \to -\infty$ and $x \to \infty$, respectively) have $n_1 = 6$ and $n_2 = 4$ dimensions (Sects.1.8.3, 2.2.2,3.5.3). Condition $n_1 + n_2 = m + 1$, which ensures existence and uniqueness of the steady shock structure, is fulfilled as long as the evolutionarity criteria are satisfied (Sect.1.7.2). However, the shock-layer equations defy straightforward integration: obviously, this could be possible only if one of the dimensionalities n_1, n_2 were equal to unity, and the other coincided with the dimensionality of the phase space. The multiscale asymptotic technique cannot be used here to reduce the dimensionality of the phase space, because of a large number of dissipative processes with matching scales (including impact ionization, electron-ion temperature relaxation, viscosity and Joule heating).

In [4.31] the integration was performed with the aid of an ingenious method which consists essentially in the following. All the variables are divided into two groups, B_y being the only one which is included in both, so that the shock-layer equations can be integrated with respect to either group of variables, as long as the variables of the other group are known functions of the variables of the group chosen for integration. Variable B_y is the independent variable. The iterative procedure starts by obtaining some initial solution, which satisfies the appropriate boundary conditions and the simplified set of shock-layer equations, reduced to an integrable form by disregarding some of the dissipative processes. Then the subsets of equations are integrated with respect to variables of each group, while the variables of the other group (as functions of B_y) are taken from the initial solution. Thus, at each step we use profiles of variables obtained in the preceding step, and so the procedure is repeated until convergence is achieved.

The calculations were carried out for $N_1 = 3.3 \cdot 10^{15} cm^{-3}$, $B_{y1} = 0.1$ T, in a wide range of ionizations: from $\alpha_1 = 0.05$, $T_1 = 10^4$K through $\alpha_1 = 0.999$, $T_1 = 2 \cdot 10^4$K (nonequilibrium ionization). The results obtained in [4.31] indicate that the structures of fast shock waves weakly depend on the initial ionization. Consequently, the transition to MHD-type structures must be assumed to occur outside the investigated range, when the initial ionization is less than a few percent. The wide region of Joule heating at the head of the shock, where for the ionizing shock wave the values of α and B_y change while all other variables remain almost constant (Sects.4.2,3.2), here appears as a minute magnetic step. Generally, the structures comply with the theory for the shock in a completely ionized plasma. At present we are dealing with partially magnetized plasma, where the Joule scale Δ_j is somewhat larger than the viscosity scale Δ_v (Sect.3.4). Accordingly, up to a certain critical Mach number ($M_a \sim 3$-4, depending on α_1) the shock structure is purely resistive, while the structures of stronger shocks contain viscous subshocks. Consistent with the results of Sect.3.4, the resistive shock width decreases as the shock velocity increases. This feature is commonly displayed by all kinds of shocks when their intensity increases from low to moderate, see e.g., (1.8.12). When the mechanism of viscous dissipations is switched on, the shock width starts increasing with velocity, as does Δ_v (Sect.3.4.2).

Here we also meet with a new quality, not encountered in Chap.3. At $\alpha_1 = 0.999$ the Joule scale is commensurate with the scale of the electronic-ionic temperature relaxation, and therefore in the resistive shock the electronic temperature is higher than the ion temperature. However, with $\alpha_1 = 0.05$ the situation is reversed: ions and atoms are hotter than electrons. In other words, the Joule heating is experienced mainly by the heavy component of the plasma! This intriguing effect is discussed in Sect.4.2.2 in the context of normal ionizing shocks since it is exactly in this geometry that it was discovered and explored. As regards the transverse shocks, it is not yet clear whether this effect arises here in full compliance with the theoretical prediction. The experiment [4.30] yields much wider profiles of electron temperature than does theoretical calculation [4.31], and so far the reason for the discrepancy is not clear. As shown below in Sect.4.42, an important role may here be attributed to the effects of nonunidimensionality of plasma flow.

4.4 Normal Ionizing Shock Waves

4.4.1 Magnetic Structures

Let us consider the magnetic structures of normal ionizing shock waves within the context of the model in Sect.4.2. From geometrical considerations, $B_{y0} = B_{z0} = 0$, i.e., the normal ionizing shock waves classify as oblique. Here we confine ourselves to studying only the strong supersonic shocks ($M_0 \gg 1$, $u_{GD} = 1/4$). Then (4.2.5,7) take the form

$$F(u_x,B_y) \equiv (4u_x - 1)(u_x - 1) + \left(\frac{5}{2}u_x - 1\right)\frac{B_y^2}{M_{a0}^2} + 2\frac{E_sB_y}{M_{a0}^2} = 0 \quad , \tag{4.4.1}$$

$$\frac{dB_y}{d\xi} = E_s + B_y(u_x - 1/M_{a0}^2) \quad . \tag{4.4.2}$$

The downstream state is determined by the intersection of curve (4.4.1) in the (u_x,B_y) phase plane with the zero-field hyperbola

$$B_y = \frac{E_s}{1/M_{a0}^2 - u_x} \quad . \tag{4.4.3}$$

We can directly point out an always-existent solution of the set (4.4.1,3): $E_s = 0$, $B_{yk} = 0$, $u_{xk} = 1/4$. This solution describes a gasdynamic ionizing shock wave in which the flow is parallel to the magnetic field and thus does not interact with it. It can be easily demonstrated, (3.2.12-14), that the gasdynamic ionizing shock in a magnetic field, consistent with our classification, pertains to type 2 ($k = 2$) for $M_{a0} > 2$ and to type 4 ($k = 4$) for $M_{a0} < 2$. This solution is peculiar in that it realizes simultaneously both GD and MHD limits.

Assuming now that $E_s \neq 0$, we notice that (4.4.1-3) are invariant under the transformation $B_y \to -B_y$, $E_s \to -E_s$. From (4.4.2) follows that $dB_y/d\xi = E_s$ at $B_y = 0$, i.e., when moving along the integral curve from $B_y = 0$ for $E_s > 0$ we always remain within the region $B_y > 0$. Therefore it suffices to consider

$E_s > 0$, $B_y > 0$. Then from (4.4.3) we find that at point k downstream

$$1/M_{a0}^2 - u_{xk} > 0 \quad .$$

Hence $M_{ak} < 1$ and, consequently, the downstream state for $E_s > 0$ may pertain only to the singular point of type 3 or type 4.

For $M_{a0} > 2$ the branch of curve (4.4.1) which contains point 0: ($u_x = 1$, $B_y = 0$) in the (u_x,B_y) phase plane lies strictly above the horizontal asymptote of hyperbola (4.4.3) and therefore has no common points with the hyperbola for $E_s > 0$. Accordingly, the present case pertains only to a GD ionizing shock of type 2 with additional boundary conditions

$$E_s = 0 \quad , \quad B_{y2} = 0 \quad . \tag{4.4.4}$$

We have yet to examine the case $0 < M_{a0} < 2$, where the switch-on ionizing shocks of types 3 and 4 with nonzero transverse magnetic field downstream are possible, including, in particular, the GD ionizing shocks complying with conditions (4.4.4). Insofar as the evolutionarity criteria do not require that two ABC exist, the GD normal ionizing shock waves for $M_{a0} < 2$ may exist only as a particular case in the one-parameter family of normal ionizing shock waves of type 4 with $E_s > 0$. Let it be emphasized that at present, as in the general case, the evolutionary shocks of type 2 correspond to isolated points, shocks of type 3 to curves (one-parameter families), shocks of type 4 to regions (two-parameter families) in the (B_{z0},E_s) parameter plane. Inasmuch as we are dealing here with only the normal ionizing shocks, we reduce by one the dimensionality of the relevant manifolds, selecting a point on the curve or a curve in the region. Therefore, the electric field E_s must be determined unambiguously for the evolutionary normal ionizing shocks of type 3, while the shocks of type 4 should comprise a one-parameter family.

If $M_{a0} < 2$, the set of equations for the boundary conditions has solutions for

$$0 \leqslant E_s \leqslant E_s^{CJ} \quad . \tag{4.4.5}$$

The lower bound of this range corresponds to the MHD limit. In this limit the shock wave is represented by a purely gasdynamic slow MHD shock for $M_{a0} < 1$, and a switch-on MHD shock for $1 < M_{a0} < 2$ (Sects.3.3.2,4,6). The upper bound of the range (4.4.5) corresponds to the case where the solutions of types 3 and 4 coincide, i.e., where the zero-field hyperbola touches the curve (4.4.1) in the (u_x,B_y) plane. As indicated above, this implies that the downstream velocity equals the velocity of the slow magnetosonic wave; for this reason such a shock is called the Chapman-Jouguet shock wave. In

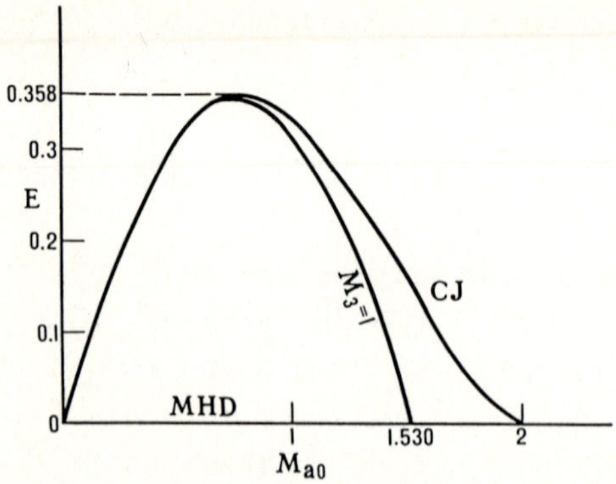

Fig.4.9. The region between the line CJ and the ray MHD (E = 0) in the (M_{a0},E) plane corresponds to the evolutionary normal ionizing shock waves of types 3 and 4, chemistry being neglected. In type 3 shock waves, whose parameters lie below the sonic line $M_3 = 1$, the isomagnetic subshock is absent

Fig.4.9 the range (4.4.5) is given in the parameter plane (M_{a0}, $E = M_{a0}E_s$). The borderline CJ of this region corresponds to the Chapman-Jouguet shock waves, in which the upstream electric induction field assumes the highest possible value for a given M_{a0}.

Since the singular point 4 is always subsonic, the structure of an ionizing shock wave of type 4 contains an isomagnetic subshock. On the contrary, the singular point 3 may be either subsonic or supersonic. Ionizing shocks of type 3_{sub}, whose parameters fall between the lines CJ and $M_3 = 1$ (Fig.4.9) contain an isomagnetic subshock, whereas the parameter region below $M_3 = 1$ pertains to ionizing shocks of type 3_{super} without isomagnetic subshock.

Note that including the energy spent on ionization and dissociation (if the test gas is a molecular one; both these processes are called "chemistry") will extend the region of parameters pertaining to the normal ionizing shock waves. For instance, in hydrogen at $p_0 = 0.2$ Torr, $B_x = 0.15 - 2.5$ T the switch-on ionizing shocks may exist in the range of $1 < M_{a0} \lesssim 3$, while the maximum dimensionless electric field E may be as high as 0.6 [4.2,32].

Figure 4.10 shows a typical magnetic structure of a normal ionizing shock wave. We see that in the normal ionizing shock of type 3_{sub} the entire variation of the magnetic field occurs upstream of the isomagnetic subshock, whereas in the shock of type 4 the magnetic field may vary both before and after the isomagnetic subshock. Accordingly, we are confronted with the problem concerning the value (range of values) of E_s, corresponding to the ionizing shock of type 3 (type 4) for a given M_{a0}, and the exact location of the isomagnetic subshock in the structure of the ionizing shock of type 4 (whether it is point a or a').

In the gasdynamic limit the answer is straightforward [4.7,8]: normal ionizing shocks of type 3 just cannot exist in this limit. The entire region

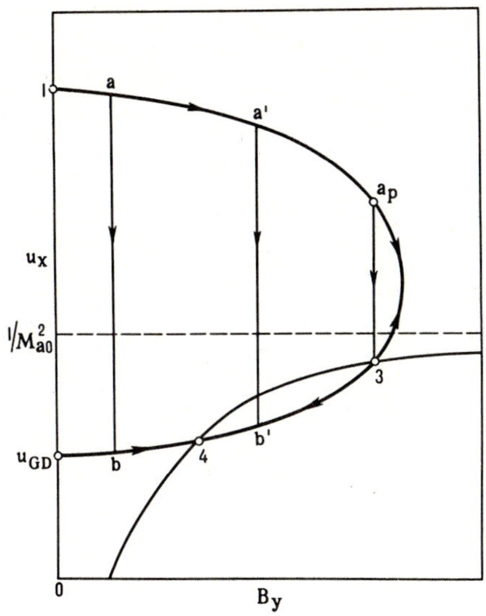

Fig.4.10. Magnetic structures of normal ionizing shock waves of types 3 and 4 for $M_3 < 1$

below the line CJ in Fig.4.9 pertains to shocks of type 4, point a coinciding with point 0. In the MHD limit there are switch-on (fast) MHD shocks (the extreme case of ionizing shocks of type 3) if $1 < M_{a0} < 2$, and purely GD slow MHD shocks (the extreme case of type 4 shocks) if $M_{a0} < 1$. To approach the problem in general, we rewrite (4.2.34) in the form

$$\Psi(B_p) \equiv \int_0^{B_p} \frac{dB_y}{E_s + B_y(u_x - 1/M_{a0}^2)} = \phi_k \quad , \tag{4.4.6}$$

where $k = 3,4$, and $u_x(B, E_s, M_{a0})$ is, as in Sect.4.3.1, the larger root of (4.4.1), quadratic in u_x. For $k = 3$, let B_3' stand for the magnetic field B_p at the end of the precursor region in the shock of type 3 (point a_p in the notation of Sect.4.2.4). For $k = 4$ we denote by B_a the magnetic field B_p in the head of the isomagnetic subshock (point a), contained in the structure of the shock of type 4. The function ϕ_k is defined by (4.2.35).

Under the conditions of Fig.4.10 $B_3' = B_{y3}$, $B_a < B_3'$. If the singular point 3 is supersonic, then the latter inequality remains valid, but (as in Sect.4.3.1) $B_3' < B_3$, so that the integral on the left-hand side of (4.4.6) converges.

The integrand in (4.4.6) is positive, and by virtue of the above-developed arguments $B_3' \geq B_a$, i.e., $\Psi(B_a) \leq \Psi(B_3')$ (in both cases the equality is possible only at $E_s = E_s^{CJ}$). This means that only one type of shock may exist for given E_s and M_{a0}:

either $\Psi(B_a) < \Psi(B_3') = \phi_3 < \phi_4$, i.e., $k = 3$,

or $\Psi(B_3') > \Psi(B_a) = \phi_4 > \phi_3$, i.e., $k = 4$.

Both types of ionizing shocks can exist simultaneously only for the degenerate case of the Chapman-Jouguet flow. Note that the purely gasdynamic solution for a shock of type 4 ($E_s = 0$, $B_{y4} = 0$) cannot be obtained from (4.4.6), since in this case both sides of (4.2.34) vanish identically, and the basic relation (4.2.34) no longer holds. Therefore this solution should be viewed as an additional solution to those of (4.4.6).

The behavior of the function Ψ_3 is analyzed along the same lines as in Sect.4.3.1, and is generally similar to the behavior of $\Psi(M_{a0}, E_s)$. In particular, when $M_{a0} > 1$ and the upper limit of the integral in (4.4.6) does not vanish as $E_s \to 0$, e.g., with $M_{a0} > 1.503$, $B_{y3}(M_{a0}, 0) = [2(M_{a0}^2 - 1)(4 - M_{a0}^2)/3]^{\frac{1}{2}}\gamma$, the main (logarithmic) singularity of $\Psi_3(M_{a0}, E_s)$ is expressed by the right-hand side of (4.3.9), where $B_2'(M_{a0}, -1)$ must be replaced by $B_3'(M_{a0}, 0)$, and $E_s + 1$ by E_s. So that both transitions $E_* \to 0$ and $R \to \infty$ lead to the same result (the transition to the MHD regime), the following condition has to be satisfied, as in Sect.4.3.1:

$$\Psi_3(M_{a0} < 1 \ , \ E_s \to 0) \to \infty \quad . \qquad (4.4.7)$$

The general form of the solutions of (4.4.6) is illustrated in Fig.4.11. In the diagram the functions $\Psi_3(M_{a0}, E_s)$, $\phi_3(M_{a0}, E_s)$, $\phi_4(M_{a0}, E_s)$ are plotted for a fixed $M_{a0} > 1$ for different values of $E = M_{a0}E_s$ for $E_* > M_{a0}E_s^{CJ} \equiv E_{CJ}$ (Fig.4.11a) and $E_* < E_{CJ}$ (Fig.4.11b). In both cases (4.4.6) defines a range of values of E for shocks of type 4: $0 \leq E_4 \leq E_{4\,max}$. As follows from the diagram, with the appropriate choice of point a the function $\Psi(B_p)$ assumes any

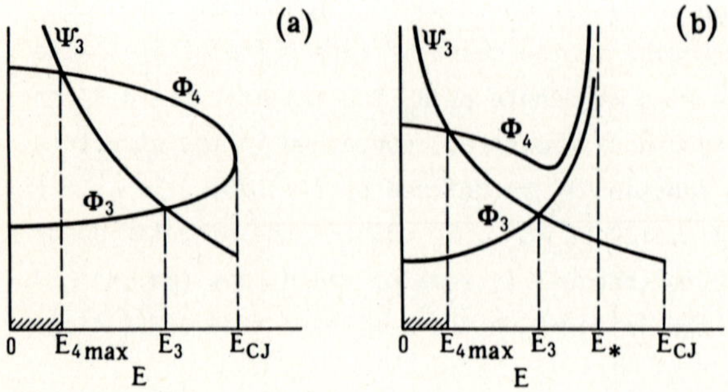

Fig.4.11a,b. The functions $\Psi_3(M_{a0}, E_s)$, $\Phi_3(M_{a0}, E_s)$, $\Phi_4(M_{a0}, E_s)$ for constant M_{a0} and variable $E = M_{a0}E_s$. The interval $0 \leq E < E_{4\,max}$ corresponds to type 4 shock waves, and the value $E = E_3$ to type 3 shocks: **(a)** $E_{CJ} < E_*$; **(b)** $E_{CJ} > E_*$

value between 0 and $\Psi_3(M_{a0},E_s)$, and a single value E_3 outside this range for shocks of type 3.

In the electrostatic breakdown limit ($R\to 0$), when $\phi_3(M_{a0},E_s)\to 0$ and $\phi_4(M_{a0},E_s)\to 0$, under the circumstances of Fig.4.11a there are no solutions for shocks of type 3 and $E<E_*$, whereas shocks of type 4 may exist in the entire range of $0\leqslant E\leqslant E_{CJ}$. For such shocks the right-hand side of (4.4.6) vanishes, and the left-hand side should also vanish: $B_a=0$. Thus, the structure of the ionizing shock of type 4 starts with the isomagnetic subshock, i.e., has the same form as in the gasdynamic limit. If $E_*<E_{CJ}$ (as in Fig. 4.11b), in the electrostatic breakdown limit we find $E_3=E_*$, $0<E_4<E_*$. Then the structure of the ionizing shock of type 3 begins with the magnetic field compression region, which may be followed by an isomagnetic subshock (providing that the parameters of the shock fall between the lines $M_3=1$ and CJ in Fig.4.9). If the condition (4.4.7) is satisfied, then similar results are obtained for $M_{a0}<1$. If the dimensionless breakdown threshold is high enough ($E_*>0.358$; Fig.4.9), the condition $E_*<E_{CJ}$ cannot be satisfied, and this extreme case is reduced to the gasdynamic limit, which allows the existence of only the ionizing shocks of type 4. Summing up the results, we may write for the electrostatic breakdown limit:

$$\begin{cases} 0\leqslant E_4 < \min(E_{CJ},\, E_*) \quad ; \\ E_3 = E_* \quad , \qquad \text{if} \qquad E_* < E_{CJ} \quad ; \\ \text{no shocks of type 3,} \qquad \text{if} \qquad E_* > E_{CJ} \quad . \end{cases} \tag{4.4.8}$$

The transition to the MHD limit occurs, as a rule, similarly both for $E_*\to 0$ as well as $R\to\infty$. In both cases the range of permissible values of E_s for the shocks of type 4 contracts to a point $E_s=0$, and E_s also tends to the same value (Fig.4.11). Of these two solutions, however, only one describes a nontrivial evolutionary MHD shock transition for each value of M_{a0}. For $1<M_{a0}<2$ the shock waves of type 4 become nonevolutionary, purely gasdynamic shocks. For $M_{a0}<1$ the shock of type 3 reduces to a null solution ($\bar{u}_{x3}=\bar{u}_{x0}$), which, obviously, cannot by itself describe an ionizing shock wave, although with $E_s\ll 1$ nontrivial ionizing shocks may exist however close to this solution.

Why should this transition to the limit give redundant (nonevolutionary from the viewpoint of MHD criteria; Sect.3.3.6) gasdynamic shocks? Formally, the answer is that the evolutionarity conditions for the ionizing shock waves do not become directly the evolutionarity criteria for the MHD limit. For instance, the trans-Alfvén shock transitions are allowed for the ionizing shock waves and forbidden for the MHD shocks.

The edge of the conducting region may be visualized to recede towards $x = -\infty$ when $R \to \infty$ or $E_* \to 0$. As soon as the width of this region exceeds a certain characteristic length, e.g., the length of the shock tube, it would be logical to switch to the MHD criteria of evolutionarity. The structures of nonevolutionary shocks, arising in the MHD limit, turn out to be ambiguous and dependent on the limiting procedure. As demonstrated in Sects. 3.3.5,4.2.1, the MHD singular points 1 and 4 are linked by an infinite number of integral curves representing such structures; any of these structures may be realized by an appropriate choice of the limiting process.

The earlier theoretical treatments of normal ionizing shock waves [4.1,7, 32] disregarded the precursor ionization, being thus confined to the gasdynamic limit, in which only the shocks of type 4 are possible. Analyzing the production of such shocks in the electromagnetic shock tube, *Kunkel* and *Gross* [4.1] concluded that only the ionizing Chapman-Jouguet shocks, considered as the extreme case of type 4 shocks (the Chapman-Jouguet hypothesis), can be observed in these shock tubes. We shall go into this matter in Sect. 5.2.5. However, the analogy with detonation theory is quite straightforward, if only we identify the velocity of the slow magnetosonic wave downstream with the speed of sound in the detonation products, and the ionizing shocks of types 3 and 4 with the weak and the strong detonation waves, respectively (Sect.1.9). From Figs.4.9,10 it becomes clear that the structure of the normal ionizing Chapman-Jouguet shock, which belongs to the shocks of type 4, in the adopted gasdynamic approximation starts with the gasdynamic subshock, followed by the magnetic field compression region, accompanied by expansion of the plasma (Fig.1.8b). The compression of plasma in such a shock must be no less than $2/(1+u_{GD}) = 1.6$.

The first detailed experimental investigation of normal ionizing shock waves [4.33] dealt with shocks in hydrogen ($p_0 = 0.2$ Torr) with $u_{x0} = 2 \div 7$ cm/μs and $B_x = 0.15$-0.25 T (a review of these and earlier works can be found in [4.2]). The results of [4.33] have confirmed the main peculiar features of shocks of this kind, predicted by the theory, in particular, the restrictedness of the range of velocities pertaining to the switch-on ionizing shocks and of the relevant range of electric fields [4.32]. The measurements of electric and magnetic fields fit in well with the theory based on the Chapman-Jouguet hypothesis.

The gasdynamic approximation was shown in Sect.4.3.1 to be utterly inadequate for describing the transverse ionizing shocks. Why is it then that it can deal successfully with the normal ionizing shocks in the same gas (hydrogen), in the same velocity range, and under more or less the same conditions? The fact is that normal shocks are less readily interpreted. It would be natural to consider the details pertaining to the possible excitation of various types of shocks within the context of the piston problem, with which we shall be concerned in Sect.5.2. Here we just point out that the gasdynamic approximation holds not only with low shock velocities, but also when the velocities are so high as to prevent the existence of the switch-on ionizing shocks, where the GD coincides with the MHD limit. This circumstance (to some extent fortuitous, since it occurs only in the normal geometry), explains why the theory based on the Chapman-Jouguet hypothesis is capable of correctly describing the asymptotic behavior at high velocities, i.e., the transition to the MHD regime.

For experimental verification of the hypothesis that the gasdynamic limiting regime is realized, apart from measuring the transverse electric and magnetic fields, one has to investigate the structure of the ionizing shock. As long as the electric field is close to E_{CJ}, the precursor ionization is insignificant, the structure begins with an isomagnetic subshock and the compression of the plasma in the shock complies with the Chapman-Jouguet hypothesis, then the conditions of applicability of the gasdynamic approximation (and thus the feasibility of the Chapman-Jouguet hypothesis) may be assumed to be satisfied. These conclusions, however, are disputed by the available experimental evidence. The magnitudes of electric field reported in [4.5] are considerably lower than E_{CJ}. The measurements of the plasma compression in the normal ionizing shocks yields consistently lower values than predicted by the Chapman-Jouguet theory (no compression at all could be detected in [4.5]). The presence of the narrow gasdynamic subshock in the head of the shock wave was not verified either [4.5,6]. At the same time, the precursor phenomena are quite considerable, and virtually determine the structure of normal ionizing shocks. Let us now take a closer look at them.

4.4.2 Tensor Conductivity and Joule Heating of Plasmas in Normal Ionizing Shocks

Weakly ionized plasma in the precursor region ahead of the normal inonizing shock is found in the crossed electric and magnetic fields. The passage of electric current produces a peculiar effect, consisting in strong Joule overheating of the heavy component of the plasma. This effect was theoretically predicted in [4.34] as early as 1956. It was first discovered in experiments with normal ionizing shocks in the electromagnetic cylindrical shock tube SUPPER II [4.5]; its theoretical analysis can be found in [4.35].

This effect arises when the movement conditions of ions in the plasma are such that the ions slip at a nonzero velocity past the neutral plasma component. Then, due to the large cross section of ion-neutral charge exchange interactions, the collisions will release a large amount of heat, redistributed between the ions and neutral particles. With low ionization ($\alpha \ll 1$) the movement of neutral particles may be described within the context of the Navier-Stokes model, whereas the ion component calls for the kinetic approach, since the ion-atom relaxation time is lower than the ion-ion relaxation time, proportional to α^{-1}. A qualitative description may also be given using the Navier-Stokes equations in which, however, the ion component is assumed to have its own mean velocity $\mathbf{u}^i$, other than the velocity of neutral particles $\mathbf{u}^a$. At the same time, there is no need to introduce the ion temperature $T_i \neq T_a$, since thermal equilibrium in the entire system of heavy particles is established in the time span of the same order as in either of its subsystems. Accordingly, we retain the designation T for the temperature of heavy particles.

To understand the circumstances which facilitate the slipping of the ions, let us turn to the three-fluid Navier-Stokes equations, describing the asymptotic stationary ionizing shock structure as $x \to -\infty$, when $\alpha \to 0$, $B_y \to 0$, $B_z \to 0$ (for reasons explained below we consider the process in three dimensions rather than in two) [4.36]. Obviously, $B_y = O(\alpha)$, $B_z = O(\alpha)$. In dimensionless variables (Sect.4.1.1) the axial mass velocity of the gas is $u_x = 1 + O(\alpha)$. As $x \to -\infty$, the ionization declines exponentially with axial distance

$$\alpha \propto \exp(x/\Delta_\alpha) \quad ,$$

where Δ_α is the characteristic scale, which generally depends on both the photoionization and on the rate of impact ionization. Here the latter is insignificant because of the low temperature of precursor electrons (see below), and hence $\Delta_\alpha = 1/N_0\sigma_{ph}$. We introduce the new variable

$$\xi = \int_{-\infty}^{x} \alpha(x')dx' \quad . \tag{4.4.9}$$

In the neighborhood of point $\xi = 0$, corresponding to $x = -\infty$, the derivatives of the velocity and the magnetic-field strength with respect to ξ are of the order of unity. From the Maxwell equations and (4.4.9) we obtain

$$u_y^i - u_y^e = -\Delta_h \frac{dB_z}{d\xi} , \tag{4.4.10}$$

$$u_z^i - u_z^e = \Delta_h \frac{dB_y}{d\xi} , \tag{4.4.11}$$

where $\Delta_h = \bar{B}_x/\mu_0 e N_0 \bar{u}_{x0}$. The equations of the transverse motion of plasma give us

$$u_y^a - u_y^i = \left(\frac{B_y}{M_{a0}^2} - u_y^i\right)(1 + O(\alpha)) , \tag{4.4.12}$$

$$u_z^a - u_z^i = \left(\frac{B_z}{M_{a0}^2} - u_z^i\right)(1 + O(\alpha)) . \tag{4.4.13}$$

From the equations of motion of electrons we get

$$u_z^e - B_z = k_e(u_y^a - u_y^e) , \tag{4.4.14}$$

$$E_s + B_y - u_y^e = k_e(u_z^a - u_z^e) , \tag{4.4.15}$$

where $k_e = 1/\Omega_e\tau_{ea}$; $\Omega_e = eB_x/m_e$; $\tau_e = [(4N_0/3)\bar{\sigma}_{ea}(8T_e/\pi m_e)^{\frac{1}{2}}]^{-1}$ is the characteristic time of electron-neutral collisions pertaining to the transfer of momentum; $\bar{\sigma}_{ea}$ is the transport cross section averaged over the Maxwellian distribution of electrons.

Adding the equations of motion of electrons and ions yields

$$\frac{d}{dx}\alpha u_y^i - \frac{1}{M_{a0}^2}\frac{dB_y}{dx} = \frac{\alpha(u_y^a - u_y^i)}{\Delta_a} , \tag{4.4.16}$$

$$\frac{d}{dx}\alpha u_z^i - \frac{1}{M_{a0}^2}\frac{dB_z}{dz} = \frac{\alpha(u_z^a - u_z^i)}{\Delta_a} , \tag{4.4.17}$$

where $\Delta_a = \bar{u}_{x0}\tau_{ia}$, $\tau_{ia} = [(4N_0/3)\bar{\sigma}_{ia}(16T/\pi m_i)^{\frac{1}{2}}]^{-1}$; $\bar{\sigma}_{ia}$ is the cross section of ion-neutral charge exchange, averaged over the Maxwellian distribution.

Equations (4.4.12-17) indicate that, as $x \to -\infty$, the transverse velocities of charged plasma species do not vanish:

$$u_y^i = \frac{k_e k_i + 1}{(k_e^2 + 1)(k_i^2 + 1)} E_s , \qquad u_z^i = \frac{k_i - k_e}{(k^2 + 1)(k^2 + 1)} E_s , \tag{4.4.18}$$

$$u_y^e = \frac{1}{k_e^2+1} E_s \quad , \qquad u_z^e = -\frac{k_e}{k_e^2+1} E_s \quad , \tag{4.4.19}$$

where $k_i = 1/\Omega_i\tau$, $\Omega_i = eB_x/m_i$, $\tau = 1/(\tau_{ia}^{-1} + \nu_i)$, $\nu_i = u_{x0}/\Delta_\alpha$.

From (4.4.18,19) we obtain Ohm's law in tensor form:

$$\mathbf{j}_\perp = \sigma_\perp \mathbf{E}_\perp + \sigma_\Lambda \frac{\mathbf{B}}{B} \times \mathbf{E}_\perp \quad , \tag{4.4.20}$$

where

$$\sigma_\perp = \sigma_0 \frac{k_e k_i (k_e k_i + 1)}{(k_e^2+1)(k_i^2+1)} \quad , \qquad \sigma_\Lambda = \sigma_0 \frac{k_e k_i (k_i - k_e)}{(k_e^2+1)(k_i^2+1)} \quad ,$$

$$\sigma_0 = \frac{e^2 n}{m_e} \tau_{ea} \quad .$$

The general expression for Ohm's law in tensor form, accounting for inelastic ionization processes with arbitrary α, can be found in [4.37].

Thus we have succeeded in demonstrating that in the normal ionizing shock in weakly ionized gas ahead of the main shock transition the electric-induction field causes transverse movement of ions relative to neutral particles, an effect that does not fit into the two-fluid model. It should be noted that similar calculations for the transverse ionizing shock bring us back to Ohm's law in scalar form $\mathbf{j}_\perp = \sigma_\perp \mathbf{E}_\perp$, i.e., our earlier findings do not call for corrections. The conductivity is a scalar quantity because of the axial electric field, produced by the polarization of plasma (Sect.3.3.4).

Let us now show that this movement overheats the heavy component. We write the equation of heat conduction for the heavy component in the precursor region in the form [4.38]

$$\frac{3}{2} N_0 u_{x0} \frac{dT}{dx} = \frac{2}{3} n N_0 \left(\frac{4T}{m_a}\right)^{1/2} \sigma_{ia} m_a u_{x0}^2 \frac{E_s^2}{(k_e^2+1)(k_i^2+1)} \quad . \tag{4.4.21}$$

Here all variables are dimensional, except E_s.

Assuming that $k_e^2 \ll k_i^2 \ll 1$ and disregarding the weak dependence of σ_{ia} on T, we integrate (4.4.21) with due regard to the fact that $n(x) \propto \exp(x/\Delta_\alpha)$:

$$n = N_0 \frac{9\sigma_{ph}}{4\sigma_{ia}} \left(\frac{\pi}{m_a}\right)^{1/2} \frac{u_{x0} B_x^2}{c^2 E_s^2} T^{1/2} \quad . \tag{4.4.22}$$

The temperature of heavy particles in the precursor region is thus found to decrease exponentially with axial distance, like the ionization, the characteristic length being

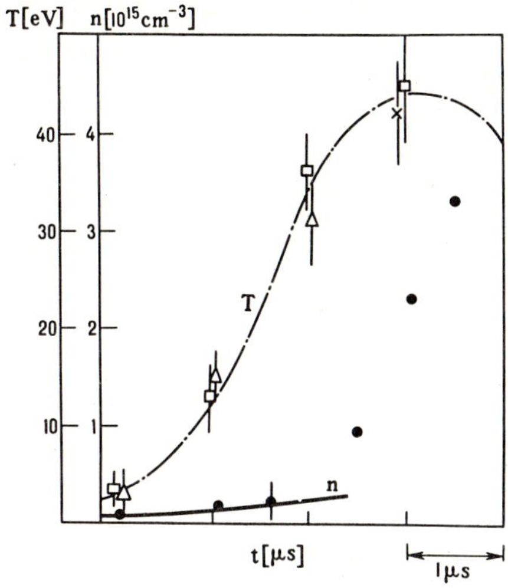

Fig.4.12. The electron density profile n in the precursor region of the structure of a normal ionizing shock wave (——), as calculated from measured T values (-·-·-) using (4.4.22). Experimental points and the interpolation of experimental data for T are from [4.5]

$$\Delta_T = \Delta_\alpha/2 = 1/2N_0\sigma_0 \quad .$$

In the experiments of [4.5] with helium ($\delta_{ph} = 5\cdot 10^{-18}\,cm^2$, $N = 3.3\cdot 10^{15}\,cm^{-3}$), the calculated value of $\Delta_T = 30$ cm fits in well with the experimental results. Substituting into (4.4.22) the magnitude of radial electric field $E = 300$ V/cm (measured with $u_{x0} = 11$ cm/μs, $B_x = 1$ T), we can find the dependence n(x) from the known T(x). As follows from Fig.4.12, the agreement with the experimental results of [4.5] for $\alpha \lesssim 0.1$ is quite fair. The temperature of the electron component can be estimated by noting that the Joule heating of electrons in the precursor region is weak, so that the heating of electrons is due mainly to elastic collisions with the neutrals. The equation for the electron temperature has the form

$$\frac{3}{2}\, nu_{x0}\, \frac{dT_e}{dx} = 8\sqrt{2}N_0 n\left(\frac{m_e T_e}{\pi}\right)^{\frac{1}{2}} \frac{\bar{\sigma}_{ea}}{m_a}\,(T - T_e) \quad , \tag{4.4.23}$$

whence [setting $\bar{\sigma}_{ea}(T) = 5\cdot 10^{-16}\,cm^2 = \mathrm{const}$] with due account of (4.4.21) we get

$$T_e = \left[\frac{4}{3}\left(\frac{2m_e}{\pi}\right)^{\frac{1}{2}} \frac{\bar{\sigma}_{ea}}{m_a u_{x0}\sigma_{\nu 0}}\right]^2 T^2 \quad . \tag{4.4.24}$$

From Fig.4.13 we see that the discrepancy between (4.4.24) and the experimental results [4.6] becomes noticeable only when $\alpha \sim 0.1$, $T_e \sim 6$ eV, when the electron energy loss on impact ionization becomes considerable.

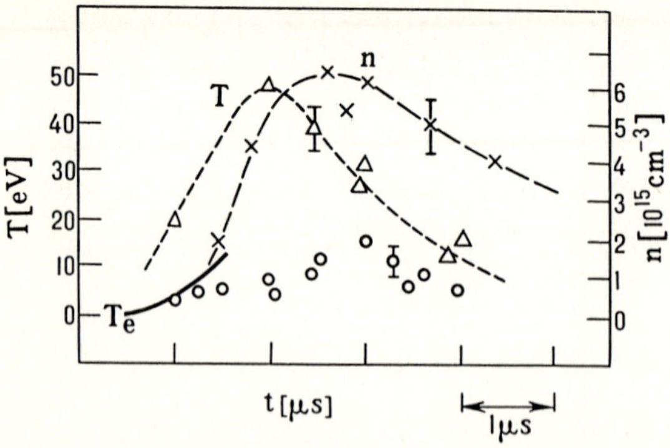

Fig.4.13. The electron temperature profile T_e in the precursor region of the structure of a normal ionizing shock wave (——), as calculated from measured T values (----) using (4.4.24). Experimental points and interpolation of experimental data for T and n (———) are from [4.6]

We see that the structures of normal ionizing shock waves, produced in electromagnetic shock tubes, are shaped by the processes of ionization and heating in the precursor region. The correct description of these processes calls for the use of kinetic equations, or, on the model level, of the three-fluid Navier-Stokes equations. The theoretical structures derived from the two-fluid equations will have little in common with the experimental structures (Sect.5.5 below). Much less adequate is the gasdynamic approximation with its total neglect of precursor phenomena. In particular, the Chapman-Jouguet hypothesis, being a corollary of the gasdynamic approximation, is unworkable.

One of the consequence of the tensor nature of the plasma's conductivity in the precursor region is the nonplanar configuration of the magnetic field. As follows from (4.4.16,17), B_y and B_z as $x \to -\infty$ vary steadily, without oscillations exhibited by the structures of switch-on MHD shocks (Sects. 3.5.3 and 4.4.3). It is straightforward that, as $x \to -\infty$,

$$\frac{B_z}{B_y} = \frac{k_i - k_e}{k_i k_e + 1} \cong \frac{\Omega_e \tau_{ea}}{(\Omega_e \tau_{ea})(\Omega_i \tau) + 1} , \tag{4.4.25}$$

and if, as typically, $\Omega_e \tau_{ea} \gg 1 \sim \Omega_i \tau$, the plane of magnetic field, i.e., the plane stretched on the vector $\mathbf{B}$ and the normal to the shock front, is rotated: as $x \to -\infty$, $(B_z/B_y) \sim 1$, and as $x \to +\infty$ $B_z \to 0$. If we assume that the unidimensional problem we are considering pertains to the cylindrical configuration of a coaxial electromagnetic shock tube with $(R_{out}/R_{in}) \ll 1$, (Sect.4.1.2 and Fig.5.1 below), then the z component of the magnetic field is directed radially. At the same time, a near-uniform magnetic field with a nonzero radial component cannot exist in cylindrical configuration, since this would be contradictory to (3.1.1). In other words, (4.4.25) is incompatible with the geometry of the coaxial shock tube. This is not surprising anyway, since (4.4.25) was derived under the assumption that both transverse components of the velocities of heavy particles are nonzero, while the boundary conditions on the walls of the shock tube allow movement only in the axial direction. Reworking the previous calculations, this time for the planar geometry of the flow ($u_z^i = u_z^a = 0$) [4.36] yields in place of (4.18) the following relation:

$$u_y^i = \frac{k_e}{k_i(k_e^2 + 1)} E_s \quad , \tag{4.4.26}$$

whence $u_y^i/u_y^e = k_e/k_i \sim (m_e/m_a)^{\frac{1}{2}} \ll 1$, i.e., the current in the plasma is conducted mainly by the electrons, while in the three-dimensional problem (in the case of magnetized plasma $k_e \ll k_i \ll 1$) the current in the z direction was conducted mainly by ions, and by both ions and electrons in the y direction ($|u_y^e - u_y^i| \ll u_y^e$). Although, in accordance with (4.4.26), some ion slip still exists, the Joule heat release in the heavy components is (m_e/m_a) times less than on the right-hand side of (4.4.21). In this model the Joule overheating of ions and atoms is precluded. The only other cause of overheating of the heavy plasma component may be related to the action of viscosity in the realm of the gasdynamic subshock. However, such subshocks were definitely not observed in the reported experiments.

We see that the problem of choosing the best approach for numerical simulation of normal ionizing shocks in electromagnetic shock tubes remains unsolved. The simplest unidimensional models cannot be extended to account simualtaneously for the realistic geometry (the existence of walls) and the Joule overheating of the heavy plasma component. It seems necessary to go beyond the limits of the unidimensional model and consider all variables to be functions not only of x,t, but also of z (or, which is the same, of r). This task is immensely complicated, and to date remains unsolved. Attempts of numerical simulation on the basis of unidimensional nonstationary equations are capable of giving only quite preliminary results, hinting at certain characteristic dynamic features of the normal ionizing shocks. A simple example is discussed in Sect.5.5 below.

4.4.3 Switch-On MHD Shocks in Partially Ionized Plasmas

In most experiments with switch-on MHD shocks the plasma contains neutral atoms [4.5,6,39]. The influence of neutral particles may be considerable even when the ionization is close to unity, since the friction of ions against neutral particles remains a major cause of Joule heating [4.35].

Let us consider the asymptotic profile of the switch-on MHD shock as $x \to -\infty$, where $B_y, B_z \to 0$. As follows from (4.2.5), $u_x = O(B_\perp^2)$ when $B_{y0} = B_{z0} = E_s = 0$, i.e., in a linear approximation with respect to B_y, B_z we may disregard the compression of the plasma ($u_x^e = u_x^i = u_x^a = 1$).

If $\alpha = \alpha_0 = \text{const}$, the dimensionless Maxwell equations retain the form (4.4.10,11) with the transcription

$$\Delta_h = \bar{B}_x/\mu_0 e\alpha N_0 \bar{u}_{x0} = c/\omega_{pi}\alpha_0^{\frac{1}{2}} M_{a0} \quad , \qquad (4.4.27)$$

where $\omega_{pi} = (e^2\alpha N_0/\varepsilon_0 m_i)^{\frac{1}{2}}$ is the plasma frequency of the ions.

The equations of conservation of the transverse component of momentum are

$$(1-\alpha_0)u_y^a + \alpha_0 u_y^i = B_y/M_{a0}^2 \quad , \qquad (4.4.28)$$

$$(1-\alpha_0)u_z^a + \alpha_0 u_z^i = B_z/M_{a0}^2 \quad . \qquad (4.4.29)$$

The equations of motion of electrons have a form similar to (4.4.14,15):

$$u_z^e - B_z = k_{ea}u_y^a + k_{ei}u_y^i - k_e u_y^e \quad , \qquad (4.4.30)$$

$$-u_y^e + B_y = k_{ea}u_z^a + k_{ei}u_z^i - k_e u_z^e \quad , \qquad (4.4.31)$$

where $k_{ea} = 1/\Omega_e\tau_{ea}$, $k_{ei} = 1/\Omega_e\tau_{ei}$, $k_e = k_{ea} + k_{ei}$. Here τ_{ei} means the characteristic time of momentum transfer across the magnetic field (τ_e in the notation of Sect.2.1.1). Under typical experimental conditions $k_{ea} \ll k_{ei}$, $k_e \approx k_{ei}$ when $\alpha_0 \sim 1$.

The equations of motion of atoms

$$\frac{du_y^a}{dx} = \frac{u_y^i - u_y^a}{\Delta_{ai}} \quad , \qquad (4.4.32)$$

$$\frac{du_z^a}{dx} = \frac{u_z^i - u_z^a}{\Delta_{ai}} \qquad (4.4.33)$$

describe the entrainment of atoms into the transverse movement due to collisions with the ions. Here $\Delta_{ai} = \bar{u}_{x0}\tau_{ai}$, $\tau_{ai} = \tau_{ia}(1-\alpha_0)/\alpha_0 \propto \alpha_0^{-1}$, see (4.4.16,17).

The set of linear first-order equations (4.4.10,11,27-33) has the solution

$$B_y, B_z \propto \exp(qx) \quad , \qquad (4.4.34)$$

which describes the sought-for asymptotic profile. If $\Delta_h \gg \Delta_{ai}$ (note that the validity of this inequality does not depend on the actual value of α_0; $\Delta_h/\Delta_{ai} = \alpha_0\tau_{ai}/\Omega_i$), then the eigenvalues $q_{1,2}$, which characterize the approach to equilibrium, as $x \to -\infty$, are given by

$$q_{1,2} = \frac{(M_{a0}^2 - 1)}{(k_e^2 + 1)M_{a0}^2\Delta_h}(\pm i + k_e) \quad ; \tag{4.4.35}$$

i.e., the fore part of the shock structure contains oscillations with the wavelength

$$\lambda = \frac{2\pi}{\mathrm{Re}\{q\}} = \frac{2\pi(k_e^2 + 1)c}{(M_{a0} - M_{a0}^{-1})\omega_{pi}\alpha_0^{\frac{1}{2}}} \tag{4.4.36}$$

and the damping length

$$\ell = \frac{1}{\mathrm{Im}\{q\}} = \frac{\lambda}{2\pi k_e} \quad . \tag{4.4.37}$$

The contribution of ion-neutral friction to the damping of oscillations must be included when the ions are magnetized ($\Omega_i\tau_{ia} \sim 1$); then k_e in (4.4.36) is replaced by a more sophisticated expression, the Hall parameter [4.35]. The formulas (4.4.36,37) have been checked experimentally for the cases when the correction indicated is negligible and $k_e \sim 1$. Comparing (4.4.36 and 37) reveals that the damping ought to be rather strong. Indeed, in experiments [4.6] only one swing of the magnetic field was clearly observed, whose wavelength complies with (4.4.36).

The calculations of Joule heating of the plasma in experiments with switch-on MHD shocks in helium [4.35] ($N_0 = 3.3\cdot10^{15}\mathrm{cm}^{-3}$, $B_x = 0.25$ T, $T_0 \sim 1.5$-4 eV) indicate that the Joule heating is experienced mainly by the electrons when $\alpha_0 > 0.3$, and by the heavy particles (as in the ionizing shock waves) when $\alpha_0 \lesssim 0.3$. These conclusions are also confirmed by experiments [4.6]. Insofar as $\Omega_i\tau_{ia} \sim m_i^{-\frac{1}{2}}$, the Joule heating of the heavy component is more pronounced in experiments with hydrogen. Indeed, in hydrogen even a small proportion of heavy neutral particles ($\alpha_0 \sim 0.85$) strongly affects the appearance of the switch-on shocks.

As in the case of normal ionizing shocks (Sect.4.2.2), physically sensible estimations of the oscillation wavelengths, Joule heating, etc., can be obtained only by sacrificing the boundary conditions on the walls of the shock tube. Attempts to include these boundary conditions in the consideration meet with difficulties similar to those mentioned in Sect.4.4.2.

4.5 Switch-Off Ionizing Shock Waves

Let us consider now another type of singular ionizing shock wave, the switch-off shock. As we have seen in Sect.3.3.2, the MHD classification places the switch-off shock wave among the slow shocks, the switch-off shock being the strongest of all slow shocks which may exist for the given initial state of the plasma. The latter circumstance accounts, in particular, for some interesting peculiarities of such shocks. Ionizing switch-off shocks are a natural counterpart of the MHD switch-off shocks; here also the upstream magnetic field is directed at an angle to the shock plane, while the downstream magnetic field is normal to the shock plane. Obviously enough, the switch-off shocks are oblique. As will be demonstrated further, there may exist supersonic switch-off shocks of types 3 and 4, and subsonic shocks of type 4. The evolutionarity criteria then require the existence of one additional boundary condition for the supersonic shocks of type 3 and subsonic shocks of type 4. Supersonic shocks of type 4 do not require any additional boundary conditions.

As in Sects.4.3.1,4.1, let us write, in dimensionless variables, the expression for the conservation laws, which define a curve in the (u_x, B_y) phase plane, and Ohm's law:

$$F(u_x, B_y) \equiv 4(u_x - 1)(u_x - u_{GD}) + \left(\frac{5}{2}u_x - \frac{1}{M_{a0}^2}\right)\frac{(B_y^2 - B_0^2)}{M_{a0}^2} = 0 \quad , \qquad (4.5.1)$$

$$\frac{dB_y}{d\xi} = B_y\left(u_x - \frac{1}{M_{a0}^2}\right) \quad , \qquad (4.5.2)$$

where $B_0 \equiv B_{y0}$. It is straightforward that for the switch-off shocks the electric field in the shock frame vanishes, while the upstream dimensionless electric field in the laboratory frame is

$$E = M_{a0}|1 - 1/M_{a0}^2|B_0 \quad . \qquad (4.5.3)$$

The vanishing of the right-hand side of (4.5.3) corresponds to two straight lines (degenerate hyperbola) in the (u_x, B_y) plane

$$B_y = 0 \quad , \qquad u_x = 1/M_{a0}^2 \quad . \qquad (4.5.4)$$

The switch-off shocks correspond to the intersection of curve (4.5.1) with the straight line $B_y = 0$.

Typical magnetic structures of supersonic switch-off shocks (in the limit $M_0 \gg 1$) are represented in Fig.4.14: $M_{a0}^2 < 2/5$ (Fig.4.14a,b); $2/5 < M_{a0}^2 < 1$ (Fig. 4.14c,d); $1 < M_{a0}^2 < 8/5$ (Fig.4.14e); $8/5 < M_{a0}^2 < 4$ (Fig.4.14f).

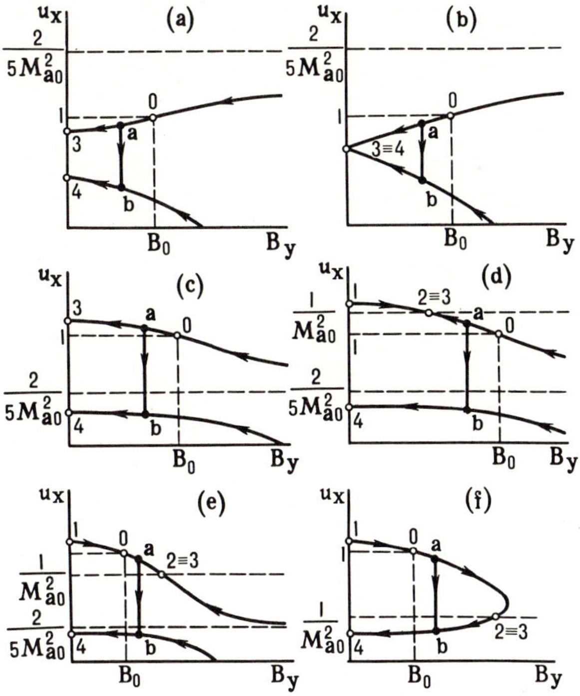

Fig.4.14a-f. Magnetic structures of supersonic switch-off ionizing shock waves. For the Chapman-Jouguet shock wave (b) the regularity of the curve (4.5.1) at the point $3 \equiv 4$ is violated

For $M_{a0} \geq 2$ the switch-off shock waves do not exist. As follows from Fig. 4.14, for $0 < M_{a0} < 2$ the switch-off shocks may only be of types 3 and 4 (the downstream signular points fall beneath the straight line $u_x = 1/M_{a0}^2$).

Note the peculiar appearance of curves representing the structure of the switch-off Chapman-Jouguet shock in the (u_x, B_y) plane. In all other cases the Chapman-Jouguet shock corresponds to the tangency of the zero-field hyperbola and the intersection of surface (4.2.5) with the plane $B_z = 0$. Presently, however, the supersonic singular point 3 and the subsonic singular point 4 may come together only on the sonic line (Fig.4.14b), i.e., when $M_3 = M_4 = 1$, which requires that

$$\frac{\partial F(u_x, B_y)}{\partial u_x} = 0 \quad . \tag{4.5.5}$$

Points 3 and 4 coincide on the straight line $B_y = 0$, where also

$$\frac{\partial F(u_x, B_y)}{\partial B_y} = 0 \quad . \tag{4.5.6}$$

If the conditions (4.5.5,6) are to hold simultaneously, the regularity of curve (4.5.1) at point $3 \equiv 4$ is violated: at this point the tangent to the curve (4.5.1) has a discontinuity (Fig.4.14b). The three equations (4.5.1, 5,6) in u_x, B_y, B_0, M_{a0} define the Chapman-Jouguet (CJ) line on the (M_{a0}, B_0) parameter plane. From Fig.4.14 it is clear that the supersonic switch-off ionizing Chapman-Jouguet shocks are possible only for $M_{a0}^2 < 2/5$. As usual, the Chapman-Jouguet condition corresponds to the highest possible value of electric field E. In the gas frame $E \leq E_{CJ}$, i.e., the solutions at $M_{a0}^2 < 2/5$ may exist only if $B_0 < B_0^{CJ}(M_{a0})$.

Using (4.5.1,4) we can readily demonstrate that the switch-off shocks of type 3 ($B_{y3} = 0$) exist only if $M_{a0} < 1$ (Fig.4.14) provided that

$$B_0 \leq B_{max}(M_{a0}) = \left[\frac{2}{3}(1 - M_{a0}^2)(4 - M_{a0}^2)\right]^{\frac{1}{2}} \quad . \tag{4.5.7}$$

The additional boundary conditions here, as in Sects.4.2-4, are given by (4.2.34), where we must set $E_s = 0$ and change the path of integration appropriately. The general analysis of solutions of these equations, performed to establish the type of shock for the given values of M_{a0} and B_0 (and the location of the starting point of the isomagnetic subshock in the solution of type 4), is similar to the analysis done in Sects.4.3,4.

The relation (4.2.34) supplies one additional boundary condition for the supersonic ionizing shocks of type 3, no additional boundary conditions being required for the shocks of type 4. As $R \to \infty$ or $E_* \to 0$ only the MHD switch-off shocks ($M_{a0} = 1$) are allowed.

For illustration let us consider the magnetic structures of switch-off shocks in the electrostatic breakdown limit $R \to 0$, in which the ABC may be written explicitly

$$\begin{cases} E_3 = E_* \quad , \\ 0 \leq E_4 < \min(E_{CJ}, E_*) \quad , \end{cases} \tag{4.5.8}$$

which corresponds to the restriction on the value of B_{y0}

$$B_0 \leq \min\left\{\frac{M_{a0}E_*}{|M_{a0}^2 - 1|} \quad , \quad B_0^{CJ}(M_{a0})\right\} \quad . \tag{4.5.9}$$

The switch-off shocks of type 4 lie within the region defined by (4.5.9), while the type 3 shocks pertain to its boundary. With large E_* (to be more precise, if $E_* > 1.14$) the shocks of type 3 do not exist. In the opposite extreme ($E_* \to 0$) the inequality (4.5.9) can be used for finding the line $M_{a0} = 1$ for any finite-valued B_0, which corresponds to MHD switch-off shocks.

Now let us consider in the electrostatic breakdown limit the subsonic ionizing switch-off shocks ($M_0 < 1$). This time we cannot hold M_{a0} for the given initial state to be a parameter independent of u_{GD}, as before. Instead

$$M_{a0}^2/M_0^2 = 5\mu_0 p_0/6\bar{B}_x^2 = \frac{5}{6}\beta_1 \quad , \tag{4.5.10}$$

where $\beta_1 = \beta_0/\cos^2\theta_0$, β_0 being the gas/magnetic pressure ratio, θ_0 the angle between the shock plane and the initial magnetic field. The parameters M_{a0}, B_0, β_1 characterize the subsonic ionizing switch-off shock wave.

The magnetic structures on the (u_x, B_y) phase plane are shown in Fig.4.15: $M_{a0}^2 < 2/5$ (a,b); $2/5 < M_{a0}^2 < 1$ (c). There are no switch-off shocks for $M_{a0}^2 > 1$. As follows from Fig.4.15, the subsonic switch-off shocks are of type 4, and belong to the class of expansion waves for $M_{a0}^2 < 2/5$ and to the class of compression waves for $M_{a0}^2 > 2/5$. As for supersonic shocks, the tangent to the curve (4.5.1) has a discontinuity at point $3 \equiv 4$, corresponding to the Chapman-Jouguet condition (Fig.4.15b).

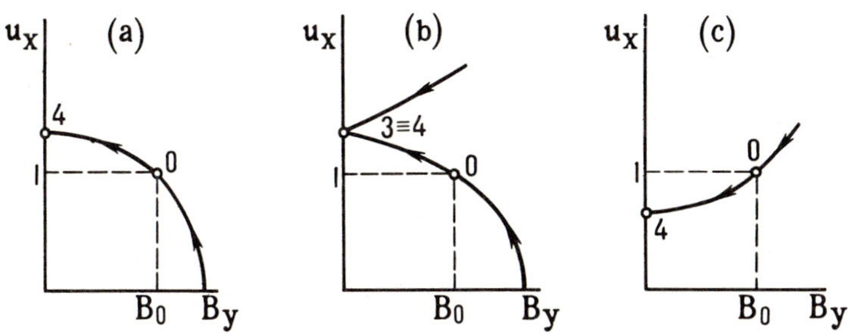

Fig.4.15a-c. Magnetic structures of subsonic switch-off ionizing shock waves: (**a**) $M_{a0}^2 < 2/5$; (**b**) $M_{a0}^2 < 2/5$, $B_{y0} = B_{y0}^{CJ}$; (**c**) $1 < M_{a0}^2 < 2/5$

The only condition required for the evolutionarity of subsonic ionizing shocks presently is given by

$$E_4 = E_* \quad . \tag{4.5.11}$$

Figure 4.16 shows curves in the (M_{a0}, B_0) plane on which condition (4.5.11) is satisfied, allowing the existence of subsonic ionizing switch-off shocks [$\beta_1 < 12/25$ (a), $12/25 < \beta_1 < 16/25$ (b); $16/25 < \beta_1 < 6/5$ (c); $6/5 < \beta_1$ (d)]. Case (b) in Fig.4.16 stands alone to the extent that with large enough values of

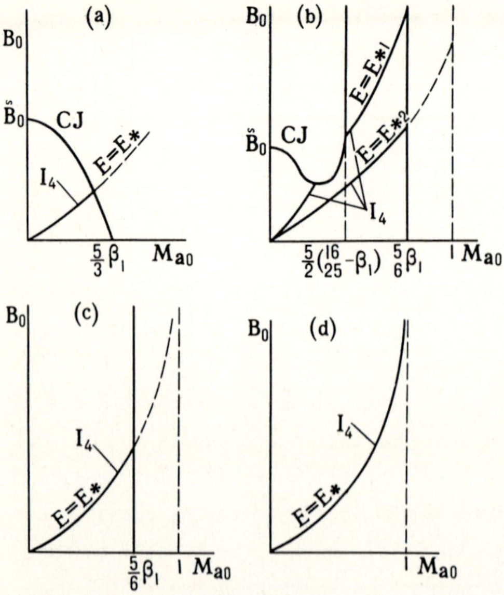

Fig.4.16a-d. Curves I_4 in the (M_{a0}, B_0) parameter plane corresponding to evolutionary subsonic switch-off ionizing shock waves in the electrostatic breakdown limit: (a) $\beta_1 < 12/25$; (b) $12/25 < \beta_1 < 16/25$; (c) $16/25 < \beta_1 < 6/5$; (d) $6/5 < \beta_1$

E_* (viz. $E_{*1} > E_{*2}$) the region of existence of the switch-off shocks separates into two.

As expected, in the limit $E_* \to 0$ only the MHD subsonic switch-off shock remains: $M_{a0}(B_0) = 1$, $\beta_1 > 6/5$.

The subsonic ionizing shocks have much in common with the deflagration waves (Sect.1.9.3). They do not contain any density jump (like the gasdynamic subshock), and heating of the gas occurs entirely at the expense of the magnetic field energy, pertaining to the switched-off transverse component. Such shocks are slow (subsonic, sub-Alfvén), but by no means weak. If $B_0^2 \gg 1$, the pressure of the strong upstream transverse field may counterbalance the high downstream kinetic pressure

$$p_4 \simeq \bar{B}_0^2/2\mu_0 \quad , \qquad \text{whence}$$

$$\bar{B}_0 = 0.062(N_4/10^{16}\text{cm}^{-3})^{\frac{1}{2}}(T_4\,[\text{eV}])^{\frac{1}{2}}[\text{T}] \quad . \tag{4.5.12}$$

The fact that subsonic ionizing switch-off Chapman-Jouguet shocks do exist in theory implies that under appropriate conditions they could be realized in electromagnetic shock tubes.

No experimental results are yet available on the switch-off ionizing shocks. We believe that investigations in this direction will contribute considerably to our understanding of the dynamics and the structures of shock waves in plasma.

5. Dynamics of Shock Waves in Magnetic Fields

As we reiterated in Chaps.2-4, the physical processes which shape the structures of shocks in gases and plasmas are strongly influenced by the conditions of generation and propagation of shock waves, all the more so in the presence of a magnetic field. To investigate these conditions we must go beyond the scope of the problem of stationary shock structure, and consider a more general problem, describing the nonstationary flow with a particular shock. This approach allows one to explore the physical possibilities for realization of shocks of various kinds, for instance, the families of ionizing shocks in a magnetic field with one and the same shock velocity and different boundary conditions. This also helps to illustrate the significance of the additional boundary conditions.

Strong shock waves in magnetic fields are produced and studied in electromagnetic shock tubes. Their design and operating principles are described in Sect.5.1.1. The simplest model, enabling the dynamics of the shock wave and the driving current sheet (magnetic piston) to be described, is discussed in Sect.5.1.2 (the so-called snowplow model). The effects of nonunidimensionality of the flow in coaxial electromagnetic shock tubes, which result in deviations from the idealized flow pattern, are considered in Sect.5.1.3.

The piston-problem's solutions (in varying statements of the problem) are used for modeling flows in electromagnetic shock tubes. The piston which drives the shock wave is considered as a rigid impenetrable diaphragm, or as a slow MHD expansion wave, which serves as a model of a magnetic piston. In Sect.5.2 we consider the piston problem in its simplest (self-similar) version, when the flow is assembled out of a set of shocks, centered expansion waves and rotational discontinuities, separated by the constant flow regions. In the MHD approximation (Sect.5.2.1) the solution of this problem supplies a physical basis to the snowplow model, and demonstrates the possible excitation of slow shock waves upon withdrawal of the magnetic piston in the electromagnetic shock tube [5.1]. The solution of the piston problem for ionizing shocks in magnetic fields (Sect.5.2.2) illustrates the linkage of the boundary conditions with the existence of the structure, discussed in Sect.4.2. The form of the solution of the piston problem depends on structural restrictions.

The piston problem on more realistic grounds is considered in Sects.5.3-5. In Sect.5.3 we give an example of numerical simulation of a nonstationary plasma flow, performed in [5.2] with a view to experiments with strong transverse MHD shocks in magnetized plasma on the shock tube at Columbia University [5.3]. Section 5.4 deals with numerical simulation of the propagation of transverse ionizing shock waves in the experiment described in [5.4]; the results are consistent with the concept of the ionizing shock structures evolved in Sects.4.2,3 [5.5]. In Sect.5.5 we describe a model of a numerical experiment, which simulates the evolution of the structure of a normal ionizing shock wave ahead of the piston. These calculations confirm the conclusions of Sect.4.4.1 regarding the results of experiments with normal ionizing shocks and the limitations of the Chapman-Jouguet hypothesis.

5.1 Electromagnetic Shock Tubes

5.1.1 Design and Operation of Electromagnetic Shock Tubes

The production of shock waves in magnetized plasma (including the strong ionizing shocks) requires a high energy density, far beyond the capacity of ordinary shock tubes. Even with the use of explosion-shock generators the shock heating cannot be raised above a few electronvolts: first, the energy released per one molecule in a chemical reaction is measured by fractions of an electronvolt; secondly, the efficiency of the hydrodynamic mechanism of energy transfer from a dense, hot driver gas to a cold rarefied test gas is limited (while the velocity of the test gas is determined by the effective speed of the driver gas which, in turn, depends on its temperature). These limitations can be removed by using, instead of the kinetic pressure exercised by the driver gas, the pressure of the magnetic field, whose energy is readily available for being converted into the kinetic and thermal energy of shock-compressed plasma. An order-of-magnitude estimate for the required magnetic-field strength can be obtained from the relation $B^2/2\mu_0 \sim 2NT$, whence for $N = 3 \cdot 10^{16} cm^{-3}$ (which corresponds, in the case of molecular hydrogen or deuterium, to the initial pressure of 0.1 Torr at room temperature) and at $T = 1$ keV we find $B \sim 3$ T, which is well within the capacity of laboratory equipment. For this reason the electromagnetic shock tubes are the major experimental tools of the physics of shock waves in magnetic fields. At this point we shall sketch some commonly used setups with the relevant geometry of currents, fields and plasma flows. A detailed account of all technicalities, dischargers, capacitor banks, vacuum and diagnostic systems, etc. was given in [5.6-9].

Most of the experimental data on the structure of shocks in collision-dominated plasma were obtained with coaxial electromagnetic shock tubes and their modifications. A coaxial electromagnetic shock tube was first described in [5.10]; its layout, which has changed little since then, is shown schematically in Fig.5.1. The test gas occupies the space between two coaxial cylindric electrodes; a third electrode runs axially in the inner cylinder. The whole unit is arranged coaxially within one or more coils. The central electrode and the coils serve to establish the bias magnetic field in the test gas: the electric current J_x, carried by the central electrode, creates the azimuthal magnetic bias B_θ transverse to the axis. (For consistency of notation, we here denote the axis of cylindric coordinate system by x rather than z, since x is the direction of propagation of the shock.) The current passed through the coils creates the axial field B_x. The axial current is supplied by the capacitor bank, connected via the discharger K_a (a similar feeder

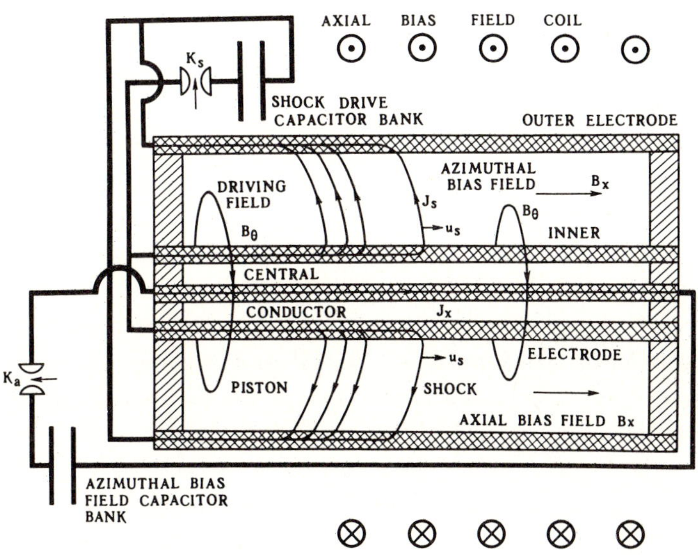

Fig.5.1. Coaxial electromagnetic shock tube

circuit of the coils is not shown in the diagram). The circuits are designed to supply steady currents and thus create a steady magnetic bias during the entire time of passage of the shock in the installation.

At the initial instant the discharger K_a connects the charged capacitor bank to the electrodes. The radial electric field creates a radial current J_p through the test gas, causing electric breakdown unless the gas is conducting. The magnetic field produced by this current has a transverse (azimuthal) direction, coinciding with the direction of magnetic field of the axial current. Under the action of the force $\mathbf{j} \times \mathbf{B}$, decomposable into axial and azimuthal nonzero components, the current-carrying plasma is accelerated in the axial direction and starts rotating about the x axis. The axially moving plasma acts as a piston, compressing and accelerating the initially unperturbed gas. A shock wave (MHD or ionizing), traveling with velocity u_s, is produced ahead of such a piston, whose velocity exceeds the velocity of propagation of waves in the unperturbed gas. If the shock produced is not purely gasdynamic and exhibits a nontrivial magnetic structure (Sects.4.2-4), it should interact with the magnetic field: electric current will be included in the shock layer. Insofar as the downstream plasma is accelerated in the direction of variation of the transverse magnetic field, and in the coaxial arrangement the mass velocity cannot possess a radial component, the current J_s across the shock is radial. It takes the same path as the current which drives the shock, modifies the azimuthal magnetic field B_θ, and the shock wave involves the plasma into rotational movement about the x axis. An exception is the transverse shock wave, in which the plasma flow is axial. However, the transverse shock waves (both MHD and ionizing) do not modify the direction of the transverse magnetic field (Sect.4.4.1,3.1), and therefore should be associated with the passage of radial current as well.

The flow of plasma in a coaxial electromagnetic shock tube is closer to the unidimensional pattern, the smaller the spacing between the outer and inner cylinders ($R_{out}-R_{in} \ll R_{in}$). If the parameter ($R_{out}/R_{in}-1$) is not small, the radial variation of the azimuthal magnetic field (proportional to r^{-1}) distorts the planeness of the shock front and the current sheet: the force acting upon the plasma is stronger towards the axis, which causes bulging of the current sheet, as shown in Fig.5.1 (see also Fig.5.5 below and Sect. 5.1.3). In practice, the gap($R_{out}-R_{in}$) cannot be reduced indefinitely (as long as R_{in} is fixed) because of the wall boundary effects, which cannot be disregarded unless the gap is large in comparison with the cyclotron radius of ions. With most installations the parameter ($R_{out}/R_{in}-1$) is more or less close to unity (0.4 for the shock tube described in [5.10]; 0.77 for the electromagnetic shock tube at Columbia University [5.7,8]; 1.25 for the coaxial shock tube SUPPER VI [5.9]). Nevertheless, the unidimensional approximation is quite suitable for describing the axial morphology of shock waves, since both the theoretical [5.11,12] and the experimental [5.13-15] data indicate that while the current sheet is considerably distorted, the deviation from the planarity of the shock front remains small. However, certain restrictions are imposed on the unidimensional model: insofar as the plasma may move in only one of the transverse directions (the azimuthal direction), radial movement being precluded by the walls, only the plane-polarized (oblique) shocks can be realized experimentally. Consequently, in the coaxial electromagnetic shock tubes the MHD and the oblique (not skew!) ionizing shock waves can be observed, whose boundary conditions comply with this requirement (Sect.4.1.2). Note that the unidimensional models often predict oscillations of both transverse components of velocity within the front of an oblique shock (Sects.3.5.3 and 4.4.3), incompatible with the actual geometry of the coaxial shock tube. Under experimental conditions the presence of the walls will probably suppress such oscillations.

In the unidimensional approximations the force acting upon the plasma in the axial direction derives from the gradient of the magnetic pressure

$$(\mathbf{j}\times\mathbf{B})_x = -\frac{\partial}{\partial x}\frac{B_x^2+B_\theta^2}{2\mu_0} \quad .$$

The differential magnetic pressure driving the current sheet and the shock wave is $(B_{\theta3}^2-B_{\theta1}^2)/2\mu_0$, where $B_{\theta1}$ is the initial azimuthal bias magnetic field, $B_{\theta3}$ is the azimuthal magnetic field at the rear wall of the shock tube (we assume that $B_x = \text{const}$). For $|B_{\theta3}| > |B_{\theta1}|$ the flow has the pattern shown in Fig.5.1. Depending on the polarity of the radial voltage applied to the electrodes, this pattern may be realized in either of two ways: with parallel and antiparallel alignment of the driving and the bias azimuthal magnetic fields. This is confirmed by the results of experiments on the coaxial electromagnetic shock tube SUPPER VI [5.15,16]. Obviously, the radial current through the plasma is stronger with the antiparallel orientation for the same value of $B_{\theta3}$, since the required variation of the magnetic field is then $|B_{\theta3}| + |B_{\theta1}|$, and not $|B_{\theta3}| - |B_{\theta1}|$ as for the parallel orientation. The possibility of obtaining MHD shocks when $|B_{\theta3}| < |B_{\theta1}|$, may seem somewhat unexpected but is demonstrated in [5.1] and discussed in Sect.5.2.1 below.

Modifications of the coaxial electromagnetic shock tube are represented by various designs of cylindric shock tubes, in which essentially the same

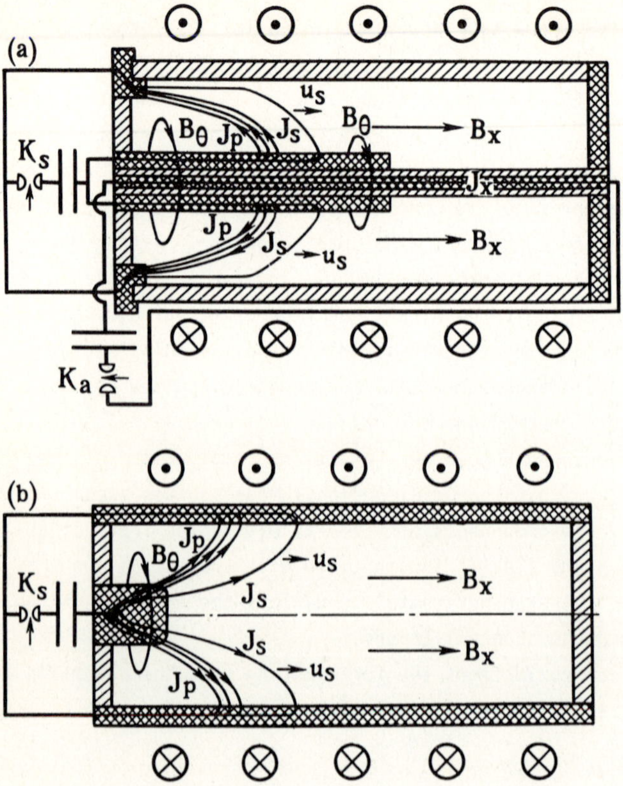

Fig.5.2a,b. Cylindrical electromagnetic shock tubes

flows are realized in slightly different configurations. The change usually consists in the truncation of one of the electrodes, either the outer (as in the *Charybdis* shock tube (Fig.5.2a) [5.1,17]), or the inner (as in SUPPER II (Fig.5.2b) [5.18]). The designations in Fig.5.2 are the same as in Fig. 5.1. The *Charybdis* shock tube has a special feature: the plasma, in which the propagation of MHD shocks is studied, is produced by the electric discharge between the outer ring electrode and the grounded opposite wall of the shock tube. In most installations the initial ionization is created by the ionizing shock wave, excited in the test gas.

Many important results on the ionizing shock waves were obtained with a different type of installation, called the inverse Z pinch [5.4,19], first proposed in [5.20] (Fig.5.3). The geometry of this installation is also cylindric; however, the electrodes are the tops of the cylinder which contains the test gas and is placed in the external axial magnetic field. The potential difference, applied to the electrodes via the discharger K, maintains an axial electric field within the cylinder (after having caused the break-

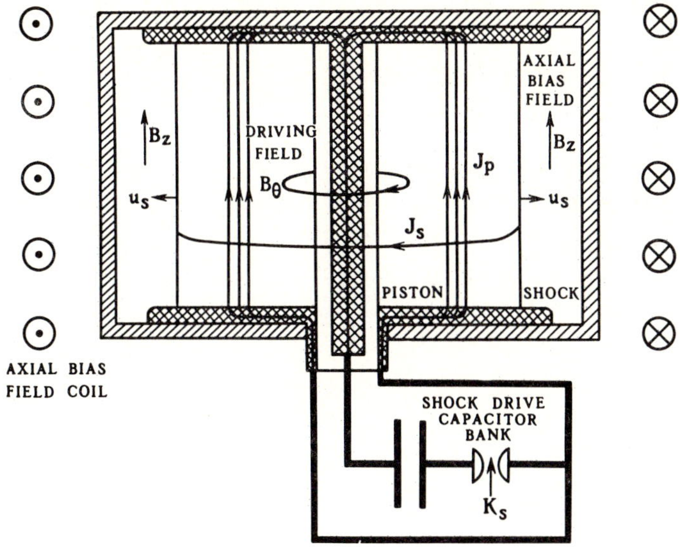

Fig.5.3. Inverse Z pinch device

down of the gas, should this be the ionizing shock wave that is studied). This field causes the axial current J_p to flow near the insulated conducting axis of the cylinder. The magnetic field B_θ of the back current, carried by the axial conductor, acts upon the nearby current sheet with the force $\mathbf{j} \times \mathrm{B}$, driving it radially away from the axis and thus making it act as a piston. A transverse shock wave (MHD or ionizing) is produced in the test gas, which interacts with the external magnetic field. The compression of axial magnetic field in the shock is associated with the passage of closed-path azimuthal current J_s through the plasma.

A definite advantage of this arrangement is the orthogonality of currents and fields, related to the shock and the piston, which facilitates the study of magnetic structure of the shock wave on the basis of magnetic-field measurements. Very advantageous is also the rather high uniformity of magnetic field throughout the entire shock surface, save for the small regions near the tops of the cylinder. The flow in the inverse Z pinch fits well into the one-dimensional approximation—this time, however, cylindric rather than planar—because the mass velocity of plasma in the transverse shocks is strictly radial. At the same time, as indicated in Sect.4.2.2, electromagnetic waves from the shock front cannot propagate freely in such geometry, since they encounter the coil producing the axial magnetic field. This impedes interpreting results of experiments with ionizing shock waves (Sect. 5.4 below).

It ought to be emphasized that the specific design of electromagnetic shock tubes determines the special status of normal and transverse shocks. For instance, the most powerful installations (shock tube at Columbia University, SUPPER VI) are not equipped with the axial magnetic bias coils, and are therefore restricted to the study of purely transverse shock waves. Only transverse shocks can be studied with the inverse Z pinch. On the other hand, the absence of a central conductor which produces the transverse bias magnetic field in the cylindric shock tube SUPPER II and in the earlier modification of *Charybdis* precluded the study of transverse shock waves. Accordingly, the experimental material accumulated so far concerns mainly these types of shocks, explaining their importance in this study.

Other types of experimental generators of high-power shocks (cylindric and noncylindric Z pinches, θ pinches, etc.) remain beyond our scope, since they are used chiefly for experimenting with hot plasmas and find limited employment in the study of shock waves. We also leave out the earlier versions of electromagnetic shock tubes (T tubes, conical tubes, etc.); descriptions can be found in [5.6] and references cited therein.

5.1.2 Elementary Theory of Electromagnetic Shock Tubes: The Snowplow Model

The snowplow model comes in handy for assessing the parameters of shocks produced in electromagnetic shock tubes. This representation of plasma flow with a current sheet, useful also for describing the dynamics of pinches, etc., in many cases reflects fairly well the performance of electromagnetic shock tubes. This model is based on the assumption that the current sheet, driven ahead by the forces of magnetic pressure, scoops up like a snowplow the entire mass of gas that falls in its way, so that the motion of the current sheet is determined, on the one hand, by the time variation of the magnetic field, and on the other, by the inertia of the entrapped mass of gas. Behind the current sheet is a vacuum space, i.e., the current sheet acts as an impenetrable piston driven by magnetic forces, whence the term "magnetic piston". The exact meaning of this approximation will be discussed in Sect.5.2. At this point we just postulate this model and derive the relevant equations.

We consider unidimensional plane-parallel movement of the plasma, i.e., the magnetic field and the velocity are nonzero in only one of the transverse directions. Physically, this corresponds to a coaxial electromagnetic shock tube with $(R_{out}/R_{in}-1) \ll 1$. Let the x axis run along the axis of the tube, the y axis run parallel to its surface, the z axis run radially; in this way we identify the coordinate axes x,y,z with the axial, azimuthal and radial directions, respectively. Since, as will be elucidated below, the snowplow can drive only the fast MHD shock waves ahead of itself, we label the upstream and the downstream states with subscripts 1 and 2, respectively (Sect. 3.3.2).

In the most simple version of the snowplow model [5.21] the gas layer, engaged with the current sheet, is considered as a whole, so that the velocity of the shock wave (which separates this layer from the undisturbed gas) can be identified with the velocity of this layer. Let x be the coordinate of the gas layer, then the relevant equation of motion has the form

$$\frac{d}{dt}\rho_1 x \frac{dx}{dt} = \Delta p_B \quad , \tag{5.1.1}$$

where ρ_1 is the upstream density, and Δp_B is the differential magnetic pressure behind the current sheet. If the initial azimuthal field, e.g., the magnetic bias created by axial current, (Sect.5.1.1), is B_1, and the radial discharge current creates azimuthal field B_p, then

$$\Delta p_B = \frac{1}{2\mu_0}\left[(B_p \pm B_1)^2 - B_1^2\right] = \frac{1}{2\mu_0}(B_p^2 \pm 2B_pB_1) = \frac{B_p^2}{2\mu_0}(1 \pm 2\lambda) \quad , \tag{5.1.2}$$

where $\lambda = B_1/B_p$, the sign before λ depends on the direction of the radial current. (If, for instance, the field B_1 is produced by an axial current, which runs in the positive direction of the x axis, i.e., $B_1 < 0$, then B_p is positive or negative together with the radial current). Substituting (5.1.2) into (5.1.1), we solve this equation with the initial condition $x(0) = 0$:

$$x(t) = \left\{2\int_0^t dt' \int_0^{t'} dt'' \frac{1}{2\mu_0\rho_1}\left[B_p^2(t'') \pm 2B_p(t'')B_1\right]\right\}^{\frac{1}{2}} \quad . \tag{5.1.3}$$

The velocity of the current sheet (and of the shock wave) is then

$$u(t) = \frac{dx(t)}{dt} = \frac{1}{x(t)}\int_0^t \frac{dt'}{2\mu_0\rho_1}\left[B_p^2(t') \pm 2B_p(t')B_1\right] \quad . \tag{5.1.4}$$

In particular, for the simplest case of instantaneous buildup of a steady driving field $B_p = \text{const}$ at $t = 0$ we find

$$x(t) = ut \quad , \tag{5.1.5}$$

$$u = \frac{b_p}{\sqrt{2}}(1 \pm 2\lambda)^{\frac{1}{2}} \quad , \quad \text{where} \quad b_p \equiv |B_p|/(\mu_0\rho_1)^{\frac{1}{2}} \quad . \tag{5.1.6}$$

In this case the velocity of the current sheet is constant.

If, however, the driving field builds up in time as, for instance, $B_p = B_{p0} \sin\Omega t$, then from (5.1.3,4) we get

$$x(t) = b_{p0}\left(\frac{t^2}{4} \pm \frac{2\lambda t}{\Omega} \mp \frac{2\lambda}{2} \sin\Omega t + \frac{\cos 2\Omega t - 1}{8\Omega^2}\right)^{\frac{1}{2}} \quad , \tag{5.1.7}$$

$$u(t) = \frac{b_{p0}^2}{2x(t)}\left[\frac{t}{2} \pm \frac{2\lambda}{\Omega} (1 - \cos\Omega t) - \frac{\sin 2\Omega t}{4\Omega}\right] \quad . \tag{5.1.8}$$

[Here b_{p0} is defined as in (5.1.6) where B_p should be replaced by B_{p0}]. Of course, for small t this solution is acceptable only if $B_{p0} > 0$. The asymptotic value of velocity is $u(t \to \infty) = b_{p0}/2$.

Provided that the current-sheet (the magnetic piston) velocity may be assumed constant, the snowplow model can account for unequal velocities of the piston u_p and the shock u_s. At present the shock-compressed gas should be considered stationary in the piston frame, i.e., the velocity of the piston u_p should coincide with the downstream velocity of the gas $\bar{u}_2$; at the same time the continuity equation implies that

$$u_p = \bar{u}_2 = u_s(1 - u_2) \quad , \tag{5.1.9}$$

where u_s is the shock-front velocity, u_2 is the dimensionless downstream velocity ($u_2 = \rho_1/\rho_2$). If x_s and x_p are the coordinates of the shock and the piston, respectively, then the mass of gas scooped up by the piston per unit of its area is defined as

$$m = \rho_2(x_2 - x_p) = \rho_1 x_s \quad . \tag{5.1.10}$$

Substituting $\rho_1 x_3$ into (5.1.1) in place of $\rho_1 x$ and u_p in place of dx/dt, we get

$$u_p u_s = \frac{b_p^2}{2} (1 \pm 2\lambda) = u^2 \quad , \tag{5.1.11}$$

where u is the snowplow velocity, calculated for the same occasion by (5.1.6). Using (5.1.9), we obtain the equation for the shock-wave velocity:

$$u_s^2[1 - u_2(u_s)] = \frac{b_p^2}{2} (1 \pm 2\lambda) \quad . \tag{5.1.12}$$

Incidentally, the same expression could have been derived straightforwardly by equating the magnetic field pressure in the vacuum region behind the current sheet $(\pm B_p + B_1)^2/2\mu_0$ to the sum of kinetic and magnetic pressures in the shock-heated plasma $p_2 + B_2^2/2\mu_0$ [5.22].

In particular, for normal MHD shocks ($B_1 = 0$, $\lambda = 0$) in dimensionless variables (Sect.3.5.1) we get

$$2M_{a1}^2(1 - u_2) = (B_p/B_x)^2 \quad .$$

In the range $1 < M_{a1} < 2$ the normal shock waves are of the switch-on type ($u_2 = 1/M_{a1}^2$), whence

$$M_{a1} = \left(\frac{1}{2}\frac{B_p^2}{B_x^2} + 1\right)^{\frac{1}{2}} \quad . \tag{5.1.13}$$

When $M_{a1} \geqslant 2$ [by virtue of (5.1.13), then $B_p \geqslant \sqrt{6}B_x$], the normal shock waves are gasdynamic and do not modify the magnetic field ($u_2 = 1/4$) whence

$$M_{a1} = \left|\frac{B_p}{B_x}\right|\sqrt{2/3} \quad . \tag{5.1.14}$$

With strictly transverse MHD shock waves ($B_x = 0$) the dimensionless variables are defined otherwise (Sect.3.4.1), and then from (5.1.12) we get

$$2M_{a1}^2[1 - u_2(M_{a1})] = \frac{B_p^2}{B_1^2}(1 \pm 2\lambda) \quad . \tag{5.1.15}$$

Here $u_2(M_{a1})$ is given by (3.4.2), where M_1^{-2} may be disregarded on the scale of M_{a1}^{-2}. In the limit of strong shock waves [when $M_{a1} \gg 1$, which is possible, in agreement with (5.1.15), only if $\lambda \ll 1$; then $u_2 = 1/4$], (5.1.15) reduces to (5.1.14), where B_x is replaced by B_1. This is quite natural, since in the limit $B_p/(B_x^2 + B_1^2) \to \infty$ the energy required for the shock compression of the initial magnetic field constitutes a small fraction of the energy of the flowing plasma. Accordingly, the shock waves are gasdynamic ($u_2 = 1/4$) no matter what the strength and orientation of the initial bias magnetic field, and their velocity depends only on the initial gas density and the pressure exerted by the magnetic piston

$$u_s = b_p\sqrt{2/3} \quad . \tag{5.1.16}$$

In Fig.5.4 the dimensionless velocities of normal and transverse MHD shocks are plotted against the ratio of the magnetic field strengths ahead of the shock and behind the magnetic piston

$$\frac{B_1}{B_3} = \left(\frac{B_{y1}^2 + B_x^2}{(B_{y1} \pm B_p)^2 + B_x^2}\right)^{\frac{1}{2}} \quad .$$

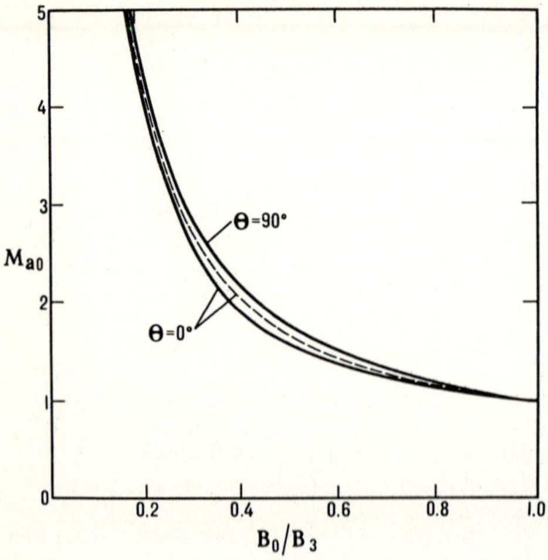

Fig.5.4. MHD shock velocity as a function of driving magnetic field in normal ($\Theta = 0^{\circ}$) and transverse ($\Theta = 90^{\circ}$) geometry calculated using the snowplow model. (---) is the result of solution of the self-similar magnetic piston problem (Sect.5.2)

For the characteristic velocity we take

$$\tilde{c}_a = \frac{(B_{y1}^2 + B_x^2)^{\frac{1}{2}}}{\mu_0 \rho_1} \quad ,$$

which allows us to use the normalization procedures adopted in Sects.3.4.1, 5.1; here M_{a1} stands for $u_s/\tilde{c}_a$.

If the azimuthal magnetic field B_p exhibits considerable time variation as the current sheet moves ahead, the snowplow model does not allow for distinguishing between the velocity of the piston and the velocity of the shock. Both are approximated by the same value u, correct to $|1-(1-u_2)^{\frac{1}{2}}| \approx u_2/2$ according to (5.1.9,11), which is about 12% with the strongest shocks. There is hardly any reason to improve this model by describing the dynamics of unsteady plasma flow between the magnetic piston and the shock front in terms of MHD equations. This is because the task would be nearly as laborious as the attempt to simulate the entire flow numerically (Sect.5.3.5), which bereaves the model of its attractive simplicity.

The simplest version of the snowplow model often needs adjustments for particular experimental conditions. For instance, in the experiments with the coaxial electromagnetic shock tube at Columbia University the velocity and displacement of the magnetic piston (with the sine-wave magnetic field) failed to comply with the initial conditions $x(0)=0$, $u(0)=0$, $t=0$ being the instant of electric discharge. Instead, the observed movement was described by the solution of the form (5.1.7,8) with the initial conditions $x(0)=x_0$, $u(0)=u_0$, i.e., additive terms $(x_0^2+2x_0u_0t)/b_{p0}^2$ and $2x_0u_0/b_{p0}^2$

appearing in the braces in (5.1.7,8), respectively. Here $x_0 = 12.5$ cm, $u_0 = 2.55 \cdot 10^8$ cm/s. In other words, the discharge accompanying the switch-on of the magnetic piston acts as an impact, which instantaneously (on the scale of the characteristic time of the driving field buildup) accelerates the plasma to a high velocity and causes considerable axial displacement.

This effect was attributed to the drift of plasma ions under the action of the strong radial electric field and the azimuthal bias magnetic field [5.21]. The high voltage applied to the plasma when the discharger is switched on grows more or less linearly over about 100 ns and creates a strong electric field in the plasma, normal to the already existing magnetic field ($B_x = 0$).

The axial movement of the initially resting ions of charge e and mass m_i in the crossed fields $\mathbf{E} = \{0, 0, E_0 t\}$, $\mathbf{B} = \{0, B_1, 0\}$ is described by

$$x(t) = \frac{v_D}{\Omega_i}\left[\frac{1}{2}(\Omega_i t)^2 + \cos\Omega_i t - 1\right] \quad , \qquad (5.1.17)$$

$$u(t) = v_D(\Omega_i t - \sin\Omega_i t) \quad , \qquad (5.1.18)$$

where $\Omega_i = eB_1/m_i$ is the cyclotron frequency of the ions, and $v_D = E_0/\Omega_i B_1$ is the characteristic drift velocity. In the experiment [5.21], performed with hydrogen, $B_1 = 0.66$ T, and the time of buildup of the electric field $t = 100$ ns $\simeq 2\pi/\Omega_i$. Substituting this value of t into (5.1.17,18) yields $x_0 = 2\pi^2 v_D/\Omega_i$, $u_0 = 2\pi v_D$, in compliance with experimental observations. According to the estimations of [5.21], at greater times the collisional effects dominate, and the plasma will move as a continuous medium described by the snowplow model.

The estimates made for the experiments with the coaxial electromagnetic shock tube SUPPER VI [5.15] indicate that the assumption of instantaneous buildup of a steady magnetic field B_p is quite adequate, while the drift of ions makes no visible contribution to the velocity and displacement of the plasma. At the same time, the observed velocities of shocks were much superior to the predictions of (5.1.12), attributed to the "leaky piston" [5.15]. This fails to scoop up the entire mass of gas in its way, and thus accelerates a smaller mass to a higher velocity. To account for this effect, the leakage coefficient ℓ is introduced into the model; by assumption, ℓ is the fraction of mass that is not compressed in the shock and not scooped up by the piston, so that the density of the gas is $\rho_2 = \rho_1(1 - \ell)/u_2$ behind the shock front, and $\rho_3 = \ell\rho_1$ behind the piston. From the equation of conservation of mass

$$\rho_3 u_p + \rho_2(u_s - u_p) = \rho_1 u_s$$

we find

$$\frac{u_s}{u_p} = \frac{1 - \ell - \ell u_2}{1 - \ell - u_2} \quad . \qquad (5.1.19)$$

If we also assume that the leaky piston is, as before, at rest with respect to the shock-compressed gas, then (5.1.11) still holds. The best-fit values of the empiric constant ℓ are found in [5.15] to be $\ell = 0.4$ and 0.6 for the parallel and the antiparallel orientation of the driving magnetic field. The concept of a leaky piston is verified by considering the two-dimensional structure of the flow in the electromagnetic shock tube, (Sect.5.1.3).

The snowplow model can be successfully used to describe current sheet dynamics in a cylindric configuration (inverse Z pinch). The relevant formulas were obtained in [5.23] and experimentally verified in [5.4,19].

5.1.3 Effects of Nonunidimensionality of the Plasma Flow in Coaxial Electromagnetic Shock Tubes

The flow of plasma in coaxial electromagnetic shock tubes, when $(R_{out}/R_{in} - 1)$ is of the order of unity, is obviously not quite unidimensional. Indeed, the azimuthal magnetic field (both the bias field, created by the passage of axial current, and the field produced by the radial current in the magnetic piston) decreases as r^{-1} with increasing distance from the axis of the tube, and so the magnetic pressure decreases as r^{-2}. Accordingly, the shock wave weakens with increasing r; this effect distorts the magnetic piston. In the context of the snowplow model this effect may be understood in the following way [5.24]. If u is the velocity of steady axial movement of the magnetic piston, then the velocity of the shock at point A (Fig.5.5) is $u \cos\Theta < u$, Θ being the angle between the normal to the piston surface and the axis of the shock tube. Equating the magnetic pressure produced by the piston to the total pressure downstream of the shock wave, we obtain a generalization of (5.1.12):

$$u_s^2 \cos^2\Theta[1 - u_2(u_s \cos\Theta)] = \frac{b_p^2}{2}(1 \pm 2\lambda)\frac{R_{in}^2}{r^2} \quad , \qquad (5.1.20)$$

where b_p is calculated for the magnitude of the magnetic field at $r = R_{in}$. The surface of the magnetic piston is assumed to be normal to the direction of travel near the inner cylinder. A rough estimate of the shape of the magnetic piston $x_p(r)$ can, in this approximation, be obtained if we neglect the

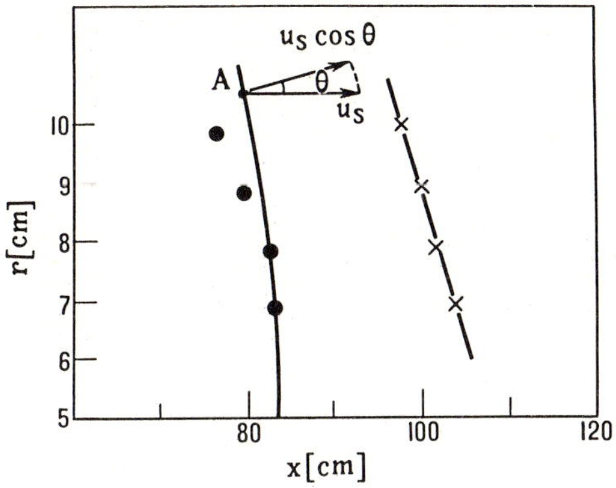

Fig.5.5. Inclination of the magnetic piston surface (• ••) and of the MHD shock front (x x x) in the coaxial electromagnetic shock tube SUPPER VI. Experimental points have been taken from [5.15]. (——) is plotted using (5.1.22)

dependence of $[1 - u_2(u_s \cos\theta)]$ on θ, and take into account that

$$\left(\frac{dx_p}{dr}\right)^2 = \frac{1}{\cos^2\theta} - 1 \quad . \tag{5.1.21}$$

Then (5.1.20,21) can be integrated to find the profile of the magnetic piston:

$$\begin{aligned} x_p(r) &= x_p(R_{in}) - \frac{1}{2} r \left[(r/R_{in})^2 - 1 \right] \\ &\quad + \frac{1}{2} R_{in} \ln\left\{ (r/R_{in}) + \left[(r/R_{in})^2 - 1 \right]^{\frac{1}{2}} \right\} \\ &\approx x_p(R_{in}) - \frac{2\sqrt{2}}{3R_{in}^{\frac{1}{2}}} (r - R_{in})^{3/2} \quad . \end{aligned} \tag{5.1.22}$$

This technique, however, is not suitable for a more detailed description of the radial structure of the flow in the coaxial electromagnetic shock tube, since that would mean pushing the model beyond its capability (Fig.5.1). A more precise assessment of the shape of the current sheet of the magnetic piston and the shock wave is possible only via a numerical solution of a two-dimensional nonstationary problem. So far a unique example of such a calculation exists, performed for the shock tube at Columbia University [5.11, 12]. The current sheet is shown to be more or less parabolic in shape, the magnetic piston being bent more strongly than the shock wave itself. The distortion of the shock front, which initially is quite prominent, flattens out with time, and finally the shock front assumes a stationary shape, whose curvature slightly exceeds the estimation by (5.1.22). These results excellently agree with experimental measurements [5.14].

The radial structure of the flow in the coaxial electromagnetic shock tube SUPPER VI was thoroughly investigated in [5.15,25]. The radial profile of the magnetic field was measured with magnetic probes [5.15]. The electron density distribution was measured by the interferometric technique [5.25]. The phase shift of interference fringes over a fixed optical length between the ports in the outer cylinder were assumed proportional to the integral electron density along this path, since at the HeNe laser frequency the permittivity of plasma is $\varepsilon = 1 - e^2 n_e / \varepsilon_0 m_e \omega^2$ [5.26]. The experiments were performed with hydrogen at the initial pressure of 50 mTorr, preionization: $\alpha_0 = 0.05$ transverse magnetic bias: 0.1 T at mid-radius, with the transverse MHD and ionizing shocks, whose velocities were 28 and 18.5 cm/μs for the parallel and the antiparallel orientations of the driving magnetic field, respectively.

The results of [5.15,25] confirm the conclusion regarding the distortion of the shock front. The inclination of the shock front to the axis ultimately assumes a steady-state value, being higher for the antiparallel shock wave. The concept of a leaky piston is confirmed by measurements of the radial electron density distribution [5.25]. Indeed, a perfect magnetic piston, which scoops up all the gas in its way, should leave behind a region of rarefaction (vacuum for the transverse shocks; Sect.5.2.1). Measurements [5.25] indicate, however, that the density drop near the walls of the shock tube is lower than at mid-radius. For the antiparallel shock waves the density increases towards the wall of the outer cylinder, which means that leakage occurs near the outer wall. On the contrary, with parallel shock waves the density increases towards both walls, and that is where the leakage occurs.

The increase in plasma density near the outer wall was explained in [5.25] by the distortion of the magnetic piston: the radial component of the force $\mathbf{j} \times \mathbf{B}$ is directed outwards, pressing the plasma to the wall. With the parallel shocks the plasma may be pressed to the wall of the inner cylinder for just the same reason, as long as the driving current takes the path through the plasma rather than through the cylinder, thus causing local perturbatin of the originally planar magnetic piston. In other words, the piston leaks just because it fails to engage the plasma layer near the inner cylinder. An alternative explanation draws attention to the possible diffusion of plasma through the piston, since the reduced conductivity of the cooler plasma near the walls raises the coefficient of magnetic field diffusion, see (3.1.7,52).

5.2 Piston Problem

5.2.1 Self-Similar Magnetic Piston Problem in Magnetohydrodynamics

In magnetohydrodynamics, just as in gasdynamics, the self-similar solution of the set of ideal MHD equations means the asymptotic solution of a complete set of equations including the dissipative processes as $t \to \infty$, when the steady shock widths Δ_s and the widths of diffusive smearing of singular non-evolutionary discontinuities $\Delta \propto t^{\frac{1}{2}}$ may be disregarded on the characteristic scale of the problem $L \propto t$ (Sect.3.3.5). As long as the length of the shock tube is large enough, these solutions can describe the actual processes that take place in the installation.

The solution of the self-similar problem is set up as a sequence of centered expansion waves and discontinuities. For the set (3.2.31-36), which describes a plane-parallel flow, the discontinuities may be of three kinds: fast and slow shock waves, and the rotational discontinuity which rotates the vector of the azimuthal magnetic field by 180°. By virtue of (3.3.27-30) the fast shocks and the expansion waves travel faster than rotational discontinuities, and the slow shocks and expansion waves move slower. Accordingly, the solution of the self-similar problem must be the following: a fast wave (either shock or expansion), a rotational discontinuity, and a slow wave (shock or expansion). The rotational discontinuity will be included only if the boundary conditions require the reversal of sign of the azimuthal magnetic field (MHD shocks and expansion waves cannot alter the sign). In the rotational discontinuity the density, pressure and axial velocity remain the same, while the azimuthal velocity varies together with the magnetic field:

$$\Delta u_y = - \frac{\Delta B_y}{\sqrt{\mu_0 \rho}} \; .$$

The magnetic piston problem is stated as follows. The undisturbed plasma found in a homogeneous magnetic field ($B_y = B_{y1}$) for $t < 0$ occupies the half-space $x > 0$, bounded at $x = 0$ by an impenetrable nonconducting diaphragm (the rear wall of the coaxial shock tube). At $t = 0$ the azimuthal magnetic field instantaneously increases by B_p (note that $|\pm B_p + B_{y1}| > |B_{y1}|$) due to the passage of electric current near the wall. The jump of the magnetic field is transmitted forward by a fast shock wave. The gas directly behind the fast shock has a nonzero velocity in the same direction and density above ρ_1. At the same time, at $x = 0$ we have to satisfy the boundary condition

$u_x = 0$ (or $\rho = 0$). The velocity and the density decrease in the expansion wave, here a slow one. As demonstrated in Sect.3.2.3, the azimuthal magnetic field in the slow expansion wave rises up to some finite value near the rear wall (if its direction is contrary to the initial orientation, the magnetic expansion wave is preceded by a rotational discontinuity, which reverses the sign of B_y), Figs.5.6-8.

In this way the slow expansion wave acts as a magnetic piston. If the gas density therein drops to zero, the magnetic piston may be considered impenetrable, scooping up all the gas in its way, leaving vacuum behind. Otherwise, the piston is leaky: some part of the shock-compressed gas penetrates through the piston, slows down and remains near the back wall. Disregarding the thickness of the region occupied by the expansion wave and the mass of gas contained therein, we arrive at the snowplow model.

An important particular case exists when this neglect is rigorously justified: the purely transverse configuration of the magnetic field ($B_x = 0$, i.e., $c_a = 0$). Then the velocity of the slow magnetosonic wave vanishes, see (3.2.11). As shown in Sect.3.3.1, the slow expansion wave then degenerates into a tangential discontinuity surface with the boundary conditions (3.2.52), the first of which implies that the discontinuity travels at the speed of the shock-compressed gas, and $u_x \neq 0$ behind the discontinuity. Then near the rear wall $\rho = 0$, $p = 0$, whence

$$\frac{(B_p + B_{y1})^2}{2\mu_0} = p_2 + \frac{B_{y2}^2}{2\mu_0} \quad , \tag{5.2.1}$$

which is equivalent to (5.1.12). The tangential discontinuity directly corresponds to the concept of the snowplow. If the initial orientation of the magnetic field is other than strictly transverse, the magnetic-piston thickness is not infinitesimal, and grows linearly with time. Nevertheless, as long as $c_s \ll c_a$, (i.e., when B_y^2/B_x^2 is large, or the pressure in the shock-compressed gas is low, $p \ll B_x^2/2\mu_0$), it expands much slower than the constant-flow downstream region. As long as the density in the expansion wave falls quickly, the snowplow approximation still works fairly well.

A complete solution of the self-similar problem was obtained in [5.27]. The calculation scheme is quite typical (Sect.1.9.3). For the chosen initial state, characterized by the dimensionless ratio B_{y1}/B_{x1}, let us consider the shock transitions produced by fast shock waves, whose velocities are varied from c_f to infinity. The downstream state for each shock transition is then taken for the initial state of the slow expansion wave. Equations (3.2.44,45,

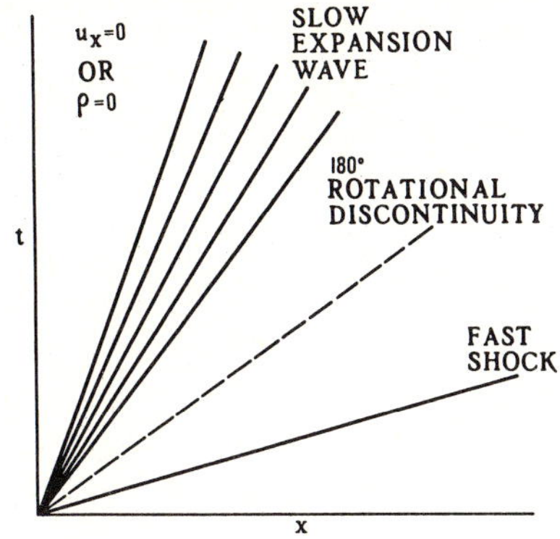

Fig.5.6. Flow patterns representing the solution of the self-similar MHD magnetic piston problem containing a fast shock wave in plane-parallel geometry. The 180° rotational discontinuity (---) may be either absent (parallel shock wave) or present (antiparallel shock wave)

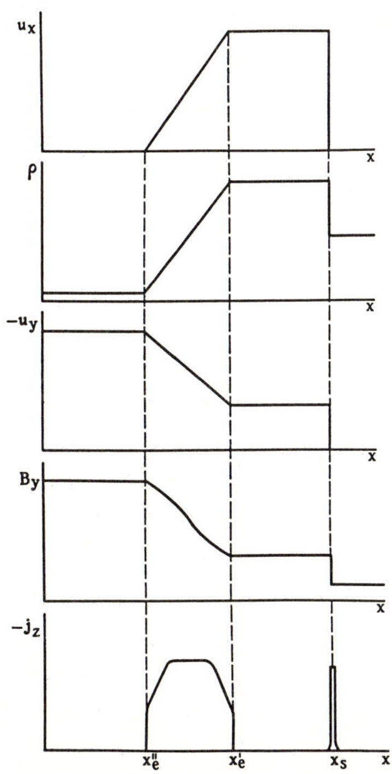

◄ Fig.5.7

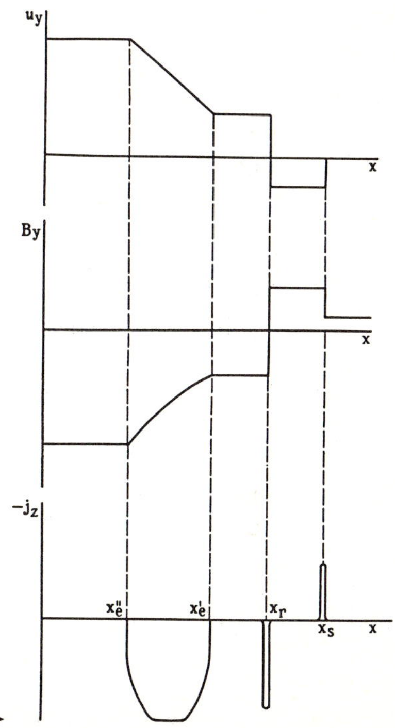

Fig.5.8 ►

Fig.5.7. Profiles of flow variables corresponding to Fig.5.6 for a parallel shock wave

Fig.5.8. Profiles of u_y, B_y, j_z corresponding to Fig.5.6 for an antiparallel shock wave (u_x and ρ profiles are the same as in Fig.5.7)

48-50) are integrated in the sense of decreasing ζ down to $\zeta=0$, unless u_x vanishes first. Upon reaching $\rho=0$ or $u_x=0$ the integration process ends, and the resulting value of the magnetic field B_3 at the rear wall is tied up with definite values of the velocity of the fast shock wave and all the other flow parameters (shock compression and heating, buildup of the azimuthal magnetic field, etc.). The velocities of shock waves as functions of the driving magnetic field for different values of $\theta=\arctan(B_{y1}/B_x)$, obtained in [5.27], fall into a narrow band between the dashed line $\theta=0$ and solid line $\theta=90^\circ$ in Fig.5.4.

Note that such self-similar solutions are mostly of methodological interest, and, as a rule, fail to supply any more detailed description of the processes in electromagnetic shock tubes than does the simple snowplow model. Actually, under experimental conditions the plasma conductivity is usually not so high as to justify the assumption of adiabaticity of the flow. Instead of being cooled down, as predicted by the ideal MHD theory, the plasma in the expansion wave, which acts as a magnetic piston, is heated to quite high temperature, Fig.5.9.

We see that the slow expansion wave in the electromagnetic shock tube is entirely similar to the gasdynamic centered expansion wave in a shock tube (Sect.1.6.4); the magnetic piston has much in common with the ordinary piston, as it transmits the high pressure forward. However, if $|B_{y1}+B_p|<B_{y1}$, the situation does not have any gasdynamic analogy. Indeed, the instantaneous reduction in the azimuthal magnetic field might be likened to the withdrawal of the magnetic piston from the shock tube. In gasdynamics the movement of the piston in the negative direction would produce an expansion wave traveling in the positive direction, without any shocks arising whatever. In an ordinary shock tube this would correspond to a situation where the pressure in the driver gas is lower than that in the test gas. After the rupture of the diaphragm the "driver" and the "test" gases would actually exchange their roles: the former would carry the shock wave, and the latter the expansion wave.

In magnetohydrodynamics, however, where two types of shocks are possible, the situation is quite different. Indeed, the decrease in the azimuthal magnetic field is transmitted forward by a fast expansion wave, which reduces the density of the gas and accelerates it in the negative direction. To satisfy the boundary condition at the rear wall the solution must also contain a slow shock wave. This cannot be an expansion wave, since in such wave the density decreases, while by the law of conservation of mass the average den-

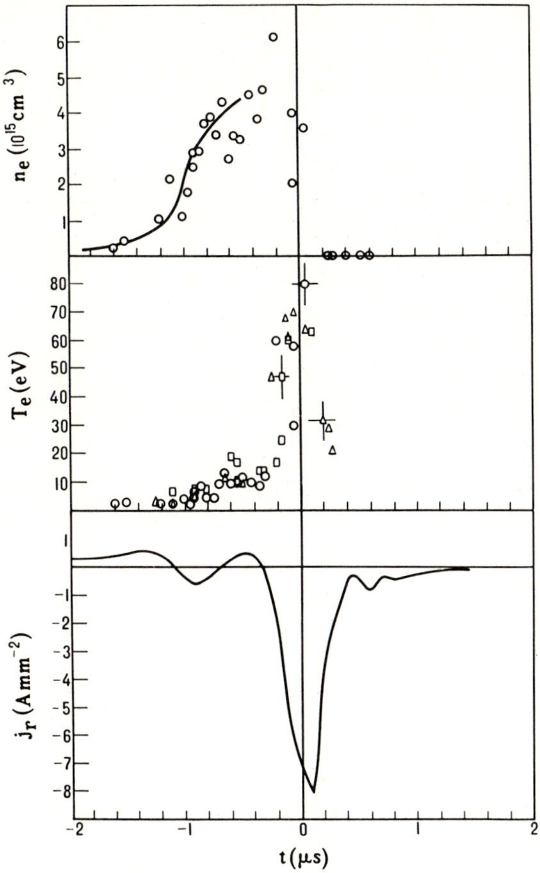

Fig.5.9. Structure of the magnetic piston region observed in the SUPPER VI shock tube (by courtesy of J. Howard)

sity between the wall and the head of the fast expansion wave should equal the initial density. Thus, the fast expansion wave can be followed only by a slow wave in which the magnitude of the axial magnetic field decreases, the gas density rises above ρ_1, and the axial velocity increases from a negative value to zero. If the azimuthal magnetic fields in the unperturbed gas and near the rear wall are directed oppositely, then the structure should include a rotational discontinuity between the fast expansion wave and the slow shock wave, Figs.5.10-12.

The solution of the self-similar problem is set up as before. Integrating (3.2.44,45,48-50) with the initial condition, reflecting the initial state, we obtain a succession of tentative downstream states 2 behind the fast expansion wave, differing in density. (As follows from Fig.3.1, the density behind the fast expansion wave is below $\tilde{\rho}$, but above a certain thres-

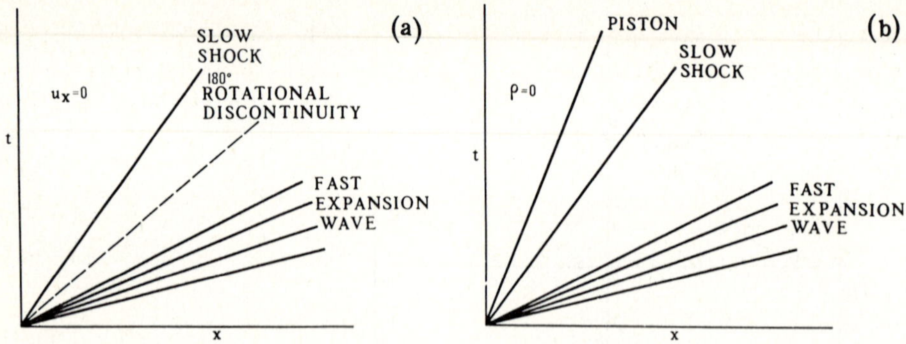

Fig.5.10a,b. Flow patterns representing the plane-parallel MHD flows containing slow shock waves: (**a**) the solution of the self-similar magnetic piston problem; (**b**) a slow shock wave driven by a boundary layer acting as a piston [5.1]

◄ Fig.5.11

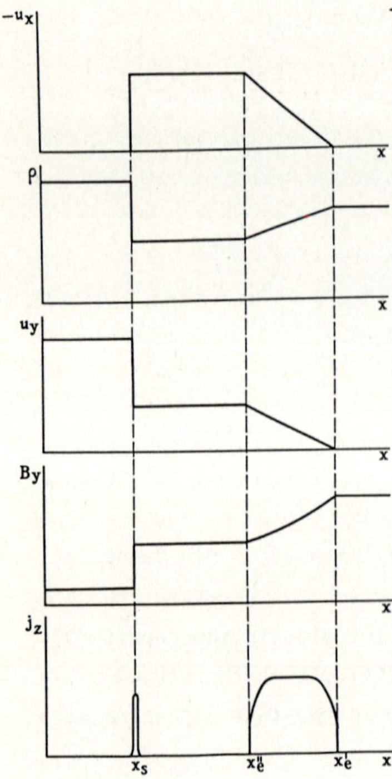

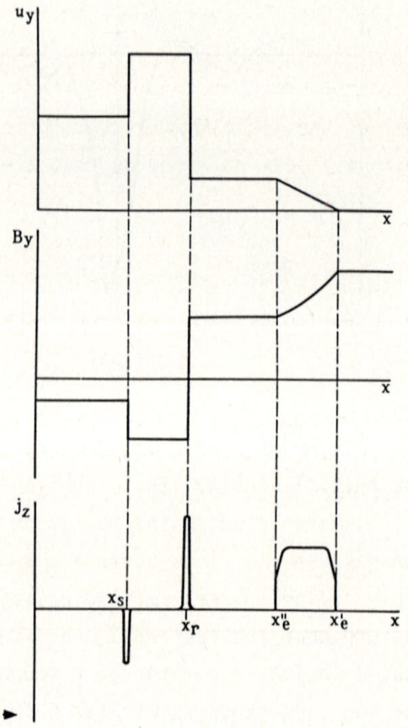

Fig.5.12 ►

Fig.5.11. Profiles of flow variables corresponding to Fig.5.10a for a parallel shock wave

Fig.5.12. Profiles of u_y, B_y, j_z corresponding to Fig.5.10a for an antiparallel shock wave (u_x and ρ profiles are the same as in Fig.5.11)

hold, which depends on the initial conditions.) From these states it is possible to proceed to the constant flow region near the back wall (where $u_x = 0$) via a slow shock wave. Its velocity in the laboratory frame u_s can be found from the continuity equation, which can be written as an equation in u_s:

$$\frac{|u_x(2)|}{u_s} = \frac{1}{u_2(u_s + |u_x(2)|, 2)} - 1 \quad , \tag{5.2.2}$$

where $u_2(\tilde{u}_s, 2)$ is the velocity variation (reciprocal compression) in the slow shock wave, propagating in the gas characterized by the parameters of state 2 with the velocity $\tilde{u}_s = u_s + |u_x(2)|$ in the gas frame. The strength of slow shock waves is limited (Sect.3.3.2): the strongest slow shock wave can be produced by just switching off the azimuthal magnetic field at the rear wall of the shock tube (the slow shock wave then classifies as a switch-off one).

Theoretical and experimental aspects of this method of obtaining slow shock waves in plasma were explored in [5.1]. The experiments revealed the existence of the following phenomenon, which does not fit into the framework of the ideal MHD theory. The boundary condition at the nonconducting rear wall does not require that the azimuthal plasma velocity vanishes, and in all self-similar solutions $u_{y3} \equiv u_y(0) \neq 0$ if $B_x \neq 0$. In real (not inviscid) plasma near the rear wall a viscous boundary layer of finite thickness δ arises, so that $u_y(0) = 0$, and $u_y(x) \to u_{y3}$ when $x/\delta \to \infty$. However, in this boudary layer the radial electric induction field

$$E_z^* = E_z - u_y B_x = E_z(0)(u_{y3} - u_y) \tag{5.2.3}$$

differs from zero, giving rise to the radial current $j_z = \sigma E_z^*$ (assuming scalar conductivity) in the boundary plasma layer. Then the boundary layer will experience axial magnetic pressure

$$\Delta p_B = -j_z B_y \cong -\sigma E_z(0) B_y(0)(u_{y3} - u_y) \quad , \tag{5.2.4}$$

the direction of the force depending on the sign of the azimuthal magnetic field near the rear wall. As follows from Figs.5.11,12, $\Delta p_B < 0$ if the azimuthal magnetic field is not reversed; then the boundary layer is pressed by the magnetic pressure against the wall, and its presence modifies the nearby flow. If the azimuthal magnetic field is reversed, then the magnetic pressure tears the boundary layer away from the rear wall and drives it ahead like a piston, leaving a vacuum behind. Rotational discontinuity then does not arise: reversal of the azimuthal magnetic field is localized to

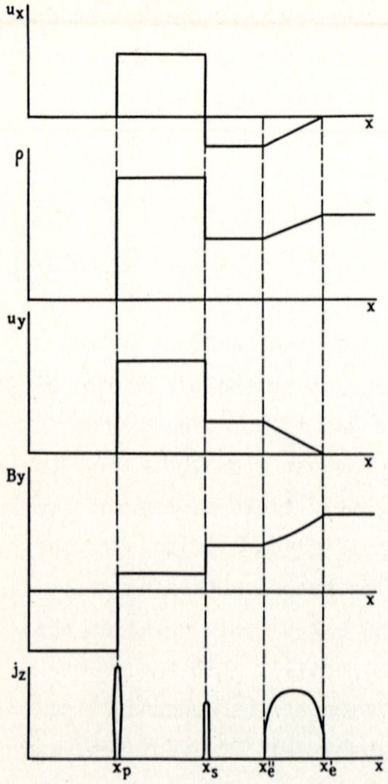

Fig.5.13. Profiles of flow variables corresponding to Fig.5.10b

the moving current layer. Accordingly, here also the flow has something in common with the snowplow model, although it defies description by the simple models of Sect.5.1.2. It can be approximated by a self-similar solution of the ideal MHD equations with appropriately modified boundary conditions: the axial flow velocity behind the slow shock wave should equal the velocity u_p of the detached boundary-layer "piston", which, in the present model, is empiric rather than calculated. Condition (5.2.2) is then replaced by (Figs.5.10b,13)

$$\frac{|u_{x2}| + u_p}{u_s - u_p} = \frac{1}{u_2(u_s + |u_{x2}|, 2)} - 1 \quad . \tag{5.2.5}$$

The experiments of [5.1], performed with hydrogen plasma (with a particle density of $4 \cdot 10^{14}\,\mathrm{cm}^{-3}$ and a temperature of about 1 eV before the application of radial electric field) reveal that when the electric field strength is not sufficient for reversing the sign of B_y, the succession of waves corresponding to Figs.5.10a,b,11 is not observed. The entire current passing through the plasma is localized to the wall boundary layer about 3 cm thick. The radial current in the wall boundary layer reduces B_y to such an extent that the amplitudes of disturbances which should have been transmitted forward by the magnetosonic waves become too small to be detected. Under these conditions the dissipative mechanisms (viscosity and conduction) gain prominence, and there is little wonder that the experimental results deviate from the predictions of the ideal MHD theory. Should the electric field be strong enough to reverse the sign of B_y, the flow pattern will comply with the diagrams shown in Figs.5.10b,13: the magnetic pressure tears the boundary layer away from the wall and drives it forward, producing waves as predicted by the MHD theory. The observed wide region of smoothly varying B_y is identified with the fast expansion wave. This is followed by an abrupt drop in B_y (slow shock wave), after which the azimuthal magnetic field is reversed in a wide current layer, identified with the piston. Calculated velocities for the head of the fast expansion wave and the slow shock wave, and variations of the azimuthal magnetic field in these waves agree excellently with experiment, which proves the viability of the theoretical model.

5.2.2 Self-Similar Piston Problem for Flows with Ionizing Shock Waves

The theory developed above is readily extrapolated to the situation when the gas is initially nonconducting. Then the sequence of waves and discontinuities, representing the solution of the self-similar problem, should include the ionizing shock wave propagated in the undisturbed gas. The ionizing shock wave can be preceded only by the electromagnetic wave, which transmits downstream the magnitudes of electric and magnetic fields, determined by the boundary conditions on the ionizing shock front. By the convention adopted in Sect.4.2.2, we do not consider the propagation of the electromagnetic wave, assuming, for instance, the values of the transverse components of magnetic field to be set in advance. In the most general context they will be identified with the values established after the passage of the electromagnetic wave. Obviously enough, we narrow down the scope of possible solutions. For instance, by moving an impenetrable piston normaly to the magnetic field and assigning definite values to the transverse components of the magnetic field on its surface, we generally obtain a nonzero transverse magnetic field, maintained by the electromagnetic wave, ahead of the ionizing shock front; in other words, the ionizing shock will not be normal. Confining our subject to only the normal ionizing shock waves, we single out of the 5-dimensional space of conceivable boundary conditions on the piston surface, which fix the values of u_x, u_y, u_z, B_y, B_z, a 3-dimensional manifold [$B_y = B_y(\mathbf{u})$, $B_z = B_z(\mathbf{u})$] complying with the conditions $B_{y0} = 0$, $B_z = 0$ downstream of the electromagnetic wave front.

The self-similar problem may be stated if we assume the downstream plasma to have a sufficiently high conductivity. In the formulation of the self-similar magnetic piston problem for a plane-parallel plasma flow the ionizing shock wave must be followed by a slow expansion wave (separated, perhaps, by a rotational discontinuity). By virtue of (3.3.27-30) the ionizing shock wave may belong to type 2 (then it may be followed by a rotational discontinuity, reversing the azimuthal magnetic field), to type 3 (then the azimuthal magnetic field retains its direction), but not, generally speaking, to type 4. The velocity of a type 4 shock wave is below c_s, so the slow expansion wave would have caught up with it. The only possible flow pattern which includes the ionizing shock of type 4 is realized when this shock is singular: $u_x(4) = c_s$. The head of the pursuing slow expansion wave travels through the gas at the same velocity c_s, so the gap between them remains the same as at $t = 0$, that is zero.

The solution of the piston problem in this case includes the ionization processes in the shock wave. This eliminates the apparent ambiguity in the boundary conditions (Chap.4) and enables the unique flow pattern to be determined. Insofar as we are here concerned mainly with the qualitative morphology, we disregard the chemistry (energy expenditure on dissociation and ionization).

This problem is most easy to deal with the case of transverse ionizing shocks, when only the type 2 ionizing shock is possible, the slow expansion wave is reduced to a tangential discontinuity, and the parameters of the shock are linked with the driving magnetic field by the snowplow relation (5.1.15). To find M_{a0} as a function of B_p, we need to know $u_2(M_{a0})$ and thus have to use an additional boundary condition. Fixing the parameters which characterize the relative contribution of photoionization and ionization due to the heating of electrons by the electric-induction field of the shock wave, and making use of the additional boundary condition (4.3.11), we can easily solve (5.1.15). The solution is plotted in Fig.5.14 for the parameter values used in Fig.4.7 (the notation was explained in the legend to Fig.4.7). We see that the velocity of the ionizing shock wave for a given B_p exhibits a weak dependence on the choice of the additional boundary condition: when $|B_1 + B_p|/B_1 \gtrsim 2$ the difference in the velocities pertaining to the GD and MHD limiting regimes is within 20%, quickly dropping down to several per cent as the strength of the driving magnetic field is further

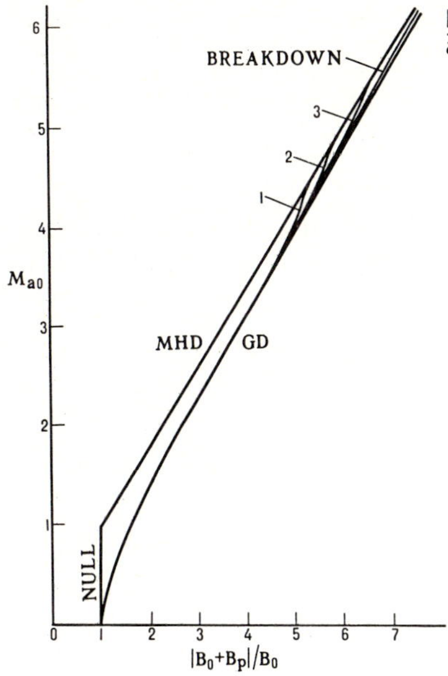

Fig.5.14. Transverse ionizing shock velocity as a function of driving magnetic field B_p

increased. The difference will be somewhat larger (but not much), if we include the chemistry.

If the azimuthal magnetic field in the undisturbed gas is zero, then ahead of the magnetic piston may be propagated the normal ionizing shock wave of type 2 (purely a gasdynamic one, in which the azimuthal magnetic field remains zero, so that the wave cannot be followed by rotational discontinuities), of type 3, and the Chapman-Jouguet shocks (an extreme case of the ionizing shocks of type 4, Sect.4.4). The choice of the additional boundary condition is essential only up to certain velocities of shocks with $M_{a0} < 2$, since with $M_{a0} \geq 2$ the normal ionizing shock waves are purely gasdynamic (of type 2), pertaining simultaneously to the GD and the MHD limits. The flow pattern for $M_{a0} < 2$ exhibits a dependence on the ionization mechanisms. When the dimensionless electric field in the ionizing shock wave is simply kept to a certain limit E_*, the additional boundary condition assumes the form (4.3.11). The piston problem is solved along the same lines as in magnetohydrodynamics (Sect.5.2.1). The solutions for several values of E_* are plotted in Fig.5.15. The line GD_2 here represents the gasdynamic shock waves existing for $M_{a0} \geq 2$.

The dash line CJ marks the gasdynamic limit: if the ionization is restricted to the viscous subshock, then in the range $0 < M_{a0} < 2$ only the type 4 normal ionizing shock waves exist, including the Chapman-Jouguet shock, which, incidentally, is the only shock that may be included in the solution of the self-similar piston problem. Therefore, the flow ahead of the magnetic pis-

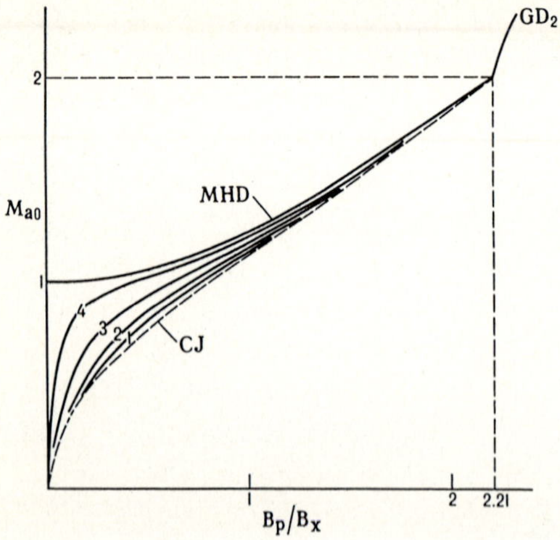

Fig.5.15. Normal ionizing shock velocity as a function of driving magnetic field B_p. Curves CJ and MHD correspond to GD and MHD limiting regimes, respectively. Curves 1, 2, 3, 4 are plotted in the electrostatic breakdown limit for $E_* = 0.3$, 0.25, 0.15 and 0.05, respectively

ton comprises the normal ionizing Chapman-Jouguet shock, immediately followed by a slow expansion wave. Thus, we have full analogy with the classical detonation theory (Sect.1.9 and [5.28,29]): the GD limit in the theory of ionizing shocks corresponds to the limit $\Delta_r/\Delta_v \gg 1$ in the detonation theory, the normal ionizing shock waves of types 3 and 4 to the strong and the weak detonation waves, respectively, the slow expansion wave to the gasdynamic expansion wave, the ignition of the chemical reaction to the appearance of conduction, i.e. switching on the interaction between the flow and the magnetic field. The entire body of argumentation, which corroborates the formation of the normal Chapman-Jouguet detonation in a free-propagating detonation wave (Sect.1.9 and [5.28,29]) is word for word transcribed to normal ionizing shock waves [5.30] in electromagnetic shock tubes.

The Chapman-Jouguet hypothesis gives a unique solution of the self-similar magnetic piston problem, although by itself it certainly does not supply the additional boundary condition at the shock front: it just defines a problem with certain initial and boundary conditions. The additional boundary condition in the present case (in the GD limit, Sect.4.4) resides in the statement that in the range of Mach numbers $0 < M_{a0} < 2$ the steady structures pertain only to the ionizing shocks of type 4 with all admissible values of the transverse electric field (from 0 through E_{CJ}).

With increasing intensity of the ionizing shock waves the GD regime goes over into the MHD regime. This transition is trivial (it occurs upon reaching $M_{a0} = 2$) as long as B_p is raised while the values of B_x and the parameters of the initial gas state are fixed. Of special interest is the transition to the MHD regime in the range $0 < M_{a0} < 2$, which may be accomplished by simultaneously increasing B_p and B_x for a fixed initial state. The higher B_x is, the higher are the velocities of shock waves in the indicated range of Mach numbers, the stronger the shock heating of the plasma, and, accordingly, the greater the contribution of precursor ionization, which is excluded from the GD limit. In the electrostatic breakdown limit the increase in B_x is accompanied by the decrease in $E_* \propto B_x^{-2}$, so the curves CJ, 1, ..., 4 may be viewed as plotted for five different values of B_x in increasing order. The transition to the MHD regime occurs, as $B_x \to \infty$ ($E_* \to 0$). As with the transverse shock waves, the shock velocity is not too sensitive to the choice of the additional boundary condition.

The GD limit in the theory of ionizing shock waves reflects the fact that the most efficient mechanism to propagate ionization (chemical reaction) is the GD shock wave. If, however, a competing mechanism of propagation offers higher velocities, it will dominate.

For example, the action of fast nongasdynamic mechanisms of ionization transfer to undisturbed gas is due primarily to radiative processes: radiant heat conduction with high plasma temperatures ($T \gtrsim 10$ eV) [5.31], and photoionization of undisturbed gas by radiation coming from downstream with less high temperatures ($T_2 \lesssim 10$ eV) [5.32]. For this reason, the laser-radiation-absorption waves are similar to weak rather than fast detonation waves: they propagate faster than the Chapman-Jouguet shock wave, and the compression of plasma is lower. The transition to the MHD regime for the normal ionizing shock waves in the range of Mach numbers $0 < M_{a0} < 2$ should occur in exactly such a fashion, insofar as the MHD switch-on shocks represent the limit of the ionizing shocks of type 3, which are similar to the weak detonation waves. The gasdynamic limit actually does exist for the relatively weak shock waves (low values of B_x), i.e., the Chapman-Jouguet hypothesis is valid. As their intensity increases, in a certain range of Mach numbers the GD regime breaks and is replaced by ionizing shocks of type 3. As B_x increases further, the range of Mach numbers extends, ultimately covering the entire region $M_{a0} < 2$. Of course, the same should apply in general to the skew and oblique ionizing shocks (of which only the oblique persist to the MHD limit).

To gain a clearer impression of the diversity of possible flow patterns with ionizing shocks, let us consider another example: the self-similar problem of a rigid conducting piston, stationed normal to the direction of the magnetic field, which at $t = 0$ is plunged into the homogeneous nonconducting gas with axial velocity U and azimuthal velocity V [5.33]. Both U and V are expressed in terms of axial Alfvén velocity c_a. The gas is assumed to become ionized in the normal ionizing shock waves. In the self-similar solution the ionizing shock of type 2 (a purely gasdynamic one; $M_{a0} \gtrsim 2$, $B_{y2} = 0$) can be followed only by the slow expansion wave, since the slow shock waves and rotational discontinuities do not propagate along the magnetic field. On the contrary, the type 3 shock which switches on the transverse magnetic field, may be succeeded by either a slow shock wave or a slow expansion wave. The slow expansion wave may bring the gas density down to zero, and then its tail will be separated from the piston by a vacuum space (cavitation). The ionizing shock wave of type 4 cannot be followed by any other shock or expansion waves or discontinuities.

From the condition of ideal conductivity of the piston it follows that on its surface (in the piston frame) not only the axial velocity of the plasma (or its density), but also the tangential electric field should vanish. If the plasma density near the piston is nonzero, then, as directly ensues from this condition, the tangential component of the relative plasma velocity vanishes on its surface (note the difference between a conducting piston and the nonconducting rear wall of the shock tube, where a viscous boundary layer may arise). Thus, the boundary condition on the surface of the piston in the absence of cavitation states that the axial and the azimuthal velocities of the plasma equal U and V, respectively. Should the density of the plasma near the piston turn to zero, the boundary condition takes the form

$$V - V_c + B_c(U - U_c) = 0 \quad . \tag{5.2.6}$$

Where U_c, V_c are dimensionless (normalized to c_a) components of plasma velocity at the point where the density becomes zero (the tail of the slow expansion wave); B_c is the magnetic field strength (normalized to B_x) at the same point. In other respects the solution of the self-similar piston problem is quite similar to what has been done in Sect.5.2.1. [We use the additional boundary condition (4.4.8)]. The results are plotted in Fig.5.16 for the MHD limit, in Fig.5.17 for the GD limit, and in Fig.5.18 for some finite value of E_*, allowing the existence of ionizing shocks of type 3.

The ray GD_2 ($U \geqslant 3/2$, $|V| = 0$) in the (U, $|V|$) plane corresponds to the formation, in front of the piston, of only the gasdynamic ionizing shock wave of type 2, not switching on the magnetic field. It is easy to demonstrate that behind such shocks $\rho > \tilde{\rho}$ (Sect.3.2.3 and Fig.3.1), so that in the next-coming slow expansion waves the decreasing density is accompanied by increasing magnetic field strength and transverse velocity. The curve bounding the region GD_2R in the (U, $|V|$) plane, which pertains to the gasdynamic ionizing shock of type 2, followed by the expansion wave (R), is determined by integration of (3.2.45,48,49) with the initial condition $U = 3/2$, $|V| = 0$.

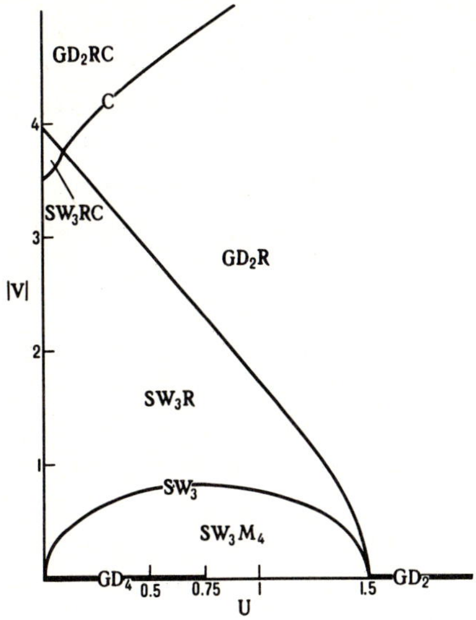

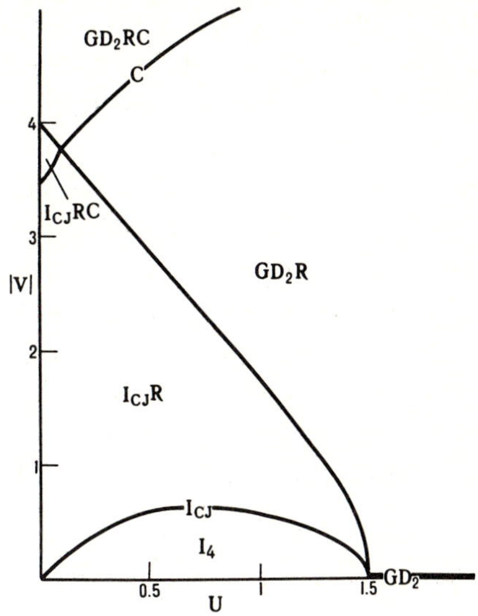

Fig.5.16

Fig.5.17

Fig.5.16. Regions in the piston velocity space (U, |V|) corresponding to different types of flow ahead of the piston in the MHD limit of the theory. Notations: GD, gasdynamic shock wave; SW, MHD switch-on shock wave; M, slow MHD shock wave; R, slow MHD expansion (rarefaction) wave; C, cavitation. The index indicates the type of downstream singular point

Fig.5.17. The same as in Fig.5.16, but in the case where the GD limit holds true. (I_{CJ}: Chapman-Jouguet ionizing shock wave; I_4: type-4 ionizing shock wave

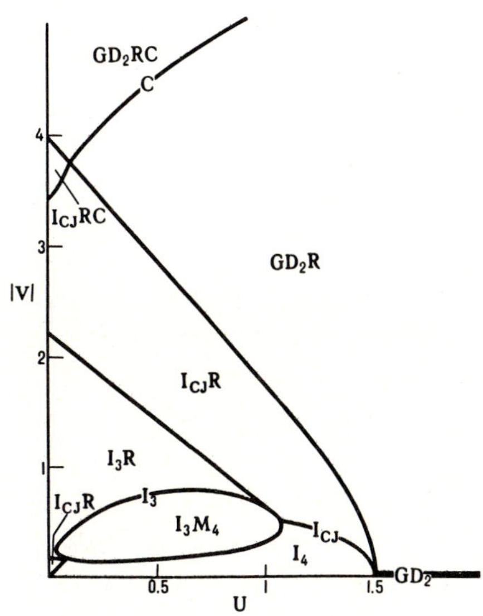

Fig.5.18. The same as in Figs.5.16, 17, but in the case where the GD limit does not hold true in a certain region of the shock parameters in the (M_{a0}, E) plane. Here $E_* = 0.15$, i.e., this region in Fig.4.9 lies below the line CJ above the straight line $E = 0.15$. I_3 is the type 3 ionizing shock wave

Curve C (cavitation) marks the density zero in the slow expansion wave. This curve demarcates the region GD_2RC (a gasdynamic wave followed by a slow expansion wave and a vacuum space), bounded also by the straight line (5.2.6), drawn through the intersection of curve C with the boundary of the region GD_2R. By virtue of arguments developed above, the regions GD_2, GD_2R, GD_2RC do not depend on the form of the additional boundary conditions and are the same in Figs.5.16-18.

In the MHD limit (Fig.5.16) the curve SW_3 corresponds to the production of the switch-on MHD shock in front of the piston (subscript 3 indicates that this shock is viewed here as the extreme case of ionizing shocks of type 3). The equation of the SW_3 curve in the (U, |V|) plane is readily obtainable from (3.5.2,4,5):

$$
\begin{aligned}
U &= M_{a0} - 1/M_{a0} \ , \\
|V| &= \left[\frac{2}{3}\,(M_{a0}^2 - 1)(4 - M_{a0}^2)\right]^{\frac{1}{2}} \ , \qquad 1 < M_{a0} < 2 \ .
\end{aligned}
\tag{5.2.7}
$$

The curve SW_3 bounds the region SW_3M_4, pertaining to the formation of the switch-on MHD shock in front of the piston, followed by a slow MHD shock wave (M_4); it would be natural to identify it with type 4 shock. The segment $(0 < U \leqslant 3/4,\ |V| = 0)$ corresponds to purely gasdynamic slow MHD shocks $(0 < M_{a0} \leqslant 1)$ in front of the piston. The segment $(3/4 < U < 3/2,\ |V| = 0)$ corresponds to the nonevolutionary gasdynamic MHD shock waves. In this piston-velocity range the diagram predicts the following configuration: the switch-on shock wave – the switch-off shock wave. The switch-on MHD shock, pertaining to singular shock waves, may also be unstable. Its instability would imply that the relevant self-similar solution does not exist, although such a solution could serve as a rough approximation (Sect.3.3.5) assuming the diffusive decay of the unstable switch-on shock.

Figure 5.17 is plotted for the GD limit, when only the ionizing shocks of type 4 are possible. The region I_4 in the (U,V) plane depicts the region of existence of such shocks in front of the piston. Curve CJ pertains to the ionizing shock of type 4 with the highest possible value of the electric field (the Chapman-Jouguet shock). Between this curve and the boundary of regions GD_2R, GD_2RC are the regions $I_{CJ}R$ (the Chapman-Jouguet shock is followed by the slow expansion wave) and $I_{CJ}RC$ (the tail of the expansion wave is separated from the piston by a vacuum space). The regions $I_{CJ}R$, $I_{CJ}RC$ in the (U, |V|) plane restrict the applicability of the Chapman-Jouguet hypothesis to the problem in question. (The self-similar magnetic piston problem implies that (3.2.45,48,49) are integrated down to the limits of the region $I_{CJ}R$ to the line $U=0$ or C). We see that here this hypothesis is not always feasible even for the GD limit, insofar as the region I_4 exists.

Figure 5.18 is plotted for an intermediate case, when ionizing shocks of type 3 are possible in a certain range of Mach numbers (curve I_3). Such shocks may be followed by either a slow expansion wave (region I_3R) or a slow MHD shock wave (region I_3M_4). Throughout the rest of the (U,|V|) plane the solution is the same as obtained in the GD limit. With increasing E_*, the curve I_3 and the regions I_3R, I_3M_4 collapse, vanishing altogether at $E_*=0.358$. When E_* is reduced to zero, curve I_3 goes into SW_3, region I_3M_4 into SW_3M_4, etc.

The self-similar piston problem in three dimensions will be solved if we rotate Figs.5.16-18 about a horizontal axis. Then, for instance, in Fig.5.18 region I_4 will become a three-dimensional space $\tilde{I}_4$, curve I_3 will describe a two-dimensional surface of revolution $\tilde{I}_3$, the ray GD_2, which lies in the axis of rotation, remains a one-dimensional manifold. The different dimensionalities of these regions correspond to the different numbers of additional boundary conditions required for the ionizing shock waves of types 2,3,4. Ionizing shocks of type 4 do not require any additional boundary conditions: under certain conditions, expressed as inequalities [namely, the point in the velocity space (U, V_y, V_z) should fall within the 3D region $\tilde{I}_4$], only this kind of shock will arise in front of the piston. Ionizing shocks of type 3 require one additional boundary condition; therefore, the dimensionality of the manifold, to which the phase point is restricted, is one less. For ionizing shocks of type 2 requiring two additional boundary conditions, the manifold has the dimension of unity. At the same time it ought to be emphasized that ionizing shocks of types 2 and 3 do not represent any particular cases, realized only when the piston velocity components are subject to some restrictions. Revolution of two-dimensional regions GD_2R, GD_2RC, I_3M_4, I_3R, I_3RC defines three-dimensional regions in the (U, V_y, V_z) space. As long as the point representing the piston velocity falls within one of these regions, ionizing shocks of types 2 or 3 will arise in front of the piston, being separated from the piston by an appropriate MHD flow.

5.3 Dynamics of Transverse Shocks in Magnetized Plasma

Theoretical estimations and experimental results using electromagnetic shock tubes indicates that the morphology of experimentally realized plasma flows is far from self-similar. Quantitative assessment of the movement of shock waves and current sheets, including the effects which destroy the self-similarity of the flow, i.e., exactly those mechanisms which shape the shock structures, such as dissipation, dispersion, etc., can be performed only numerical simulation. Systematic studies in this direction were initiated in the early seventies [5.2,34], concerned with the dynamics of MHD shocks in plasma. The description of the flow was based on the two-fluid transfer equations with classical kinetic coefficients.

The unsteady plane-parallel unidimensional flow of the completely ionized simple plasma ($\gamma_e = \gamma_i = 5/3$, $Z = 1$) in a transverse magnetic field is described by the equations of continuity, of motion and heat balance and the equation of induction, which represents Maxwell equations in the quasi-steady approximation. Under the assumption of quasi-neutrality of the plasma ($n_e = n_i = N$), one continuity equation is sufficient; then the axial velocities of the electrons and the ions are the same ($u_x^e = u_x^i = u_x$). In the transverse configuration we may exclude the azimuthal motion (rotation) of the plasma, insofar as it becomes involved in this movement only through viscosity, while the off-diagonal components of the viscous stress tensor in magnetized plasma are as small as $(\Omega_i\tau_i)^{-1}$. In view of all this, the necessary equations in the five variables N, u_x, T_e, T_i, B_y are

$$\frac{\partial N}{\partial t} + \frac{\partial}{\partial x} N u_x = 0 \quad , \tag{5.3.1}$$

$$m_i N\left(\frac{\partial u_x}{\partial x} + u_x \frac{\partial u_x}{\partial x}\right) + \frac{\partial}{\partial x}\left(NT_e + NT_i + \frac{B_y^2}{8\pi} - \frac{1}{3}\eta_0^i \frac{\partial u_x}{\partial x}\right) = 0 \quad , \tag{5.3.2}$$

$$\frac{3}{2} N\left(\frac{\partial T_i}{\partial t} + u_x \frac{\partial T_i}{\partial x}\right) + NT_i \frac{\partial u_x}{\partial x} + \frac{\partial}{\partial x} \kappa_i \frac{\partial T_i}{\partial x} - \frac{1}{3}\eta_0^i \left(\frac{\partial u_x}{\partial x}\right)^2 - Q_\Delta = 0 \quad , \tag{5.3.3}$$

$$\frac{3}{2} N\left(\frac{\partial T_e}{\partial t} + u_x \frac{\partial T_e}{\partial x}\right) + NT_e \frac{\partial u_x}{\partial x} + \frac{\partial}{\partial x} \kappa_e \frac{\partial T_e}{\partial x} + Q_\Delta - \frac{1}{\mu_0^2 \sigma_\perp}\left(\frac{\partial B_y}{\partial x}\right)^2 = 0 \quad , \tag{5.3.4}$$

$$\frac{\partial B_y}{\partial t} + \frac{\partial}{\partial x}\left(u_x B_y - \frac{1}{\mu_0 \sigma_\perp} \frac{\partial B_y}{\partial x}\right) = 0 \tag{5.3.5}$$

(the notation is the same as in Sect.3.1.2). The stationary solutions of this set of equations have already been analyzed in Sect.3.4. In particular, we have demonstrated that in a magnetized plasma ($\Omega_i\tau_i \gg 1$) the terms representing ion and electron heat conduction, as well as the Joule heating of the electrons, on the right-hand sides of (5.3.3,4) are negligible. The dynamics of the plasma and the heating are governed by the isotropic component of ion viscosity; the electron component of the plasma is heated adiabatically due to compression and heat transfer in elastic collisions with the ions.

The set of equations similar to (5.3.1-5) has been solved numerically in [5.2]. The performance test was accomplished by solving the same equations with zero magnetic field; this procedure yielded adequate results for the steady shock structures in plasma without external magnetic field [5.35,36]. The calculations used Lagrangian coordinates, the boundary condition being set in the layer $\zeta = 0$, associated with certain gas particles. In the absence of a magnetic piston the velocity was assigned the value $u_x(\zeta = 0)$, i.e., it was the rigid piston problem that was actually being solved. The magnetic piston in the calculations of the dynamics of the MHD shocks was simulated by applying the transverse magnetic field B_{y3}, parallel to B_{y1}, at point $\zeta = 0$ and time $t = 0$. This, of course, does not embrace the magnetic piston problem in the entire space region near the rear wall of the shock tube. The boundary condition employed is equivalent to the assumption that the magnetic piston scoops up all the gas in front of itself, leaving a vacuum behind the layer $\zeta = 0$, as is the case in the self-similar problem. However, this simple approach proved quite efficient for describing the dynamics of the shock wave and the current sheet. The boundary conditions for all other variables are stated in the form

$$\frac{\partial}{\partial\zeta} = 0 \quad \text{at} \quad \zeta = 0 \quad ,$$

which is also consistent with the results of the solution of the self-similar problem. The plasma in the initial state is assumed to be homogeneous, with the particle density $N_1 = 2.6 \cdot 10^{15}\,\mathrm{cm}^{-3}$, which corresponds to the initial pressure $p_1 = 50$ mTorr at room temperature. To justify the use of the plasma transfer coefficients in the entire range of variation of flow parameters, the initial temperature was taken to be $T_1 = 1$ eV. Although in the shock tube at Columbia University the gas is initially nonconducting, i.e., the shocks formally classify as ionizing ones, the shock waves used for simulation are so strong as to satisfy the criteria of transition to the MHD limit with a broad margin (Sect.4.2.5), and therefore the calculated results show practically no dependence on the initial temperature as long as $T_1 \ll T_2 \sim$

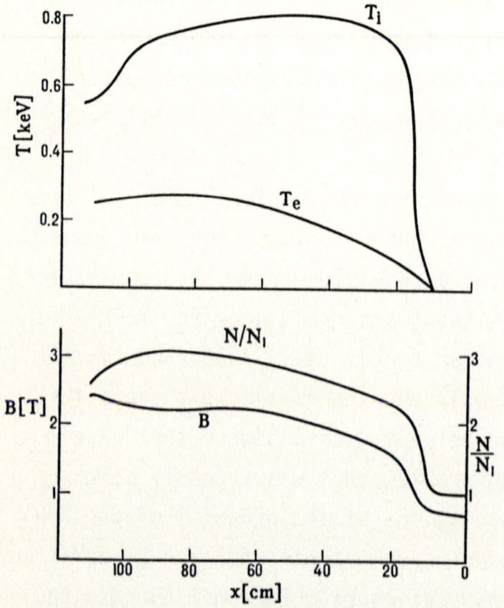

Fig.5.19

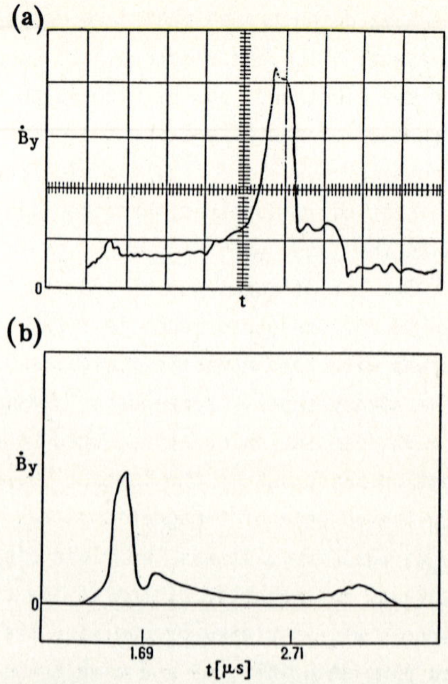

Fig.5.20

Fig.5.19. Profiles of flow variables in a nonstationary transverse MHD shock wave [5.2] (by courtesy of the American Institute of Physics)

Fig.5.20a,b. Profiles of $\dot{B}_y$, observed in experiments [5.3,8] (**a**), and calculated [5.2] (**b**) (by courtesy of the American Institute of Physics)

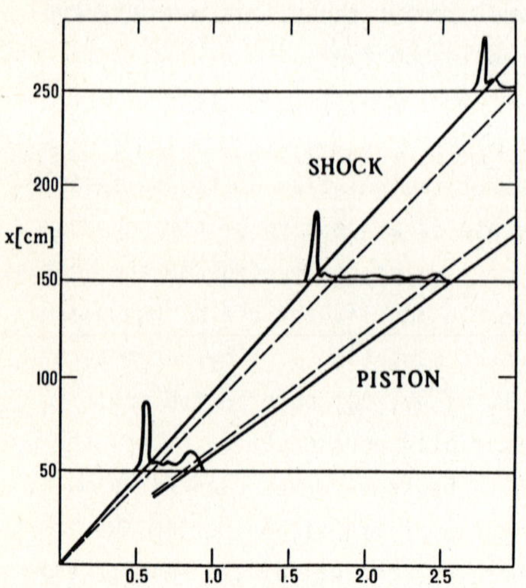

Fig.5.21. The x-t diagrams for a transverse MHD shock wave and a magnetic piston. (——) are results of calculations [5.2], (---) are calculated using the snowplow model (by courtesy of the American Institute of Physics)

$10^2 \div 10^3$eV. To smooth out the discontinuity in the velocity, which initially appears due to the discontinuity of the boundary condition [$B_y(0,t) = B_{y3}(0,t)$], the first steps of integration use an auxiliary numerical viscosity, which is later replaced by the actual ion viscosity.

The results of calculations are shown graphically in Figs.5.19-21 and 3.14. Figure 5.19 represents the profiles of flow variables in the shock wave, which fit in well with the theory of steady structures of the strong MHD shocks (Sects.3.4.2,5). Figure 5.20 shows the calculated (b) and the measured (a) [5.3,8] profiles of $\dot{B}_y$, proportional to the outputs of magnetic probes, which verify the excellent agreement between theory and experiment. The calculated and measured shock widths are plotted in Fig.3.14. The pertinent length scale was chosen on the basis of measurements of $\dot{B}_y$. The agreement here is also quite fair, although the calculated value is about twice the measured one. Recall that the experimental path length of the shock (the same as used in the calculations) did not exceed 3 m (the length of the shock tube at Columbia University), and therefore the strongest shocks, whose velocities correspond to steady structures more than 3 m wide ($100 < u_s < 300$ cm/μs), do not have time to evolve into steady structures. Accordingly, the experimental results should conform with the theoretical steady structure for $u_s \lesssim 100$ cm/μs, just as they really do. Successful numerical simulation of the evolution of the shock wave when u_s exceeds 100 cm/μs ($100 < u_s < 300$ cm/μs) indicates the correct assessment of the shock-shaping mechanisms and proves the adequacy of the Navier-Stokes approximation.

Figure 5.21 shows the space-time diagram of the moving shock front and current sheet of the magnetic piston. The two regions are quite distinct. We see that the calculated profiles of $\dot{B}_y$ do not resemble self-similar ones. In particular, the current sheet (the magnetic piston), represented in the self-similar solution by a tangential discontinuity, has a distinct diffusive structure. This nothwithstanding, the movement of the shock wave and the piston is found to be adequately approximated by the self-similar solution (Fig.5.21), here represented by the snowplow model. For the used strength of driving magnetic field ($B_{y3}/B_{y1} = 3.33$) the shock velocity in this model is found to be 85 cm/μs, merely 10% below the result yielded by the exact numeric solution.

Equations used in [5.2] included also the inertia of electrons and energy loss by bremsstrahlung. In accordance with the estimates made in Sect.3.4.2, these effects are quite immaterial. The Joule heating of electrons is also small.

The dynamics of oblique MHD shocks in plasma ($B_x \neq 0$) was explored in [5.34]. The main characteristic feature of such shocks is the presence of the region of heating due to electron heat conduction. The electron heat flux in magnetized plasma

$$q_e = \left(\frac{B_y^2}{B_y^2 + B_x^2}\kappa_{e\perp} + \frac{B_x^2}{B_y^2 + B_x^2}\kappa_{e\parallel}\right)\nabla T \sim \frac{\kappa_{e\parallel}}{1 + (B_y/B_x)^2}\nabla T$$

is no longer small if the inclination of the magnetic field to the x axis $\theta = \arctan(B_y/B_x)$ is not close to 90°. For this reason the width of an oblique shock is by a factor of $(m_i/m_e)^{1/2}$ greater than the width of a transverse shock (Sects.3.4,5). It follows that such shocks, which would raise the plasma temperature to the order of 10^3eV, cannot be obtained with a shock tube of reasonable size. From this we may conceive the importance of theoretical and experimental investigation of the transverse strong shock waves.

In the experiments with coaxial electromagnetic shock tubes the nonunidimensionality of the flow, arising from the inhomogeneity of the azimuthal magnetic field, becomes essential. The numerical simulation of the dynamics of MHD shocks in plasma on the basis of two-dimensional two-fluid nonstationary Navier-Stokes equations [5.11,12] demonstrated excellent consistency with the experimental results [5.14] and proved the viability of this approach in describing collision-dominated MHD shock waves.

5.4 Evolution of the Initial Ionizing Discontinuity in the Transverse Magnetic Field

The morphology of plasma flows, observed in electromagnetic shock tubes with ionizing shock waves, is more sophisticated and deviates more strongly from the idealized self-similar pattern than the MHD flow with the initially conducting plasma. This is not at all surprising, since the basic criterion of the adequacy of self-similar approximations (the relative smallness of the space and time scales of the shock-shaping collisional and radiative processes) in this case is almost never satisfied. The finite conductivity of the plasma, which starts rising from zero at the shock front, leads to a situation when the scale Δ_j (the diffusion length of magnetic field) becomes considerable, and, as a rule, not small in comparison with the size of the installation. This Δ_j characterizes also the Joule heating of the plasma, violating its adiabaticity. The width of the shock layer depends on the finite-valued downstream rate of ionization relaxation and finite-width precusor region in the head of the shock. As a result, the shock front is by no means thin in comparison with the width of the current sheet (the magnetic piston), and sometimes is even comparable with the size of the installation, which implies the importance of nonstationary effects. All the sophisticated flow pattern may be adequately described only by numerical simulation. Comparison of the results of numerical simulation with the predictions of the theory of stationary shock structures allows the consistency of both approaches to be checked and throws light on the peculiar features of experimental flows arising from their nonstationarity.

Especially difficult is the numeric simulation of the magnetic piston, because of the highly intricate configuration of currents and fields in this region. It appears that the approach should include the analysis of both stages of formation of the magnetic piston: the relatively short-lived breakdown of the gas and the production of discharge plasma, when a strong electric field exists and the charged particles drift in crossed fields (Sect.5.1.2), and the gasdynamic stage of acceleration of the plasma by magnetic-pressure forces. These two stages are essentially different. First, they are characterized by different time scales, the first stage being at least an order of magnitude shorter than the second. Secondly, the underlying physical processes are dissimilar to such an extent as to require essentially different approaches. Therefore the initial development of the discharge should be described by the kinetic equations. The movement of neutral particles, which starts when the breakdown is well advanced, can be accounted for by the three-fluid Navier-Stokes equations, whereas the movement of highly ionized discharge plasma is described by the two-fluid equations. The analysis of the initial breakdown stage should supply the values of flow variables, which serve for the initial conditions for the two-fluid equations describing the subsequent movement of the plasma.

Failing this, the statement of the problem of numerical simulation of the flows with ionizing shock waves requires certain speculative assumptions, i.e., is to some extent arbitrary. This is of little consequence as long as we are concerned not with the flow on the whole, but rather with the evolution of the magnetic structure of the ionizing shock wave to its final asymptotic state. Then it is possible to vary the initial and boundary conditions, assuming that the role of the magnetic piston consists in maintaining differential pressure, required for driving the shock wave at a constant velocity. For instance, the magnetic piston can be replaced by a rigid conducting piston, plunged into the homogeneous nonconducting gas at $t=0$ with the velocity

U_0. This formulation of the problem seems especially natural when we deal with transverse ionizing shock waves, since then the conducting piston in the self-similar problem is entirely equivalent to the magnetic piston (Sect.5.2).

Let us now consider the numerical solution of the problem of dynamics of the transverse ionizing shock wave in hydrogen, formulated in this manner [5.5]. At $t = 0$ immediately in front of the piston the gasdynamic shock wave arises in which the magnetic field is constant by virtue of inequality $Pm \ll 1$. This holds in the experiments with ionizing shock waves (recall that the MHD calculations described in Sect.5.3 were performed for that range of plasma parameters where, on the contrary, $Pm \gg 1$). The width of the gasdynamic shock being large compared to the characteristic length of ionization, the increase in the number of electrons within the confines of the shock may be disregarded. When the ionization in front of the shock wave is above 10^{-3}, the scale of the electron heat conduction Δ_{Te}, which is smaller than the other characteristic scales of the problem, is still larger than the shock width, so the temperature within the gasdynamic shock changes little. As follows from these arguments, the analysis of the magnetic structure of ionizing shock does not require resolving the structure of the gasdynamic subshock. It can be substituted by a zero-thickness discontinuity surface, containing jumps of the heavy-component temperature, pressure, density and velocity, with appropriate boundary conditions (viscous subshock) separating the flow into regions 1 and 2. Assume that the gas in region 1 before the subshock is partly ionized molecular hydrogen ($\gamma = 7/5$, $J = 15.4$ eV). The gas suffers complete dissociation in the subshock, and in region 2 becomes atomic ($\gamma = 5/3$, $J = 13.6$ eV). Then the boundary conditions on the viscous discontinuity surface may be written in the form

$$n_1(u_1 - U) = n_2(u_2 - U) \quad , \tag{5.4.1}$$

$$M_1N_1(u_1 - U) = M_2N_2(u_2 - U) \quad , \tag{5.4.2}$$

$$M_1N_1(u_1 - U)^2 + N_1T_1 + n_1T_e = M_2N_2(u_2 - U)^2 + N_2T_2 + n_2T_e \quad , \tag{5.4.3}$$

$$M_1N_1 \frac{(u_1 - U)^3}{2} + \frac{7}{2} N_1T_1(u_1 - U) + E_d(u_1 - U)$$
$$= M_2N_2 \frac{(u_2 - U)^3}{2} + \frac{5}{2} N_2T_2(u_2 - U) \quad . \tag{5.4.4}$$

Here the subscripts 1 and 2 pertain to regions 1 and 2, $M_{1,2}$ are the masses of heavy particles (molecules in region 1 and atoms in region 2), N is the particle density, T temperature, n the density of electrons, E_d is the energy of dissociation of hydrogen, U is the velocity of viscous subshock.

In regions 1 and 2 the flow is described by the two-fluid hydrodynamic equations, which here are actually reduced to the one-fluid two-temperature model. As long as the ion-atom charge exchange cross section is large, the velocities of the ion and atom components may be considered equal. It must be emphasized that this approximation, which is justified here, is not, however, always acceptable (Sect.4.4). The equations of continuity (5.3.1), motion (5.3.2) and induction (5.3.5) still hold, except that the term NT_e in (5.3.2) should be replaced by nT_e, and the viscous term dropped altogether. The variable n is included in the equation of ionization kinetics

$$\frac{\partial n}{\partial t} + \frac{\partial}{\partial x}(nu_x) = \dot{n}_e + \dot{n}_{ph} - \frac{D_e}{\Lambda^2} n \quad . \tag{5.4.5}$$

Here $\dot{n}_e$ and $\dot{n}_{ph}$ stand for the ionization rates due to collisional and radiative processes, respectively; the last term on the right-hand side accounts for the diffusive electron losses.

The equations of heat balance for the plasma components take the form

$$\frac{\partial}{\partial t}\left(\frac{3}{2} nT_e\right) + \frac{\partial}{\partial x}\left(\frac{3}{2} nT_e u_x\right) + nT_e \frac{\partial u_x}{\partial x} - \frac{1}{\mu_0^2 \sigma_\perp}\left(\frac{\partial B_y}{\partial x}\right)^2$$

$$+ Q_\Delta + J\dot{n}_e - T_{\nu *}\dot{n}_{ph} = 0 \quad , \tag{5.4.6}$$

$$\frac{\partial}{\partial t}\left(MN \frac{u_x^2}{2} + \frac{1}{\gamma - 1} NT\right) + \frac{\partial}{\partial x}\left(MN \frac{u_x^3}{2} + \frac{\gamma}{\gamma - 1} NTu_x\right)$$

$$+ u_x \frac{\partial}{\partial x}\left(nT_e + \frac{B_y^2}{2\mu_0}\right) - Q_\Delta = 0 \quad . \tag{5.4.7}$$

The expressions for Q_Δ and $\sigma_\perp$ used here differ from those quoted in Sect. 3.1.2 for the completely ionized plasma by the inclusion of neutral particles:

$$Q_\Delta = \frac{2m_e}{M}(T_e - T)n(\nu_{ea} + \nu_{ei}) \quad ,$$

where

$$\nu_{ea} + \nu_{ei} = \left(\frac{8T_e}{\pi m_e}\right)^{\frac{1}{2}}\left[(N - n)\bar{\sigma}_{ea} + n\frac{2\pi}{3}\left(\frac{e^2}{T_e}\right)^2 \ln\lambda\right] \quad .$$

Here $\bar{\sigma}_{ea}$ is the Maxwell-averaged cross section of the electron-neutral elastic scattering; $\ln\lambda$ is the Coulomb logarithm. Further,

$$\sigma_\perp = \frac{e^2 n}{m_e(\nu_{ea} + \nu_{ei})} \quad .$$

In the equation of ionization kinetics (5.4.5)

$$\dot{n}_e = \alpha_e n(N - n) - \beta_e n^3 \quad .$$

The coefficients of impact ionization α_e and triple recombination β_e are linked, in accordance with the principle of detailed balancing, by

$$\frac{\alpha_e}{\beta_e} = \frac{2g_+}{g_0}\left(\frac{2\pi m_e T_e}{h^2}\right)^{3/2} \exp(-J/T_e) \quad , \qquad (5.4.8)$$

where g_+, g_0 are the statistical weights of ions and atoms (molecules), respectively; J is the ionization energy. The values of α_e and β_e were chosen from the relevant experimental data with due account of (5.4.8).

The contribution of photoionization and photorecombination is given by

$$\dot{n}_{ph} = \alpha_\nu(N - n) - \beta_\nu n^2 \quad , \qquad (5.4.9)$$

where

$$\alpha_\nu = \frac{8\pi J^2}{c^2 h^3} T_\nu \sigma_{ph} \exp(-J/T_\nu) \quad ; \quad \beta_\nu = \frac{g_0}{2g_+}\left(\frac{8T_e}{\pi m_e}\right)^{\frac{1}{2}} \frac{J^3 \sigma_{ph}}{m_e^2 c^2 T_e} \quad .$$

Here σ_{ph} is the photoionization cross section near the threshold $h\nu = J$; T_ν is the effective radiation temperature, found from the density of radiation U_ν in accordance with

$$U_\nu^{(eq)}(T_\nu) = U_\nu \quad , \qquad (5.4.10)$$

where $U_\nu^{(eq)}(T)$ is defined by Planck's formula

$$\frac{1}{\exp(h\nu/T_\nu(\tau)) - 1} = \int_0^\infty \frac{Ei(|\tau' - \tau|)}{\exp(h\nu/T(\tau')) - 1} d\tau' \quad .$$

Here τ is the optical coordinate

$$\tau(x) = \int_{x_p}^{x} \kappa(x')dx'$$

where $\kappa(x') = [N(x') - n(x')]\sigma_{ph}$, x_p is the coordinate of the piston; $Ei(z) = \int_1^\infty e^{uz} du/u$ is the exponential integral. Here we disregard the radiation flux propagating head-on to the shock and arising from photorecombination.

The effective radiation temperature at a given frequency is a convenient parameter, because under the present circumstances $J/T \gg 1$. Accordingly, the number of ionizing quanta decreases rapidly as we depart from the threshold, as does the photoionization cross section. Therefore, photoelectrons are produced mainly by the quanta in a narrow energy range near J. The effective temperature of radiant electron heating is defined similarly

$$U_{\nu*}^{(eq)}(T_{\nu*}) = U_{\nu*} \quad , \tag{5.4.11}$$

since the heating of electrons as a function of frequency exhibits a maximum at a certain frequency $\nu = \nu^*$.

The calculations were performed for hydrogen under the conductions corresponding to experimental realization of the inverse Z pinch (initial pressure: $p_1 = 0.25$ Torr, axial magnetic field 0.042-0.439 T, the diffusion length of electrons along the cylindric chamber axis with height $h = 15$ cm is taken equal to $\Lambda = h/\pi = 5$ cm) [5.4]. The piston velocity U_0 was chosen so as to obtain the experimentally observed shock velocities u_s (identified with the viscous subshock velocity U).

Typical (x,t) plots of the viscous subshock are shown in Fig.5.22. The curves are almost linear; velocity variations do not exceed 15%. With fixed piston velocity, the velocity of the shock increases with increasing strength of the axial magnetic field.

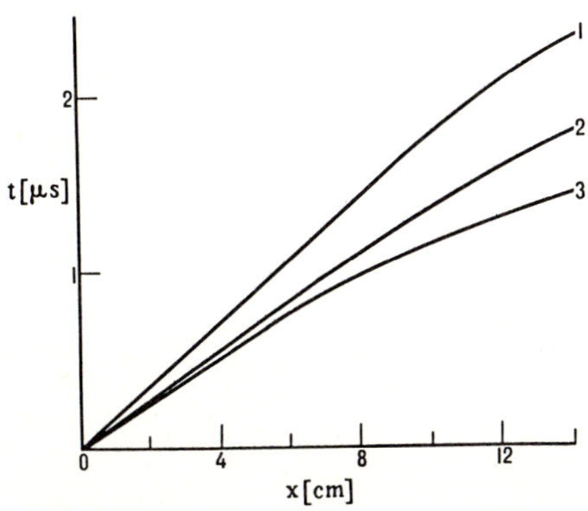

Fig.5.22. Typical x-t diagrams for gasdynamic subshocks. 1: $B_{y0} = 0.172$ T, $U_0 = 4.9$ cm/μs; 2: $B_{y0} = 0.349$ T, $U_0 = 6$ cm/μs; 3: $B_{y0} = 0.439$ T, $U_0 = 7$ cm/μs

Figure 5.23 shows typical profiles of velocity u_x and magnetic field B_y. An abrupt change in the gas velocity directly behind the viscous subshock is due to the avalanche multiplication of electrons, the recombination in this region being immaterial. Figures 5.24,25 show the profiles of temperature and ionization α. The latter is defined as the ratio of the number of

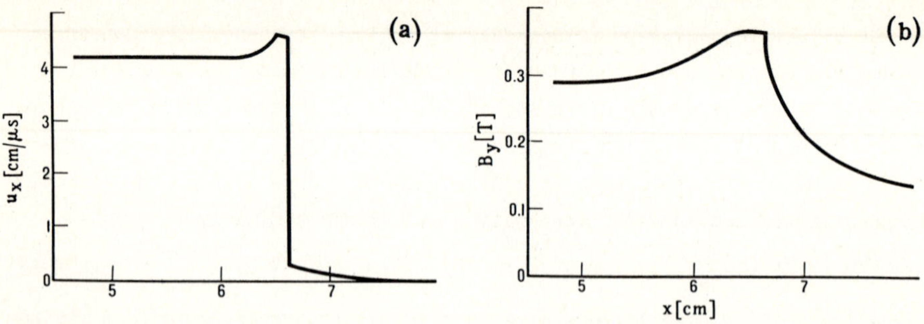

Fig.5.23. Velocity (**a**) and magnetic field (**b**) profiles for $B_{y0} = 0.128$ T, $U_0 = 4.8$ cm/μs

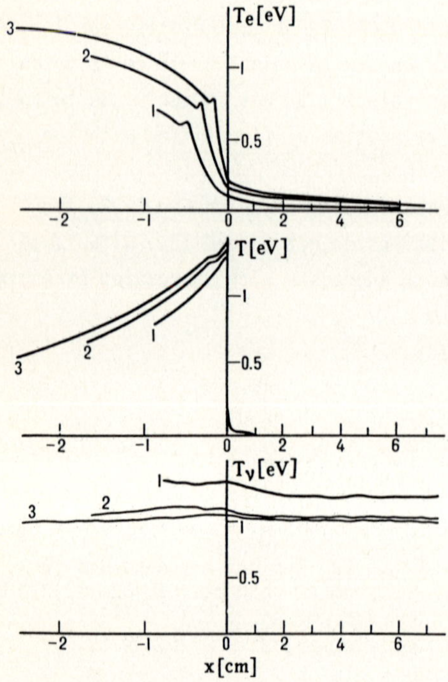

◄ Fig.5.24. Temperature profiles for $B_{y0} = 0.172$ T, $U_0 = 4.9$ cm/μs. 1: $t = 0.5$ μs, $x_p = 3.27$ cm, $U = 6.69$ cm/μs; 2: $t = 1$ μs, $x_p = 6.58$ cm, $U = 6.8$ cm/ s; 3: $t = 1.5$ μs, $x_p = 9.9$ cm, $U = 6.77$ cm/μs, x_p being the piston coordinate

Fig.5.25. Profiles of the degree of ionization for the same values as in Fig.5.24

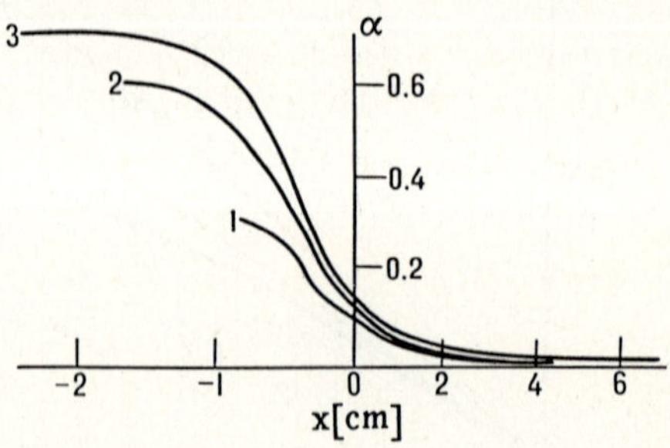

electrons to the total number of ions, including the ions bound in atoms, molecules and molecular ions, so that the relative number of molecular ions ahead of the shock is 2α. By virtue of (4.1,2), α is continuous across the viscous subshock.

Figure 5.26 represents the readings of the magnetic probes B_y, together with its time derivative $\dot{B}_y$ and the electric field in laboratory frame E_{lab}. Similar probe signals were observed in [5.4].

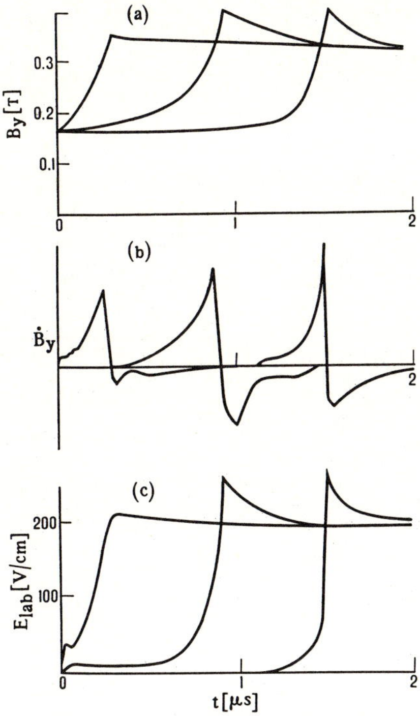

Fig.5.26a-c. The "probe signals": (**a**) B_y; (**b**) $\dot{B}_y$, arb. units; (**c**) E_{lab}, taken at 2, 6 and 10 cm from where gasdynamic shock is formed

As the ionization before the viscous subshock increases, the magnetic field is further compressed, reducing the electric induction field and slowing down the advance of the breakdown. This results in steady profiles of magnetic field, as illustrated in Fig.5.27. In accordance with the theory (Sect.4.3.1 and [5.37]), the numerical calculation indicates that the time taken to achieve the steady state is smaller, the higher the initial transverse magnetic field B_{y1}.

Figure 5.28 shows the compression of the magnetic field as a function of B_{y1}. The experimental points are abstracted from [5.4]. The theory is found to agree quite well with the experiment.

In the steady-state regime the electric induction field ahead of the shock has the value of 7-10 V/cm, in compliance with the predictions of the theory of steady-state shock structure of ionizing shock waves (Sect.4.3.1). This accounts for the good agreement between theory and experiment (Fig.4.8). Numerical simulation on the basis of the two-fluid transfer equations with due account taken of radiative processes allows one to describe all the principal features of experimental transverse ionizing shocks. The calculations

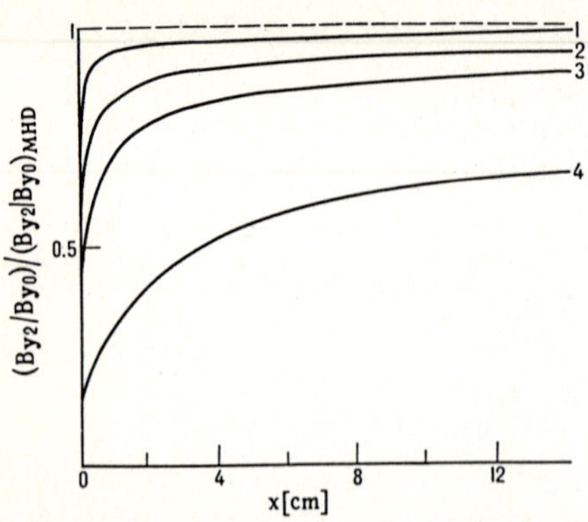

Fig.5.27

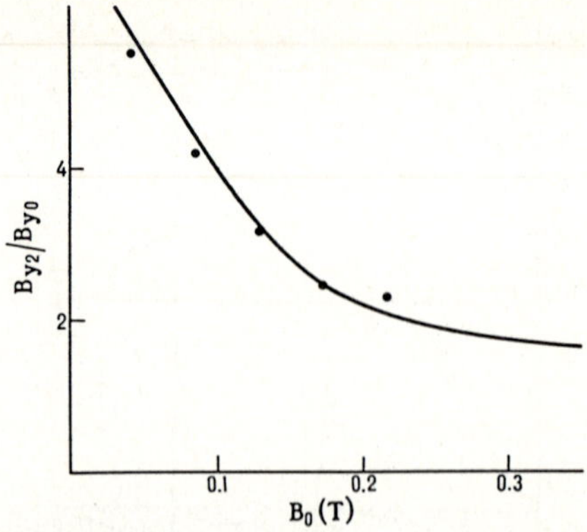

Fig.5.28

Fig.5.27. The ratio of compression of the magnetic field in the ionizing shock wave B_{y2}/B_{y0} to that in a MHD shock $(B_{y2}/B_{y0})_{MHD}$, having the same velocity U, as a function of the ionizing shock front coordinate. 1: $B_{y0} = 0.439$ T, $U_0 = 7$ cm/μs; 2: $B_{y0} = 0.214$ T, $U_0 = 5$ cm/μs; 3: $B_{y0} = 0.128$ T, $U_0 = 4.8$ cm/μs; 4: $B_{y0} = 0.042$ T, $U_0 = 4.6$ cm/μs

Fig.5.28. Compression of the magnetic field in a transverse ionizing shock wave propagated at x = 14 cm from where it is formed, as a function of the initial magnetic field. Experimental points are from [5.4]

neglect photoionization, and predict, as expected (Sects.4.2.5,3.1) a very low amount of compression of magnetic field, and their results deviate widely from experiment.

In concluding this section, let us emphasize that one should be rather cautious in interpreting the experiments with transverse ionizing shock waves produced on the inverse Z pinch installations. Indeed, the electromagnetic waves cannot propagate freely in installations of this kind, since they strike conducting surfaces (the coils for producing axial magnetic bias, Fig.5.3). In the quasi-stationary approximation the presence of these coils gives rise to a situation when the movement of the conducting fluid (the shock and the magnetic piston) from the axis to the walls of the cylinder reduces the magnetic flux confined between the coils and the shock front. This induces electric current in the coils, which tends to counterbalance the reduction in the magnetic flux. Accordingly, the change in the magnetic field in the region before the pressure jump (the gasdynamic isomagnetic subshock) in the transverse ionizing shock wave adds up from the buildup of the axial field, due to the movement of the shock, and the compression of magnetic field within the shock itself. These contributions are quite distinct in the oscillograms of magnetic field [5.4]. At first the field increases slowly (the "ramp"); then follows a relatively narrow region of rapid buildup across the shock, terminated by the pressure jump. The main variation of the magnetic field in these experiments was restricted to the shock wave. This justifies the use of the theory developed in Sect.4.3.1 and the results of numerical simulation described in this section for interpreting the results of these experiments.

5.5 Shaping the Structure of the Normal Ionizing Shock

Let us now consider the nonstationary problem concerning shaping the ionizing shock-wave structure similarly to the method in Sect.5.4 [5.38]. At $t=0$ a rigid piston, whose surface is normal to the axial magnetic field, is plunged into the undisturbed nonconducting gas at a constant velocity. (Actually, for stability of the computative scheme, the calculations of [5.38] included a small transverse magnetic field, $B_{y0}/B_x = 0.01 \div 0.3$; the ratio becomes smaller, the higher B_x.) A time-constant transverse magnetic field B_d is assigned to the piston surface. The flow is assumed to be planar, i.e., $B_z = u_z = 0$. The ion slip in relation to atoms is disregarded: Ohm's law is used in its simplest scalar form (see below).

Let us mark the difference between the present statement of the problem and the self-similar piston problem (Sect.5.2.2). The self-similar problem fixes only the velocity of the conducting piston, whereas the magnetic field on its surface is defined as an eigenvalue, which characterizes the flow as a whole. Conversely, we may fix the value of B_y on the piston surface (Sect. 5.3 and [5.2]), but then the eigenvalue will be the piston velocity (with the boundary condition $\partial u_x/\partial x = 0$ on the piston surface, or on the surface $\zeta = 0$ in Lagrangian coordinates). In [5.38] the piston was assumed to be not perfectly conducting $(j_z(0) \propto (dB_y/dx)_0 \neq 0)$, enabling the calculations to be simplified. An a priori value was assigned to B_d, without trying to find its value for given velocity U self-consistently.

Numerical solution was performed for helium with the initial parameters $p_0 = 0.1$ Torr, $T_0 = 300$ k, $N_0 = 3.2 \cdot 10^{15}\,\mathrm{cm}^{-3}$. As in Sect.5.4, at $t=0$ the moving piston produces a gasdynamic shock wave, which is further viewed as the internal discontinuity in the flow velocity, pressure and temperature of the heavy component (viscous discontinuity), which separates the flow into regions 1 and 2. Its velocity U is identified with the velocity of the shock wave. The boundary conditions on the discontinuity surface differ from

(5.4.1-4), because now we have a monatomic gas both before and after the discontinuity surface, so we set $M_1 = M_2 = M$ in (5.4.2-4). The left-hand side of (5.4.4) is now identical to the right-hand side, except for the subscript 1 to the variables N, u ($\equiv u_x$), T. The equations of continuity (5.3.1), axial movement (5.3.2), ionization kinetics (5.4.5) and electron energy balance (5.4.6) have the same form as in Sect.5.4. The azimuthal motion of the plasma is described by

$$M \frac{\partial}{\partial t} N u_y - \frac{\partial}{\partial x}\left(M N u_x u_y - \frac{B_x B_y}{\mu_0}\right) = 0 \quad . \tag{5.5.1}$$

The equations of plasma energy balance and induction become

$$\begin{aligned} &\frac{\partial}{\partial t}\left[N\left(M\frac{u_x^2 + u_y^2}{2} + \frac{3}{2}T\right) + n\left(\frac{3}{2}T_e + J\right) + \frac{B_y^2}{2\mu_0}\right] \\ &\quad + \frac{\partial}{\partial x}\left[N u_x\left(M\frac{u_x^2 + u_y^2}{2} + \frac{5}{2}T\right) + n u_x\left(\frac{5}{2}T_e + J\right)\right. \\ &\quad \left. + \frac{B_x}{\mu_0}\left(u_x B_y - u_y B_x - \nu_m \frac{\partial B_y}{\partial x}\right)\right] = 0 \quad , \end{aligned} \tag{5.5.2}$$

$$\frac{\partial B_y}{\partial t} + \frac{\partial}{\partial x}\left(u_x B_y - u_y B_x - \nu_m \frac{\partial B_y}{\partial x}\right) = 0 \quad . \tag{5.5.3}$$

Calculations by the method of an effective ionization potential (Sect.2.2.1 and [5.39]) accounted for the possibility of double ionization of helium; in most cases, however, the gas was singly ionized ($J = J_{eff} = J_1 = 24.5$ eV).

At the initial instants the ionization in region 2 increases rapidly (mainly due to photoionization), and the magnetic field B_y steadily decreases with increasing distance from the piston. When the ionization becomes noticeable ($\alpha \sim 10^{-3} - 10^{-2}$), the profile of B_y changes its appearance. The magnetic field near the piston starts to increase, and the profile of B_y exhibits a maximum, indicating the switch on of the transverse magnetic field in the shock wave. At the same time the temperature profile exhibits a deepening depression, which indicates the formation of the expansion wave between the piston and shock wave. The expansion wave is followed by a plateau on which the electric and magnetic field remain constant, and the temperature, density and axial velocity are small (for instance, $u_x < 10^{-2} U$). This configuration is quite similar to the situation realized in the self-similar magnetic piston problem, the magnetic field on the plateau B_p acting as the driving magnetic field. The region between the piston and the plateau does not influence the flow pattern between the piston and the shock, since the time of transfer of the disturbance from the piston to the plateau is by two or three orders of magnitude greater than the characteristic time scale of the problem. The interpretation of B_p as the driving magnetic field is corroborated by the fact that with different combinations of piston velocity U_0 with the magnetic field B_d on its surface, resulting in the same shock velocity U, the profiles of flow variables are also the same, although the time of formation of the expansion wave depends on U_0. With low piston velocities the expansion wave may fail to appear altogether. When it does exist, the driving magnetic field $B_p \gtrsim 0.08$-0.09 T.

Thus, while solving the problem of a rigid piston, we rather unexpectedly come to a typical flow pattern in front of the magnetic piston, which certainly does not conform to self-similar solutions (Sect.5.2.2). The region of hot plasma between the viscous jump and the expansion wave is here optically thin, and the region adjoining the rigid piston is detached from the viscous subshock. Therefore the degree of ionization in front of the viscous subshock is very low, and the precursor phenomena, which shape the structures of real normal ionizing shocks (Sect.4.4.2) are immaterial at present.

The ionization in the case in question starts behind the viscous subshock, i.e., in these numeric examples we are dealing with an artificially conjectured situation, corresponding to the GD limit in the theory of ionizing shock waves. It is exactly in this sense that these calculations are called model calculations. In this limit, in accordance with the idealized theory (Sect.5.2.2), the normal ionizing Chapman-Jouguet shocks ought to be realized in front of the magnetic piston (CJ).

Figure 5.29 shows the profiles of temperatures T and T_e (a) and particle density (b) for $B_x = 0.1$ T, $B_p = 0.01$ T, $U = 6$ cm/μs. The origin of coordinates corresponds to the instant when the viscous subshock passes a probe located at $\ell = 180$ cm from the point of origination of the shock wave. Figure 5.30 shows the profiles of the magnetic and electric fields E (a) and B_y (b) for the same conditions, with $\ell = 60$ cm and $\ell = 180$ cm. These profiles distinguish the shock front and the expansion wave, where the density and temperature go down and the magnetic field goes up. We see that the ionizing shock front is not detached from the head of the expansion wave, as should be expected for the Chapman-Jouguet flow. In particular, the profile of B_y does not exhibit the two-step structure, observed experimentally when the shock wave detaches from the magnetic piston [5.40]. Figure 5.29a explains the smallness of pre-

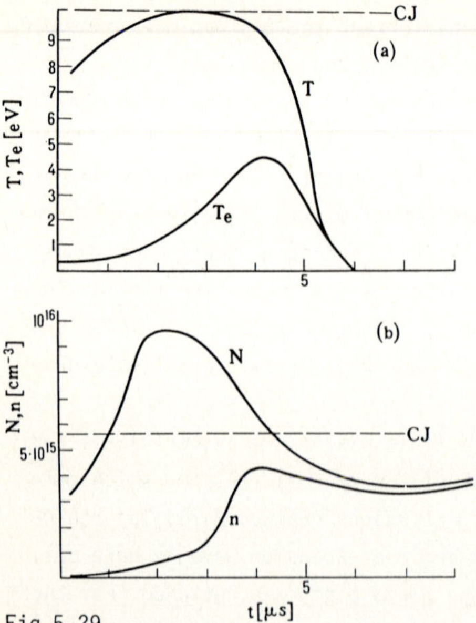

Fig.5.29

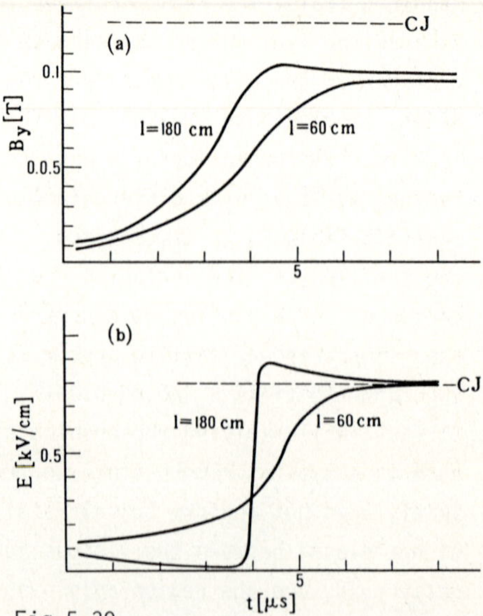

Fig.5.30

Fig.5.29a,b. Profiles of (**a**) temperatures T and T_e and (**b**) number densities N and n in a normal ionizing shock wave

Fig.5.30a,b. Profiles of magnetic (**a**) and electric (**b**) fields in normal ionizing shock waves

cursor ionization: the magnetic piston in these examples is cold, while under real experimental conditions [5.40,41] it had the temperature of several electronvolts and emitted strong ionizing radiation.

Figure 5.30 illustrates the stationarity of the calculated shock structures: the separation between the viscous subshock and the expansion wave remains more or less constant (about 40 cm) as the shock wave travels from $\ell = 60$ cm to $\ell = 180$ cm. The profiles of E and B_y become more articulate, approaching the asymptotic state. This observation conforms with the general theory of the shaping of the steady shock structure [5.42]: this process takes the time required for the shock wave to travel a distance equal to several shock widths.

The numerical solution of the two-fluid hydrodynamic equations under these circumstances then predicts the formation of the ionizing shock wave, whose structure begins with a viscous subshock and which does not detach from the magnetic piston. In the idealized self-similar problem (Sect.5.2.2) this corresponds to the narrow Chapman-Jouguet shock wave, adjoining a wider slow MHD expansion wave. The profiles of flow variables, reproduced in Figs. 5.29,30, hardly conform to this configuration. Here the shock wave and the expansion wave have commensurate widths. The jumps of density and temperature across the shock front differ from the predictions of the Chapman-Jouguet hypothesis (dashed lines CJ in Figs.5.29,30). Recall that under the circumstances of Fig.5.29a the equilibrium downstream temperature (not the temperature maximum) is about 5 eV, i.e., almost one-half of T_{CJ}. The compression of plasma (Fig.5.29a) is stronger than in the Chapman-Jouguet shock wave, but the electron density is lower because of the lower temperature. The downstream values of the electric and magnetic field, however, comply with Chapman-Jouguet hypothesis (Fig.5.30).

Figures 5.31,32 compare calculated results with the measured values [5.43]. As apparent, the numerical curves generally agree with the experiment; the predicted electric field behind the shock wave (identified with the plateau

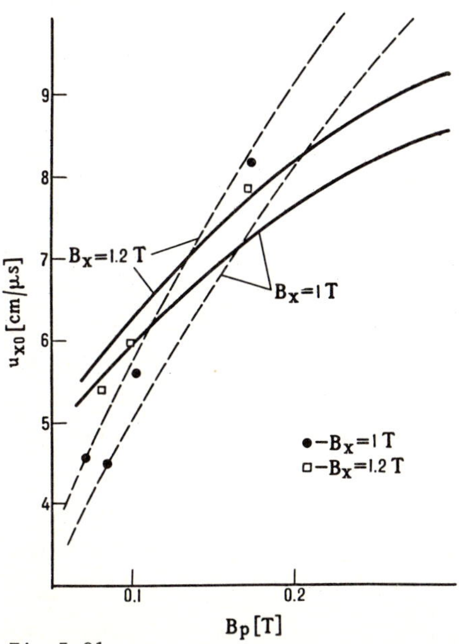

Fig.5.31

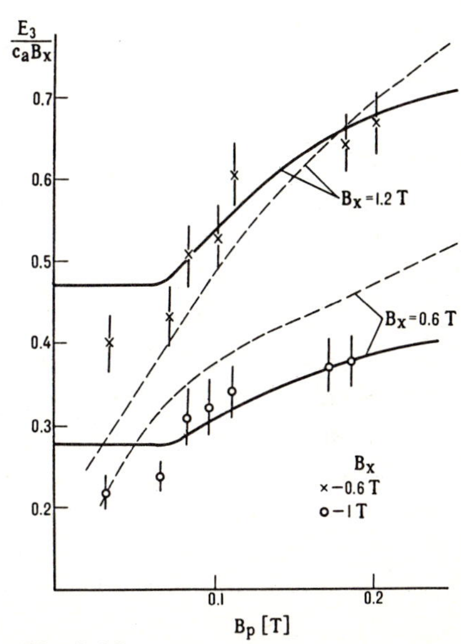

Fig.5.32

Fig.5.31. Normal ionizing shock velocity as a function of driving magnetic field B_p. (——) denote numerical calculations, (---) Chapman-Jouguet shock waves; experimental points are from [5.43]

Fig.5.32. The downstream electric field E_3 as a function of driving magnetic field B_p. (——) denote numerical calculations, (---) Chapman-Jouguet shock waves; experimental points have been taken from [5.43]

value) fits the experiment much better than does the Chapman-Jouguet hypothesis. Note that the agreement is achieved despite the fact the calculated shock structures differ from the observations. In particular, the viscous subshocks failed to be observed in the experiments of [5.40,41], and the plasma compression, contrary to the theoretical forecast, is not detected. Hence the conclusion [5.22,38] that the velocity of the normal ionizing shock wave and the electric field – as functions of the driving magnetic field – are insensitive to the particularities of the shock structure. Therefore, the agreement between the results of numeric calculations, the predictions made on the basis of Chapman-Jouguet hypothesis and the experimental findings do not yet prove that these calculations correctly render the main physical processes underlying the shaping of the shock structures. In a sense, it might be taken to indicate quite the opposite: correctly described are perhaps just those parameters of shock waves which are insensitive to the pertinent physical processes, and the agreement between theory and experiment has nothing to do with the adequacy of description of these processes. On the contrary, the compression and heating of the plasma are extremely sensitive to the structures and to the choice of additional boundary conditions. So far these processes have defied description within the framework of any simple model.

Summing up, we must admit that so far there is no theory capable of adequately describing the propagation of ionizing shock waves in electromagnetic shock tubes. We do know the basic physical processes which shape the structures of these shocks: first of all, the photoionization of neutral gas and the Joule heating of the heavy component due to the ion slip. It is possible to correlate the profiles of flow variables on the basis of the known current profiles J_r [5.41]. However, the description of the flow on the whole, and, in particular, the problem of zero or almost-zero compression in the normal ionizing shocks in helium [5.40,41], still remain subjects for future research.

References

Chapter 1

1.1 L.D. Landau, E.M. Lifshitz: *Statistical Physics*, Part 1 (Pergamon, Oxford 1980)
1.2 L.D. Landau, E.M. Lifshitz: *Fluid Mechanics* (Pergamon, New York 1959)
1.3 E.M. Lifshitz, L.P. Pitaevskii: *Physical Kinetics* (Pergamon, New York 1981)
1.4 D. Burnett: Proc. London Math. Soc. **39**, 385 (1935): **40**, 382 (1935)
1.5 M.N. Kogan: *Dinamika razrezhennogo gaza* (rarefied gas dynamics), (Nauka, Moscow 1967)
1.6 M.N. Kogan, V.S. Galkin, O.G. Fridlender: Usp. Fiz. Nauk **119**, 111 (1976)
1.7 S.M. Dikman, L.P. Pitaevskii: Pis'ma Zh. Eksp. Teor. Fiz. **78**, 1752 (1980)
1.8 H. Grad: Commun. Pure Appl. Math. **2**, 331 (1949)
1.9 V.M. Zhdanov: *Yavleniya perenosa v mnogokomponentnoi plazme* (transport phenomena in multi-component plasma) (Energoizdat, Moscow 1982)
1.10 H.M. Mott-Smith: Phys. Rev. **82**, 885 (1951)
1.11 I.E. Tamm: Tr. FIAN **29**, 239 (1965) [in Russian]
1.12 S. Ziering: Phys. Fluids **4**, 765 (1961)
1.13 C. Muckenfuss: Phys. Fluids **5**, 1325 (1962)
1.14 H. Salwen, C. Grosch, S. Ziering: Phys. Fluids **7**, 180 (1964)
1.15 V.V. Struminskii, V.Yu. Velikodnyi: Dikl. Akad. Nauk SSSR **266**, 28 (1982)
1.16 H. Grad: Commun. Pure Appl. Math **5**, 257 (1952)
1.17 L.H. Holway: Phys. Fluids **7**, 911 (1964)
1.18 P.A. Thompson, T.W. Strock, D.S. Lim: Phys. Fluids **26**, 48 (1983)
1.19 H.W. Liepmann, R. Narasimha, M.T. Chanine: Phys. Fluids **5**, 1313 (1962)
1.20 G.B. Whitham: *Linear and Nonlinear Waves* (Wiley, New York 1974)
1.21 A. Jeffrey, T. Taniuti: *Nonlinear Wave Propagation* (Academic, New York 1964)
1.22 V.I. Karpman: *Nelineinye volny v dispergiruyushchikh sredakh* (nonlinear waves in dispersive media) (Nauka, Moscow 1973)
1.23 E. Hopf: Commun. Pure Appl. Math. **3**, 201 (1950)
1.24 J.D. Cole: Q. Appl. Math. **9**, 225 (1951)
1.25 D.T. Blackstock: J. Acoust. Soc. Am. **39**, 6, 1019 (1966)
1.26 R.D. Fay: J. Acoust. Soc. Am. **3**, 222 (1931)
1.27 G.I. Barenblatt: *Podobie, avtomodel'nost', promezhutochnaya asimptotika* (similarity, self-similarity, intermediate asymptotics) (Gidrometeoizdat, Leningrad 1978)
1.28 Ya.B. Zel'dovich: *Teoriya udarnykh voln i vvedenie v gazodinamiku* (theory of shock waves and introduction to gas dynamics) (Izdatel'stvo AN SSSR, Moscow 1946)
1.29 Al.A. Borisov, A.A. Borisov, S.S. Kutateladze, V.E. Nakoryakov: Pis'ma Zh. Eksp. Teor. Fiz. **31**, 619 (1980)
1.30 H.W. Liepmann: "Cryogenic Fluid Mechanics", in *Recent Developments in Shock Stube Research*, ed. by D. Bershader and W. Griffith (Stanford U. Press, Stanford CA 1973) p.11

1.31 H.W. Liepmann, J.C. Cumming, V.C. Rupert: Phys. Fluids **16**, 332 (1973)
1.32 V.V. Pukhnachev: Dokl. Akad. Nauk SSSR **149**, 798 (1963)
1.33 G.I. Barenblatt, Ya.B. Zel'dovich: Prikl. Mat. Mekh. **21**, 856 (1957)
1.34 Ya. B. Zel'dovich, Yu.P. Raizer: *Physics of Shock Waves and High-Temperature Hydrodynamic Phenomena*, Vols. 1, 2 (Academic, New York 1966,1967)
1.35 G.Ya. Liubarskii: Prikl. Mat. Mekh. **25**, 1041 (1961)
1.36 A.G. Kulikovskii: Prikl. Mat. Mekh. **32**, 1125 (1968)
1.37 A.F. Andreev: Zh. Eksp. Teor. Fiz. **59**, 1819 (1970)
1.38 A.F. Andreev, A.E. Meierovich: Zh. Eksp. Teor. Fiz. **64**, 1640 (1973)
1.39 S.I. Anisimov, V.I. Matsaev: "Nonlinear stability of weak shock waves in dissipative media" Max-Planck-Gesellschaft, Projektgruppe für Laserforschung, Preprint PLF 14 (1979)
1.40 S.P. D'yakov: Zh. Eksp. Teor. Fiz. **27**, 288 (1951)
1.41 V.M. Kontorovich: Zh. Eksp. Teor. Fiz. **33**, 1525 (1957)
1.42 V.M. Kontorovich: Akust. Zh. **5**, (1959)
1.43 G.R. Fowles: "Shock Wave Stability", in Proc. Topical Conf. on Shock Waves, Menlo Park, CA (1981)
G.R. Fowles, A.F.P. Howning: Phys. Fluids **28**, 1982 (1984)
1.44 I.I. Glass: *Shock Waves and Man* (University of Toronto Institute for Aerospace Studies, Toronto 1974)
1.45 G.R. Fowles: Phys. Fluids **24**, 220 (1981)
1.46 D.D. Joseph: *Stability of Fluid Motions*, Vols. 1, 2, *Springer Tracts Nat. Phil., 27, 28* (Springer, Berlin, Heidelberg 1976)
1.47 J.T. Montgomery: Phys. Fluids **18**, 148 (1975)
1.48 N.F. Sather: Phys. Fluids **16**, 2106 (1973)
1.49 Ya.B. Zel'dovich, A.S. Kompaneets: *Teoriya detonatsii* (theory of detonation) (Izdatel'stvo AN SSSR, Moscow 1955)
1.50 Ya.B. Zel'dovich, G.I. Barenblatt, V.B. Librovich, G.M. Makhviladze: *Matematicheskaia teoriya goreniya i vzryva* (mathematical theory of combustion and detonation) (Nauka, Moscow 1980)
1.51 F. Bartlmä: *Gasdynamik der Verbrennung* (Springer, Wien 1975)
1.52 W. Fickett, W.C. Davies: *Detonation* (University of California Press, Berkeley 1979)
1.53 Ya.B. Zel'dovich, D.A. Frank-Kamenetskii: Zh. Fiz. Khim. **12**, 100 (1938)
1.54 J.Y.S. Mar, V. Makios: Phys. Fluids **16**, 2160 (1973)
1.55 P. Rosenau, S. Frankental: Phys. Fluids **19**, 1889 (1976)

Chapter 2

2.1 Ya.B. Zel'dovich: Zh. Eksp. Teor. Fiz. **32**, 1126 (1957)
2.2 V.D. Shafranov: Zh. Eksp. Teor. Fiz. **32**, 1453 (1957)
2.3 J.D. Jukes: J. Fluid Mech. **3**, 275 (1957)
2.4 O.W. Greenberg, H.K. Sen, Y.M. Trève: Phys. Fluids **3**, 379 (1960)
2.5 O.W. Greenberg, Y.M. Trêve: Phys. Fluids **3**, 769 (1960)
2.6 V.S. Imshennik: Zh. Eksp. Teor. Fiz. **42**, 236 (1962)
2.7 V.S. Imshennik: Zh. Vychisl. Mat. Mat. Fiz. **2**, 206 (1962)
2.8 M.Y. Jaffrin, R.F. Probstein: Phys. Fluids **7**, 1659 (1964); Zh. Prikl. Mekh. Tekh. Fiz. **6**, 6 (1964)
2.9 M.S. Greywall: Phys. Fluids **16**, 561 (1973)
2.10 M.S. Greywall: Phys. Fluids **18**, 1439 (1975)
2.11 M.S. Greywall: Phys. Fluids **19**, 2046 (1976)
2.12 V.S. Imshennik: Fiz. Plazmy **1**, 202 (1975)

2.13 K. Abe: Phys. Fluids **18**, 1125 (1975)
2.14 S.I. Braginskii: In *Reviews of Plasma Physics*, ed. by M.A. Leontovich, Vol. 1 (Consultants Bureau, New York 1965)
2.15 Ya.B. Zel'dovich, Yu.P. Raizer: *Physics of Shock Waves and High-Temperature Hydrodynamic Phenomena*, Vols. 1, 2 (Academic, New York 1966, 1967)
2.16 V.P. Silin, A.A. Rukhadze: *Elektromagnitnye svoistva plazmy i plazmopodobnykh sred* (electromagnetic properties of plasma and plasma-like media) (Atomizdat, Moscow **1961**)
2.17 L.D. Landau, E.M. Lifshitz: *Statistical Physics*, Part 1 (Pergamon New York 1980)
2.18 E.A. Mason: In *Processes in Gases and Plasma*, ed. by A.R. Hochstim (Academic, New York 1969) p.50
2.19 N.N. Magretova, N.T. Pashchenko, Yu.P. Raizer: Zh. Prikl. Mekh. Tekh. Fiz. **5**, 11 (1970)
2.20 M.J. Seaton: In *Atomic and Molecular Processes*, ed. by D.R. Bates (Academic, New York 1962) p.414
2.21 I. Amdur, E.A. Mason: Phys. Fluids **1**, 370 (1958)
2.22 M.Y. Jaffrin: Phys. Fluids **8**, 606 (1965)
2.23 Y. Enomoto: J. Phys. Soc. Jpn. **35**, 1228 (1973)
2.24 C.E. Moore: Nat. Bur. Stand. Rep. NSRDS-NBS 34 (Washington 1970)
2.25 H. Petschek, S. Byron: Ann. Phys. (NY) **1**, **270** (1957)
2.26 A.J. Kelly: J. Chem. Phys. **45**, 1723 (1966)
2.27 H.W. Drawn, P. Felenbok: *Data for Plasmas in Local Thermodynamic Equilibrium* (Gauthier-Villars, Paris 1965)
2.28 V. Shanmugasundaram, S.S.R. Murty: J. Plasma Phys. **20**, 419 (1978)
2.29 C.D. Mathers: J. Plasma Phys. **24**, 121 (1980)
2.30 M. Pinègre, P. Valentin: Comptes Rendus **283**, Série B, 445 (1976)
2.31 J.H. Clarke, C. Ferrari: Phys. Fluids **8**, 2121 (1965)
2.32 D.L. Chubb: Phys. Fluids **11**, 2363 (1968)
2.33 H.F. Nelson, R. Goulard: Phys. Fluids **12**, 1605 (1969)
2.34 S.S.R. Murty: Phys. Fluids **12**, 1830 (1969)
2.35 J.H. Clarke, M.J. Onorato: J. Appl. Mech. **37**, 783 (1970)
2.36 H. Honma, S. Nakadaira: Trans. Jpn. Soc. Aero. Space Sci. **14**, 102 (1971)
2.37 G. Kamimoto, K. Teshima: Trans. Jpn. Soc. Aero. Space Sci. **15**, 29, 124 (1972)
2.38 W.H. Foley, J.H. Clarke: Phys. Fluids **16**, 1612 (1973)
2.39 A.R. Vinolo, J.H. Clarke: Phys. Fluids **16**, 1612 (1973)
2.40 H.F. Nelson: Phys. Fluids **16**, 2132 (1973)
2.41 V. Shanmugasundaram, S.S.R. Murty: J. Plasma Phys. **23**, 43 (1980)
2.42 R.W. Rutowski, D. Bershader: Phys. Fluids **7**, 568 (1964)
2.43 K.P. Horn, H. Wong, D. Bershader: J. Plasma Phys. **1**, 157 (1967)
2.44 F.Y. Su, D.B. Olfe: Phys. Fluids **15**, 263 (1972)
2.45 G. Larcher, C. Thenard, P. Maillot, P. Valentin: Comptes Rendus **271**, Série B, 765 (1970)
2.46 M. Pinègre, P. Valentin: Comptes Rendus **280**, Série B, 215 (1975)
2.47 M. Pinègre, P. Valentin: Comptes Rendus **283**, Série B, 421 (1976)
2.48 M. Pinègre: Thèse, University of Rouen (1976)
2.49 L.M. Biberman, G.E. Norman, K.N. Ulyanov: Opt. Spektrosk. **10**, 565 (1961)
2.50 W.M. Robinson: Ph.D. Thesis, California Institute of Technology (1969)
2.51 G.W. Paxton, R.G. Fowler: Phys. Rev. **128**, 993 (1962)
2.52 R. Klingbeil, D.A. Tidman, R.F. Fernsler: Phys. Fluids **15**, 1969 (1972)
2.53 R.G. Fowler: Adv. Electron. Electron. Phys. **41**, 1 (1976)
2.54 E.E. Sanmann, R.G. Fowler: Phys. Fluids **18**, 1433 (1975)
2.55 E.J. Morgan, R.D. Morrison: Phys. Fluids **8**, 1608 (1965)
2.56 H. Wong, D. Bershader: J. Fluid Mech. **26**, 459 (1966)
2.57 P.E. Oettinger, D. Bershader: AIAA J. **5**, 1625 (1967)
2.58 M.I. Hoffert, H. Lien: Phys. Fluids **10**, 1769 (1967)

2.59 G. Bekefi: *Radiation Processes in Plasmas* (Wiley, New York 1966)
2.60 M. Pinègre, P. Valentin: Comptes Rendus **279**, Série B, 551 (1974)
2.61 T.I. McLaren, R.M. Hobson: Phys. Fluids **11**, 2152 (1968)
2.62 I.I. Glass, W.S. Liu: J. Fluid Mech. **84**, 55 (1978)
2.63 B.T. Whitten: Ph.D. Thesis, University of Toronto, Institute for Aerospace Studies (1977)
2.64 R.E. Duff: Phys.Fluids **2**, 207 (1959)
2.65 A. Roshko: Phys. Fluids **3**, 835 (1960)
2.66 M. Mirels: Phys. Fluids **6**, 1201 (1963)
2.67 H. Mirels: AIAA J. **2**, 84 (1964)
2.68 H. Mirels: Phys. Fluids **9**, 1907 (1966)
2.69 M. Merilo, E.J. Morgan: J. Chem. Phys. **52**, 2192 (1970)
2.70 V.H. Blackman, G.B.F. Niblett: *Fundamental Data Obtained from Shock Tube Experiments* (Pergamon, London 1961)
2.71 N.R. Jones, M. McChesney: Nature **209**, 1080 (1966)
2.72 H. Oguchi: Bull. Inst. Space Aero. Sci., Univ. of Tokyo **3**, 367 (1967)
2.73 L.V. Kochmanova, Ts.G. Breido, V.L. Goryachev, G.S. Sukhov: Zh. Tekh. Fiz. **40**, 600 (1970)
2.74 R. Shreffler, R. Christian: J. Appl. Phys. **18**, 1008 (1950)
2.75 Yu.A. Zatsepin, E.G. Popov, M.A. Tsikulin: Zh. Eksp. Teor. Fiz. **54**, 112 (1968)
2.76 M.A. Tsikulin, E.G. Popov: *Izluchatel'nye svoistva udarnykh voln v gazakh* (radiative properties of shock waves in gases) Nauka, Moscow 1977)
2.77 R.W. Griffiths, R.J. Sandeman, H.G. Hornung: J. Phys. D. (Appl. Phys.) **9**, 1681 (1976)
2.78 J.A. Johnson III, J.P. Santiago, Lin I: Phys. Lett. **83A**, 443 (1981)
2.79 H.L. Swinney, J.P. Gollub: Phys. Today **31**, 41 (1978)
H.L. Swinney, J.P. Gollub (eds): *Hydrodynamic Instability and the Transition to Turbulence*, 2nd ed., Topics Appl. Phys., Vol. 45 (Springer, Berlin, Heidelberg 1985)
2.80 A.M. Yaglom: Ann. Rev. Fluid Mech. **11**, 505 (1979)
2.81 G. Speziale: Phys. Fluids **23**, 459 (1980)
2.82 S. Tsuge: Phys. Fluids **17**, 22 (1974)
2.83 Yu.L. Klimontovich: *The Statistical Theory of Nonequilibrium Processes in a Plasma* (MIT Press, Cambridge, MA 1967)
2.84 J.A. Johnson III, Lin I, R. Ramaich: "Reduced Molecular Chaos and Flow Instability", Proc. of IUTAM Conference on Stability in the Mechanics of Continua, Numbrecht (1981)

Chapter 3

3.1 L.D. Landau, E.M. Lifshitz: *Electrodynamics of Continuous Media* (Pergamon, New York 1982)
3.2 S.I. Braginskii: In *Reviews of Plasma Physics*, ed. by M.A. Leontovich, Vol. 1 (Consultants Bureau, New York 1965) p.205
3.3 E.P. Sirotina, S.I. Syrovatskii: Zh. Eksp. Teor. Fiz. **39**, 746 (1960)
3.4 T.H. Stix: *The Theory of Plasma Waves* (McGraw-Hill, New York 1962)
3.5 K.O. Friedrichs, H. Kranzer: "Notes on Magnetohydrodynamics, VIII, Non-Linear Wave Motion" (Institute of Mathematical Sciences, New York University, July 31, 1958)
3.6 J. Bazer: Astrophys. J. **128**, 686 (1958)
3.7 W.B. Ericson, J. Bazer: Astrophys. J. **129**, 758 (1959)

3.8 F. Hoffman, E. Teller: Phys. Rev. **80**, 692 (1950)
3.9 S.I. Syrovatskii: Zh. Eksp. Teor. Fiz. **24**, 622 (1953)
3.10 P. Germain: Rev. Mod. Phys. **32**, 951 (1960)
3.11 J.A. Shercliff: Rev. Mod. Phys. **32**, 980 (1960)
3.12 J.E. Anderson: *Magnetohydrodynamic Shock Waves* (MIT Press, Cambridge, MA 1963)
3.13 A.I. Akhiezer, G.Ya. Liubarskii, R.V. Polovin: Zh. Eksp. Teor. Fiz. **35**, 731 (1958)
3.14 S.I. Syrovatskii: Zh. Eksp. Teor. Fiz. **35**, 1466 (1958)
3.15 R.V. Polovin: Usp. Fiz. Nauk **72**, 33 (1960)
3.16 A. Jeffrey, T. Taniuti: *Non-Linear Wave Propagation* (Academic, New York 1964)
3.17 Z.B. Roikhvarger, S.I. Syrovatskii: Zh. Eksp. Teor. Fiz. **66**, 1338 (1979)
3.18 S.F. Pimenov: Zh. Eksp. Teor. Fiz. **84**, 1703 (1983)
3.19 C.K. Chu, R.T. Taussig: Phys. Fluids **10**, 249 (1967)
3.20 C.K. Chu, R.A. Gross: "Shock Waves in Plasma Physics", in *Advances in Plasma Physics*, ed. by A. Simon, W.B. Thompson, Vol. 2 (Interscience, New York 1969)
3.21 A.L. Velikovich, M.A. Liberman: Usp. Fiz. Nauk **129**, 377 (1979)
3.22 W. Marshall: Proc. R. Soc. **A233**, 367 (1955)
3.23 C.S.S. Ludford: J. Fluid Mech. **5**, 67 (1959)
3.24 H.S. Green: Phys. Fluids **2**, 341 (1959)
3.25 H.L. Helfer: Astrophys. J. **117** (1953)
3.26 H. Sen: Phys. Rev. **102**, 5 (1956)
3.27 G.S. Golitsyn, K.P. Staniukovich: Zh. Eksp. Teor. Fiz **33**, 1417 (1957)
3.28 Ya.B. Zel'dovich: Zh. Eksp. Teor. Fiz. **32**, 1126 (1957)
3.29 V.D. Shafranov: Zh. Eksp. Teor. Fiz. **32**, 1453 (1957)
3.30 V.S. Imshennik: Zh. Eksp. Teor. Fiz. **42**, 236 (1962)
3.31 A.G. Kulikovskii, G.A. Liubimov: Prikl. Mat. Mekh. **23**, 868 (1959)
3.32 S.B. Pikel'ner: Zh. Eksp. Teor. Fiz. **36**, 1536 (1959)
3.33 W. Geiger, H.J. Kaeppeler, B. Mayser: Nucl. Fusion Suppl. part 2, 403 (1962)
3.34 P.N. Hu: Phys. Fluids **9**, 89 (1966)
3.35 H. Grad, P.N. Hu: Phys. Fluids **10**, 2597 (1967)
3.36 P.N. Hu, G. Grad: Phys. Fluids **15**, 402 (1972)
3.37 R.T. Taussig: Phys. Fluids **16**, 384 (1973)
3.38 V.A. Dixon, L.C. Woods: Plasma Phys. **18**, 627 (1976)
3.39 A.L. Velikovich, M.A. Liberman: Fiz. Plazmy **2**, 334 (1976)
3.40 A.L. Velikovich, M.A. Liberman: Zh. Eksp. Teor. Fiz. **71**, 1390 (1976)
3.41 A.L. Velikovich, M.A. Liberman: Izv. Akad. Nauk SSSR, Mekh. Zhidk. Gaza no. 3, 134 (1977)
3.42 R.Z. Sagdeev: In *Reviews of Plasma Physics*, ed. by M.A. Leontovich, Vol. 2 (Consultants Bureau, New York 1966)
3.43 V.I. Karpman: *Nelineinye volny v dispergiruyushchikh sredakh* (nonlinear waves in dispersive media) (Nauka, Moscow 1973)
3.44 L.C. Woods: Plasma Phys. **11**, 967 (1969)
3.45 D. Sherwell, R.A. Cairns: J. Plasma Phys. **20**, 265 (1978)
3.46 J.P. Barach: Phys. Fluids **22**, 837 (1979)
3.47 R.A. Gross, B. Miller: In *Methods of Experimental Physics*, Vol. 9 part A, ed. by H.R. Griem, R.H. Lovberg, (Academic, New York 1970) p.169
3.48 R.A. Gross: Nucl. Fusion **15**, 729 (1975)
3.49 F.L. Sandel, M. Niimura, S.N. Robertson, R.A. Gross: Phys. Fluids **18**, 1075 (1975)
3.50 P. Moriette: Phys. Fluids **15**, 51 (1972)
3.51 E. Halmoy: Phys. Fluids **14**, 2134 (1971)

3.52 D.H. McNeill: Phys. Fluids **18**, 44 (1975)
3.53 S.H. Robertson, Y.G. Chen: Phys. Fluids **18**, 917 (1975)
3.54 S.H. Schneider: Phys. Fluids **15**, 805 (1972)
3.55 E.Y. Sommer, Jr., J.P. Barach: Phys. Fluids **14**, 2102 (1971)
3.56 J.W.M. Paul, L.S. Holmes, M.J. Parkinson, J. Sheffield: Nature **208**, 133 (1965)
3.57 D.L. Morse: Plasma Phys. **15**, 1262 (1973)
3.58 J.W.M. Paul, G.C. Goldenbaum, A. Iiyoshi, L.S. Holmes, R.A. Hardcastle: Nature **216**, 363 (1976)
3.59 B.P. Leonard: Phys. Fluids **9**, 917 (1966)
3.60 R.J. Bickerton, L. Lehamon, R.V.W. Murphy: J. Plasma Phys. **5**, 177 (1971)
3.61 R.Kh. Kurtmullaev, V.S. Masalov, K.I. Mekler, V.I. Semenov: Zh. Eksp. Teor. Fiz. **60**, 400 (1971)
3.62 B.P. Leonard: J. Plasma Phys. **9**, 917 (1966)
3.63 V.G. Ledenev: Izv. Vuzov, Radiofizika **18**, 1594 (1975)
3.64 Yu.A. Berezin: *Chislennoe issledovanie nelineinykh voln v razrezhennoi plasme* (numerical investigation of nonlinear waves in rarefied plasma) (Nauka, Novosibirsk 1977)
3.65 M.A. Liberman: Zh. Eksp. Teor. Fiz. **75**, 1652 (1978)
3.66 J.W.M. Paul, C.C. Daughney, L.S. Holmes, P.T. Rumsby, A.D. Craig, E.L. Murray, D.D.R. Summers, J. Beaulieu: In: Proc. 4th Intern. Conf. on Plasma Physics and Controlled Nuclear Fusion, Madison (1971) Vol. 3, Vienna Int. p.251
3.67 A.D. Craig, J.W.M. Paul: J. Plasma Phys. **9**, 161 (1973)
3.68 A.D. Craig: J. Plasma Phys. **12**, 129 (1974)
3.69 A.D. Craig: Plasma Phys. **17**, 1111 (1975)
3.70 L. Bighel, A.R. Collins, R.C. Cross, Phys. Lett. **47A**; 333 (1974)
3.71 N.F. Cramer: J. Plasma Phys. **14**, 333 (1975)
3.72 L. Bighel, A.R. Collins, N.F. Cramer: J. Plasma Phys. **18**, 77 (1977)
3.73 A.D. Craig: J. Plasma Phys. **12**, 149 (1974)

Chapter 4

4.1 W.B. Kunkel, R.A. Gross: In *Plasma Hydromagnetics*, ed. by D. Bershader (Stanford Press, Stanford, CA 1962) p.58
4.2 C.K. Chu, R.A. Gross: In *Advances in Plasma Physics*, ed. by A. Simon, W.B. Thompson, Vol. 2 (Interscience New York, 1969)
4.3 G.A. Liubimov: Dokl. Akad. Nauk SSSR **126**, 291 (1959)
4.4 A.G. Kulikovskii, G.A. Liubimov: Dokl. Akad. Nauk SSSR **129**, 52, 525 (1959)
4.5 L. Bighel, N.F. Cramer, D.D. Millar, R.A. Niland: Phys. Lett. **44A**, 449 (1973)
4.6 L. Bighel, A.R. Collins, N.F. Cramer: J. Plasma Phys. **18**, 77 (1977)
4.7 B.P. Leonard: J. Plasma Phys. **7**, 133, 157, 177 (1972)
4.8 B.P. Leonard: J. Plasma Phys. **10**, 13 (1973)
4.9 M.D. Cowley: J. Plasma Phys. **1**, 37 (1967)
4.10 M.A. Liberman, A.L. Velikovich: J. Plasma Phys. **26**, 29 (1981)
4.11 B.P. Leonard: J. Plasma Phys. **17**, 69 (1977)
4.12 M.A. Liberman: Zh. Eksp. Teor. Fiz. **77**, 124 (1979)
4.13 M.A. Liberman: Usp. Fiz. Nauk **127**, 528 (1979)
4.14 M.A. Liberman, A.L. Velikovich: Plasma Phys. **20**, 439 (1978)
4.15 A.L. Velikovich, M.A. Liberman: Zh. Eksp. Teor. Fiz. **74**, 1650 (1978)
4.16 M.A. Liberman: In: Proc. 15th Conf. on Phenomena in Ionized Gases, Minsk (1981) Vol. 1, p.237

4.17 A. von Engel: *Ionized Gases* (Clarendon, Oxford 1955)
4.18 Yu.P. Raizer: *Osnovy sovremennoi fiziki gazorazryadnykh protsessov* (fundamentals of modern physics of gas discharge processes) (Nauka, Moscow 1980)
4.19 A.M. Maksimov, V.E. Ostashev: Teplofiz Vys. Temp. **13**, 644 (1975)
4.20 C.K. Chu: Phys. Fluids **7**, 1349 (1964)
4.21 R.T. Taussig: Phys. Fluids **8**, 1616 (1965)
4.22 R.T. Taussig: Phys. Fluids **9**, 421 (1966)
4.23 R.T. Taussig: Phys. Fluids **10**, 1145 (1967)
4.24 B.P. Leonard: Phys. Fluids **12**, 1816 (1969)
4.25 B.P. Leonard: Phys. Fluids **13**, 833, 3063 (1970)
4.26 M.I. Hoffert: Phys. Fluids **11**, 77 (1968)
4.27 M.A. Liberman, A.L. Velikovich: J. Plasma Phys. **26**, 55 (1981)
4.28 C.F. Stebbins, G.C. Vlases: J. Plasma Phys. **2**, 633 (1968)
4.29 S.H. Robertson, Y.G. Chen: Phys. Fluids **18**, 917 (1975)
4.30 A.R. Collins, C.D. Mathers: Phys. Fluids **21**, 1939 (1978)
4.31 C.D. Mathers: J. Plasma Phys. **24**, 121 (1980)
4.32 R.A. Gross, L. Levine, F. Geldon: Phys. Fluids **9**, 1033 (1966)
4.33 L. Levine: Phys. Fluids **11**, 2479 (1968)
4.34 T.G. Cowling: Mon. Not. R. Astron. Soc. **116**, 114 (1956)
4.35 N.F. Cramer: J. Plasma Phys. **14**, 333 (1975)
4.36 M.A. Liberman, V.S. Synakh, A.L. Velikovich, V.V. Zakajdakhov: Plasma Phys. **24**, 519 (1982)
4.37 C.D. Mathers, N.F. Cramer: Aust. J. Phys. **31**, 171 (1978)
4.38 M.Y. Jaffrin: Phys. Fluids **8**, 606 (1965)
4.39 A.D. Craig, J.W.M. Paul: J. Plasma Phys. **9**, 161 (1973)

Chapter 5

5.1 A.D. Craig: J. Plasma Phys. **12**, 149 (1974)
5.2 S.H. Schneider: Phys. Fluids **15**, 51 (1972)
5.3 P. Moriette: Phys. Fluids **15**, 51 (1972)
5.4 C.F. Stebbins, G.C. Vlases: J. Plasma Phys. **2**, 633 (1968)
5.5 M.A. Liberman, V.S. Synakh, A.L. Velikovich, V.V. Zakajdakhov: Plasma Phys. **22**, 317 (1980)
5.6 R.A. Gross, B. Miller: In *Methods of Experimental Physics*, Vol. 9, part A, ed. by H.R. Griem, R.H. Lovberg (Academic, New York 1970) p.169
5.7 R.A. Gross: In *Recent Developments in Shock Tube Research*, ed. by D. Bershader, W. Griffith (Stanford U. Press, Stanford CA 1973) p.72
5.8 E. Halmoy: Phys. Fluids **14**, 2134 (1971)
5.9 A.R. Collins: Ph.D. Thesis, University of Sydney (1977)
5.10 R.M. Patrick: Phys. Fluids **2**, 589 (1959)
5.11 H.C. Lui: Columbia University Plasma Laboratory Report No.60 (1973)
5.12 H.C. Lui: Ph.D. Thesis, Columbia University, (1973)
5.13 S.H. Robertson, Y.G. Chen: Phys. Fluids **18**, 917 (1975)
5.14 R.A. Gross: In: Proc. 8th Bi. Symp. on Fluid Dynamics, Bialowieza, Poland (1975) p.9
5.15 A.R. Collins, C.D. Mathers: Phys. Fluids **21**, 1939 (1978)
5.16 L. Bighel, J. Howard: Phys. Lett. **69A**, 39 (1978)
5.17 A.D. Craig, J.W.M. Paul: J. Plasma Phys. **9**, 161 (1973)
5.18 M.H. Brennan, I.G. Brown, D.D. Millar, C.N. Watson-Munro: J. Nucl. Energy **5C**, 82 (1963)

5.19 G.C. Vlases: J. Fluid Mech. **16**, 82 (1963)
5.20 O.A. Anderson: Phys. Fluids **1**, 489 (1958)
5.21 R.T. Taussig, Y.G. Chen, R.A. Gross: Phys. Fluids **16**, 212 (1973)
5.22 M.A. Liberman: Zh. Eksp. Teor. Fiz. **77**, 124 (1979)
5.23 C. Greifinger, J.D. Cole: Phys. Fluids **4**, 527 (1961)
5.24 F.J. Fishman, H. Petschek: Ühys. Fluids **5**, 631 (1962)
5.25 J. Howard, B.W. James, A.R. Law: Phys. Fluids **27**, 277 (1984)
5.26 G. Bekefi: *Radiation Processes in Plasmas* (Wiley, New York 1966)
5.27 N.H. Kemp, H.E. Petschek: Phys. Fluids **2**, 599 (1959)
5.28 Ya.B. Zel'dovich, A.S. Kompaneets: *Teoriya detonatsii (*theory of detonation) (Tzdatel'stvo AN SSSR, Moscow 1955) (California Press, University of Berkeley 1979)
5.30 W.B. Kunkel, R.A. Gross: In *Plasma Hydromagnetics*, ed. by D. Bershader (Stanford U. Press, Stanford CA 1962) p.58
5.31 Yu.P. Raizer: *Lazernaya iskra i rasprostranenie razriadov* (laser spark and discharge propagation) (Nauka, Moscow 1974)
5.32 V.I. Fisher: Zh. Eksp. Teor. Fiz. **79**, 2142 (1980)
5.33 M.A. Liberman, A.L. Velikovich: J. Plasma Phys. **26**, 55 (1981)
5.34 S.H. Schneider: Columbia University Plasma Laboratory Report No.55 (1971)
5.35 R.T. Taussig: Phys. Fluids **16**, 384 (1973)
5.36 A.L. Velikovich, M.A. Liberman: Zh. Eksp. Teor. Fiz. **71**, 1390 (1976)
5.37 A.L. Velikovich, M.A. Liberman: Zh. Eksp. Teor. Fiz. **74**, 1650 (1978)
5.38 M.A. Liberman, V.S. Synakh, A.L. Velikovich, V.V. Zakajdakhov: Plasma Phys. **24**, 519 (1982)
5.39 Ya.B. Zel'dovich, Yu.P. Raizer: *Physics of Shock Waves and High Temperature Hydrodynamic Phenomena*, Vol.1 (Academic, New York 1966)
5.40 L. Bighel, A.R. Collins, N.F. Cramer: J. Plasma Phys. **18**, 77 (1977)
5.41 L. Bighel, N.F. Cramer, D.D. Millar, R.G. Niland: Phys. Lett. **44**A, 449 (1973)
5.42 M.J. Lighthill: In *Surveys in Mechancs*, ed. by G.K. Batchelor and R.M. Davies (Cambridge U. Press, Cambridge 1956)
5.43 L. Bighel, C.N. Watson-Munro: Nucl. Fusion **12**, 193 (1972)

Subject Index